信息科学与技术丛书

Java 服务端研发知识图谱

何　为　迟文恒　编著

机械工业出版社

本书覆盖内容较广，从研发基础、框架、组件、部署、工具几个方面分别讲述了 Java 后台研发涉及的知识，各种技术选取最常用和实用的部分，可以让读者花较少的时间获取精要的内容。

基础篇包含 Java 语言的使用和特性、Maven 工程管理、Svn 和 Git 代码管理、Linux 服务器命令。服务框架篇包含 Spring 框架治理、Spring MVC、SpringBoot、框架演进、Spring Cloud 微服务框架。组件篇包含 MySQL 数据库及操作、MongoDB 存储、Redis 缓存、Zookeeper 配置及注册发现原理、FastDFS 文件存储、ElasticSearch 搜索、定时任务、RabbitMQ 消息队列、ELK 日志展示及分析。部署篇包含 Docker 镜像技术、Jenkins 持续集成、Harbor 镜像仓库、Rancher 容器管理。工具篇包含 Swagger 接口文档编写及测试工具、JMeter 测试工具、VisualVM 分析工具等。

图书在版编目（CIP）数据

Java 服务端研发知识图谱 / 何为，迟文恒编著. —北京：机械工业出版社，2018.12
（信息科学与技术丛书）
ISBN 978-7-111-61011-3

Ⅰ.①J… Ⅱ.①何… ②迟… Ⅲ.①JAVA 语言－程序设计 Ⅳ.①TP312.8

中国版本图书馆 CIP 数据核字（2018）第 220353 号

机械工业出版社（北京市百万庄大街 22 号　邮政编码 100037）
责任编辑：车　忱
责任校对：张艳霞
责任印制：张　博

三河市宏达印刷有限公司印刷

2019 年 1 月第 1 版・第 1 次印刷
184mm×260mm・26.75 印张・655 千字
0001－3000 册
标准书号：ISBN 978-7-111-61011-3
定价：98.00 元

凡购本书，如有缺页、倒页、脱页，由本社发行部调换

电话服务　　　　　　　　　　　　　　网络服务
服务咨询热线：（010）88361066　　　机 工 官 网：www.cmpbook.com
读者购书热线：（010）68326294　　　机 工 官 博：weibo.com/cmp1952
　　　　　　　（010）88379203　　　教育服务网：www.cmpedu.com
封面无防伪标均为盗版　　　　　　　　金 书 网：www.golden-book.com

业 界 推 荐

箭速迭代一直是我秉承的团队管理理念，在产品和项目的实践过程中，箭速迭代的好处就是能最快地进行试错，从而才能最快地找到正确的道路。在我的产品实践过程中，非常重视技术的实用性，快速地使用成熟好用的技术才能快速地抢占市场。这本书的编写理念和我的产品理念非常契合，希望大家能够从本书中有所得。

——刘春河　赤子城创始人、CEO

这是一本朴实且全面的 Java 服务端研发技术类图书，书中覆盖了应用类服务端研发所用到的各种技术及使用场景，并且恰当地圈定了写作的深度和范围，各块内容介绍精炼又不失完整性，技术覆盖全面又重点突出，能帮助读者快速搭建业务需要的框架和组件，加速业务的开发和落地。

——解爽　北京文云易迅科技有限公司 CEO，
前美国西部数据存储事业部亚太区总裁

我从工作伊始就一直在带领以 Java 作为服务端开发语言的技术团队，曾系统地实践了以 Spring Cloud 作为微服务框架的分布式系统。此前，我一直在感叹业内缺乏一本从实践角度总结的框架性教材。而今天这本书，恰好从 Java 基础、常见组件使用、服务构建、微服务框架介绍与使用乃至 CI/CD 等角度，把一个服务端研发人员会接触的方方面面都系统地进行了讲解，算是了了我的一个心愿。强烈推荐相关从业者阅读此书，通过学习本书可以得到许多 Java 服务端编程的知识，同时可以从作者这里汲取经验。此书是一本不可多得的服务端研发参考书。

——朱清　创始人&CEO @ 链博科技&VENA NETWORK，
SPRING CLOUD 中国社区联合创始人

作者兼历经百战的架构功力与乐于分享的技术情怀于一身，在本书中将分布式微服务框架做了淋漓尽致的阐述。基于多年来对技术的追求和积累，沉淀了非常宝贵的金石之言。学以致用、用以促学、学用相长是本书的写作精髓。强烈推荐给从事分布式开发的程序员、架构师们作为必读书籍细细品鉴阅读。

——王士磊　聚分享平台事业部总经理

出 版 说 明

随着信息科学与技术的迅速发展，人类每时每刻都会面对层出不穷的新技术和新概念。毫无疑问，在节奏越来越快的工作和生活中，人们需要通过阅读和学习大量信息丰富、具备实践指导意义的图书来获取新知识和新技能，从而不断提高自身素质，紧跟信息化时代发展的步伐。

众所周知，在计算机硬件方面，高性价比的解决方案和新型技术的应用一直备受青睐；在软件技术方面，随着计算机软件的规模和复杂性与日俱增，软件技术不断地受到挑战，人们一直在为寻求更先进的软件技术而奋斗不止。目前，计算机和互联网在社会生活中日益普及，掌握计算机网络技术和理论已成为大众的文化需求。由于信息科学与技术在电工、电子、通信、工业控制、智能建筑、工业产品设计与制造等专业领域中已经得到充分、广泛的应用，所以这些专业领域中的研究人员和工程技术人员越来越迫切需要汲取自身领域信息化所带来的新理念和新方法。

针对人们了解和掌握新知识、新技能的热切期待，以及由此促成的人们对语言简洁、内容充实、融合实践经验的图书迫切需要的现状，机械工业出版社适时推出了"信息科学与技术丛书"。这套丛书涉及计算机软件、硬件、网络和工程应用等内容，注重理论与实践的结合，内容实用、层次分明、语言流畅，是信息科学与技术领域专业人员不可或缺的参考书。

目前，信息科学与技术的发展可谓一日千里，机械工业出版社欢迎从事信息技术方面工作的科研人员、工程技术人员积极参与我们的工作，为推进我国的信息化建设做出贡献。

机械工业出版社

前　　言

期望您能把这本书当成在 Java 服务端领域探索的一张微缩地图。

计算机自发明以来至今不到百年时间，但是其发展速度却是超乎想象的。这种快速的发展不仅表现在纯科研的方面，还表现在技术的普及及应用方面。首先，计算机硬件在长达半个多世纪的时间内，遵循摩尔定律[①]的发展规则；其次，计算机软件从最初的纸带打孔编程已经发展到现在的各种高级语言、框架等；最后，在技术应用上，近几年刚刚成熟的移动互联网，已经改变了千千万万人的生活。

计算机如此迅猛的发展速度，当然与广大从业者的不断努力和探索是分不开的。目前计算机的飞速发展仍处于人类的控制之下，还没有出现独立的苗头，但是计算机在某些特定领域已经可以战胜人类了。不久前计算机在围棋上战胜了人类，这其实并没有那么让人吃惊，因为笔者认为在所有条件和规则已知，并且有明确胜负标准（棋类作为代表）的前提下，人类与计算机对局已经毫无还手之力。

计算机现在的发展已经如此健全和强大，对于计算机从业者来说可能并不是一个好消息，因为从业者要学习大量的计算机知识。这也是笔者写作本书想解决的问题：面对那么多知识，应该如何学习以及如何最快地学习。所以本书尽量涵盖了与 Java 相关的语言要点、服务框架、功能组件体系以及其他辅助工具，目的就是通过最精炼的篇幅，讲述某一技术领域最常用的部分，而此部分会让读者快速地理解、接收并运用到实际工作中。

全书共分为五篇，每篇内容如下：

第一篇主要讲解 Java 语言，以及工程构建、代码管理和基本的服务器命令，以这些内容作为本书的起点和基础。

第二篇主要讲解 Spring 框架治理、服务框架 Spring MVC 和 Spring Boot、服务架构的演进以及微服务框架 Spring Cloud。

第三篇主要讲解在服务中使用的数据库、缓存、定时任务、消息队列、全局搜索等功能组件的使用方法和使用场景。

第四篇主要讲解镜像技术的用法，使用镜像技术快速搭建功能环境的服务组件，并且讲解使用 Jenkins 构建工程以及服务部署相关的内容。

第五篇主要讲解在日常工作中，为了提高工作质量和效率所使用的研发工具。

本书采用循序渐进的方式，讲述 Java 服务端研发所涉及的几个领域。希望读者阅读本书后，对相关内容进行实践和总结，从而在脑海中绘制出属于自己的技术版图。

书中包含大量代码，为了避免分散读者的注意力，书中省略了部分重复的和不重要的代码。如果读者想查看完整的代码可以下载本书附带的源代码进行了解。

[①] 当价格不变时，集成电路上可容纳的元器件的数目，约每隔 18~24 个月便会增加一倍，性能也将提升一倍。

编写技术类书籍是一件非常辛苦的事情，与日常研发不同，编写技术类书籍不仅要会用涉及的技术，还要了解其原理，并且要以读者能够理解的方式讲述出来，同时还要保证技术使用的正确性以及描述的准确性。在编写此书的过程中，两位作者一直秉承着实用且精简的原则，经过几轮的代码复查和文档复查才终于结稿。在此特别感谢默默支持着我们的家人，朋友，感谢曾经一起工作奋战过的同事冯剑、侯金砖、尹波，感谢机械工业出版社车忱编辑，感谢曾经支持过我们的所有人。谢谢大家！

目　　录

业界推荐
出版说明
前言

第一篇　基　础　篇

第1章　Java 概要 ········· 2
1.1　Java 环境搭建 ········· 2
    1.1.1　Java 基础环境搭建 ········· 2
    1.1.2　Eclipse 的安装 ········· 3
    1.1.3　第一个 Java 程序 ········· 3
1.2　基本类型与运算 ········· 3
    1.2.1　基本类型概述 ········· 3
    1.2.2　操作符 ········· 6
    1.2.3　类型转换与越界 ········· 10
1.3　流程控制 ········· 12
    1.3.1　If-else ········· 13
    1.3.2　Switch ········· 13
    1.3.3　For ········· 14
    1.3.4　While ········· 14
    1.3.5　break 与 continue ········· 15
    1.3.6　Return ········· 16
1.4　对象 ········· 16
    1.4.1　什么是对象 ········· 17
    1.4.2　方法 ········· 17
    1.4.3　初始化 ········· 19
    1.4.4　This 与 Static ········· 21
    1.4.5　访问权限 ········· 22
    1.4.6　垃圾回收 ········· 22
1.5　继承和多态 ········· 23
    1.5.1　Object ········· 23
    1.5.2　组合 ········· 24
    1.5.3　继承 ········· 25
    1.5.4　多态 ········· 26
    1.5.5　接口 ········· 27
    1.5.6　抽象类 ········· 29
1.6　容器 ········· 29
    1.6.1　数组 ········· 29
    1.6.2　List ········· 31
    1.6.3　Set ········· 32
    1.6.4　Map ········· 34
1.7　泛型 ········· 36
    1.7.1　泛型的基本使用 ········· 36
    1.7.2　通配符 ········· 37
    1.7.3　泛型接口 ········· 38
    1.7.4　自定义泛型 ········· 38
1.8　异常 ········· 39
    1.8.1　运行时异常 ········· 39
    1.8.2　检查性异常 ········· 41
    1.8.3　自定义异常 ········· 42
1.9　I/O ········· 43
    1.9.1　控制台 I/O ········· 43
    1.9.2　查看文件列表 ········· 44
    1.9.3　文件 I/O ········· 45
    1.9.4　序列化 ········· 46
    1.9.5　网络 I/O ········· 48
1.10　并发 ········· 50
    1.10.1　多线程的实现 ········· 50
    1.10.2　线程冲突 ········· 52
    1.10.3　锁 ········· 53
1.11　反射与注解 ········· 54
    1.11.1　反射 ········· 55
    1.11.2　注解 ········· 57
1.12　JUnit ········· 59
    1.12.1　JUnit 的集成 ········· 59
    1.12.2　JUnit 的基本使用 ········· 59

第2章　Maven ········· 62
2.1　Maven 安装和配置 ········· 62
    2.1.1　Maven 环境的搭建 ········· 62
    2.1.2　在 Eclipse 中配置 Maven 的 settings 文件 ········· 63
2.2　Maven 使用 ········· 63

2.2.1 在 Eclipse 中创建第一个 Maven 项目 ················ 63
2.2.2 认识 pom 文件 ················ 64
2.2.3 运行 Maven 项目 ················ 66
2.3 Maven 坐标和依赖 ················ 67
 2.3.1 什么是坐标 ················ 67
 2.3.2 什么是 Maven 依赖 ················ 68
 2.3.3 Maven 依赖的 scope 范围 ················ 69
 2.3.4 Maven 的依赖调解原则 ················ 70
 2.3.5 Maven 仓库使用 ················ 70
2.4 Maven 生命周期和插件 ················ 71
 2.4.1 Maven 生命周期 ················ 71
 2.4.2 Maven 插件 ················ 72
 2.4.3 生命周期与插件的关系 ················ 72
2.5 Maven 聚合和继承 ················ 73
 2.5.1 聚合应用的场景 ················ 73
 2.5.2 Maven 的继承 ················ 74
 2.5.3 Maven 中 dependencyManagement 的使用 ················ 74
 2.5.4 Maven 中的 pluginManagement 的使用 ················ 75

第 3 章 代码管理 ················ 77
3.1 Svn ················ 77
 3.1.1 Svn 客户端的安装 ················ 77
 3.1.2 Svn 基本使用 ················ 77
 3.1.3 Svn 解决冲突 ················ 79
 3.1.4 Svn 分支 ················ 81
3.2 Git ················ 81
 3.2.1 Git 客户端安装 ················ 81
 3.2.2 Git 基本使用 ················ 81
 3.2.3 Git 分支管理 ················ 83
 3.2.4 Git 标签 ················ 83
 3.2.5 在 Git 中配置 SSH ················ 84
 3.2.6 用 Git stash 暂存代码 ················ 85

第 4 章 Linux 命令 ················ 86
4.1 Linux 简介 ················ 86
4.2 Linux 常用命令 ················ 86
4.3 Linux 文件管理 ················ 88
 4.3.1 Linux 文件操作命令 ················ 88
 4.3.2 Linux 文件权限 ················ 89
4.4 Linux 启动服务 ················ 90

第二篇 服务框架篇

第 5 章 Spring ················ 94
5.1 Spring 概述 ················ 94
 5.1.1 核心模块 ················ 94
 5.1.2 预备知识 ················ 94
5.2 构建第一个 Spring 工程 ················ 95
5.3 IoC ················ 97
 5.3.1 IoC 和 DI 基本原理 ················ 97
 5.3.2 IoC 的配置使用 ················ 97
 5.3.3 Bean 定义 ················ 99
 5.3.4 Bean 的作用域 ················ 100
 5.3.5 Bean 的生命周期 ················ 102
 5.3.6 注解实现 IoC ················ 107
 5.3.7 注解的作用域 scope ················ 109
 5.3.8 自动装配 ················ 109
 5.3.9 @Autowired 与@Qualifier ················ 109
5.4 Aop ················ 111
 5.4.1 AOP 的核心概念 ················ 111
 5.4.2 AOP 的代理机制 ················ 112
 5.4.3 基于 Schema 的 AOP 使用 ················ 112
 5.4.4 基于@AspectJ 的 AOP 使用 ················ 115
5.5 集成 Logback ················ 116
 5.5.1 SLF4J 简介 ················ 116
 5.5.2 Logback 概述 ················ 116
 5.5.3 Logback 的集成 ················ 117
 5.5.4 输出日志到文件 ················ 119
5.6 集成 MyBatis ················ 120
 5.6.1 数据准备 ················ 120
 5.6.2 添加 Spring 与 Mybatis 集成相关依赖 ················ 121
 5.6.3 编写相关配置文件 ················ 121
 5.6.4 使用 generator 生成单表增删改查代码 ················ 123

第 6 章 Spring MVC ················ 127
6.1 Spring MVC 概述 ················ 127
 6.1.1 MVC ················ 127
 6.1.2 HTTP 请求处理流程 ················ 127

6.1.3　Servlet 与 Tomcat 的关系·················128
6.1.4　Spring MVC 的执行流程··············128
6.2　构建第一个 Spring MVC 项目······129
6.2.1　添加依赖···130
6.2.2　配置相关文件·································130
6.2.3　基本页面展示·································132
6.3　Spring MVC Restful 实现··················132
6.3.1　REST 概述·····································133
6.3.2　创建 REST 风格的 Controller·····133
6.4　Spring MVC 拦截器························137
6.4.1　拦截器···137
6.4.2　自定义拦截器·································138
6.4.3　拦截器执行规则·····························139
6.5　Spring MVC 异常处理器················141
6.5.1　Spring MVC 异常处理方式·········141
6.5.2　实现自定义异常处理类···············141
6.6　Spring MVC 上传和下载文件········144
6.6.1　MultipartFile 对象·························145
6.6.2　上传文件···145
6.6.3　下载文件···148

第 7 章　Spring Boot ···············150
7.1　构建第一个 Spring Boot 工程········150
7.1.1　IDE 搭建及特性·····························150
7.1.2　工程目录···152
7.2　起步依赖··153
7.3　配置··154
7.3.1　自动配置···154
7.3.2　设置配置值·····································155
7.3.3　配置优先级·····································155
7.3.4　多环境配置·····································156
7.3.5　自定义类的注入·····························157
7.4　使用 Thymeleaf 构建页面···············159
7.4.1　Thymeleaf 基本使用·······················159
7.4.2　添加页面逻辑·································161
7.5　使用 JPA 构建持久化存储···············164
7.5.1　JPA 基本使用··································164
7.5.2　定义 JPA 扩展接口·························167
7.6　Actuator··168
7.6.1　Actuator 的基本使用·······················169
7.6.2　端点的保护·····································170
7.7　部署··171

7.8　参数校验··172
7.8.1　前台完成基本参数校验···············172
7.8.2　前后台配合完成数据校验···········173
7.9　MyBatis 的框架整合及数据校验······176
7.9.1　整合 MyBatis···································176
7.9.2　后台接口请求校验·························181
7.9.3　规范数据返回·································182
7.10　添加日志及记录请求信息············186
7.10.1　添加日志模块·······························186
7.10.2　AOP 实现接口信息打印·············188

第 8 章　服务架构···················190

第 9 章　Spring Cloud············195
9.1　Eureka··195
9.1.1　Eureka 基础使用·····························195
9.1.2　配置服务注册信息·························199
9.1.3　基于 Host 的高可用 Eureka·········201
9.1.4　基于 IP 的高可用 Eureka·············202
9.2　Ribbon 与 Feign···································204
9.2.1　Ribbon··204
9.2.2　Feign···208
9.3　Hystrix 与 Turbine·······························210
9.3.1　Hystrix 基本使用·····························211
9.3.2　Feign 与 Hystrix 结合·····················214
9.3.3　Hystrix 相关配置·····························215
9.3.4　Hystrix 作为限流工具·····················217
9.3.5　Turbine 聚合展示·····························219
9.3.6　Turbine 通过总线聚合信息···········221
9.4　Zuul··223
9.4.1　Zuul 的基本使用·····························223
9.4.2　Zuul 的配置······································225
9.4.3　Filter 基本使用································227
9.4.4　简单的鉴权服务·····························229
9.4.5　Filter 使用其他服务进行鉴权······231
9.4.6　Zuul 的其他使用方法·····················233
9.5　Config··234
9.5.1　配置 Config 服务端·························234
9.5.2　服务通过 Config 获取配置···········237
9.5.3　添加加密···238
9.5.4　通过 Config 服务名读取配置·······239
9.5.5　配置动态刷新·································240

IX

9.5.6 批量刷新配置 240
9.6 Sleuth 与 Zipkin 242
　9.6.1 Sleuth 信息采集 242
　9.6.2 Zipkin 数据聚合展示 243
9.6.3 数据解读 246
9.6.4 通过消息中间件收集信息 246
9.6.5 数据保存 249

第三篇　组　件　篇

第 10 章　MySQL 252
10.1 MySQL 基本介绍和使用场景 252
　10.1.1 MySQL 概述 252
　10.1.2 MySQL 常用存储引擎 252
　10.1.3 MySQL 使用场景 252
10.2 MySQL 基本操作 253
　10.2.1 MySQL 创建和删除数据库 253
　10.2.2 DDL 基本操作 253
　10.2.3 DQL 基本操作 255
　10.2.4 DML 基本操作 255
　10.2.5 DCL 基本操作 256
10.3 事务处理 258
　10.3.1 事务概述 258
　10.3.2 事务处理方法 258
10.4 MyBatis 插入获取主键 261
10.5 MyBatis 多表查询 263
10.6 查询优化 265
　10.6.1 优化查询的方向 265
　10.6.2 EXPLAIN 分析 266
　10.6.3 小结 268
10.7 数据库主从复制原理 268

第 11 章　MongoDB 269
11.1 MongoDB 基本介绍和使用场景 269
　11.1.1 MongoDB 概述 269
　11.1.2 MongoDB 使用场景 270
11.2 MongoDB 基本操作 270
　11.2.1 MongoDB 基本命令 270
　11.2.2 MongoDB 图形化工具 271
11.3 SpringBoot 集成 MongoDB 272
　11.3.1 整合 MongoDB 272
　11.3.2 操作数据 273
　11.3.3 缓存商品详情页面功能 275

第 12 章　Redis 278
12.1 基本的 Redis 操作 278
12.2 Redis 常用命令和可视化工具 279
　12.2.1 Redis 命令 280
　12.2.2 可视化工具 280
12.3 Redis 的五种数据格式的操作 281
　12.3.1 String 操作 281
　12.3.2 List 操作 283
　12.3.3 Hash 操作 284
　12.3.4 Set 操作 285
　12.3.5 ZSet 操作 287
12.4 Redis 事务处理 288
　12.4.1 批量操作 288
　12.4.2 对值进行监控 289
12.5 Redis 分布式锁 290
12.6 Redis 实现秒杀 292

第 13 章　Zookeeper 295
13.1 Zookeeper 介绍 295
13.2 基本操作 296
　13.2.1 Zookeeper 客户端操作 296
　13.2.2 Java 客户端操作 Zookeeper 298
　13.2.3 订阅子节点变化 301
　13.2.4 订阅节点的数据内容变化 303
13.3 服务注册与发现 304
　13.3.1 服务注册 304
　13.3.2 服务发现 306

第 14 章　FastDFS 308
14.1 FastDFS 基本介绍 308
　14.1.1 FastDFS 概述 308
　14.1.2 FastDFS 上传和下载过程 308
14.2 Spring Boot 集成 FastDFS 309
　14.2.1 文件上传 310
　14.2.2 文件下载 314

第 15 章　ElasticSearch 316
15.1 ElasticSearch 基本介绍 316
　15.1.1 ElasticSearch 概述 316
　15.1.2 分片与副本的关系 316

- 15.1.3　ElasticSearch 主要特性 …………… 317
- 15.2　ElasticSearch 基本用法 …………… 317
  - 15.2.1　索引操作 …………… 318
  - 15.2.2　索引映射 mappings …………… 318
  - 15.2.3　ElasticSearch 之 Head 插件 …………… 320
  - 15.2.4　ElasticSearch 中文插件集成 …………… 322
  - 15.2.5　ElasticSearch 中文检索示例 …………… 325
- 15.3　SpringBoot 集成 ElasticSearch …………… 328
  - 15.3.1　整合 ElasticSearch …………… 328
  - 15.3.2　ElasticSearch 操作数据 …………… 329
- 15.4　SpringBoot 集成 Java Rest Client …………… 331

第 16 章　定时任务 …………… 335
- 16.1　Spring Boot 定时任务 …………… 335
  - 16.1.1　单线程定时任务 …………… 335
  - 16.1.2　多线程定时任务 …………… 336
  - 16.1.3　用定时任务实时统计 …………… 337
- 16.2　Cron 配置 …………… 340
- 16.3　ElasticJob 介绍 …………… 341
- 16.4　简单任务 …………… 343
- 16.5　流式任务 …………… 345

第 17 章　RabbitMQ …………… 348
- 17.1　队列传递字符串 …………… 348
  - 17.1.1　消息队列基本配置 …………… 348
  - 17.1.2　发送方配置及使用 …………… 349
  - 17.1.3　接收方配置及使用 …………… 350
  - 17.1.4　多对多实现 …………… 351
- 17.2　队列传递对象 …………… 353
  - 17.2.1　发送方配置及使用 …………… 353
  - 17.2.2　接收方配置及使用 …………… 354
- 17.3　队列传递 Json 数据 …………… 354
  - 17.3.1　发送方配置及使用 …………… 354
  - 17.3.2　接收方配置及使用 …………… 355
- 17.4　Topic 模式 …………… 356
  - 17.4.1　Topic 模式讲解 …………… 356
  - 17.4.2　发送方配置及使用 …………… 357
  - 17.4.3　接收方配置及使用 …………… 358
- 17.5　Fanout 模式 …………… 359
  - 17.5.1　发送方配置及使用 …………… 359
  - 17.5.2　接收方配置及使用 …………… 360

第 18 章　ELK …………… 361
- 18.1　Logstash 使用 …………… 361
  - 18.1.1　Logstash 概要介绍 …………… 362
  - 18.1.2　文件搜集及 ElasticSearch 存储 …………… 362
  - 18.1.3　使用 Json 格式日志 …………… 363
  - 18.1.4　使用 filter 处理数据 …………… 364
- 18.2　Kibana 使用 …………… 365

第四篇　部　署　篇

第 19 章　Docker …………… 370
- 19.1　Docker 基础环境搭建 …………… 370
  - 19.1.1　Docker 环境安装 …………… 370
  - 19.1.2　Docker 环境卸载 …………… 371
  - 19.1.3　镜像加速 …………… 371
- 19.2　Docker 常用命令 …………… 371
  - 19.2.1　针对镜像的命令 …………… 372
  - 19.2.2　针对容器的命令 …………… 373
  - 19.2.3　使用 Dockerfile 创建镜像 …………… 374
- 19.3　Docker 搭建功能组件 …………… 377

第 20 章　项目构建 …………… 380
- 20.1　Jenkins 基本介绍 …………… 380
- 20.2　Jenkins 基本设置 …………… 380
  - 20.2.1　Jenkins 的安装 …………… 380
  - 20.2.2　Jenkins 初次使用配置 …………… 381
  - 20.2.3　Jenkins 环境变量配置 …………… 383
  - 20.2.4　Jenkins 日志级别设置 …………… 384
  - 20.2.5　安装常用插件 …………… 384
- 20.3　构建 Maven 项目 …………… 386
  - 20.3.1　Maven 构建设置 …………… 387
  - 20.3.2　服务的执行 …………… 388
- 20.4　Harbor 镜像管理 …………… 390
  - 20.4.1　Harbor 安装 …………… 390
  - 20.4.2　生成镜像并保存 …………… 391
- 20.5　Rancher 容器管理 …………… 393
  - 20.5.1　Rancher 的安装及主机添加 …………… 393
  - 20.5.2　Rancher 启动单一容器 …………… 394
  - 20.5.3　Rancher 启动批量容器 …………… 396
  - 20.5.4　服务更新 …………… 398

第五篇 工 具 篇

第 21 章 常用工具·····················402
21.1 Swagger·····················402
21.1.1 Swagger 基本配置·············402
21.1.2 使用 Swagger 编写接口文档······403
21.1.3 Swagger 测试演示··············404
21.2 JMeter······················406
21.2.1 JMeter 的环境搭建·············406
21.2.2 测试计划····················407
21.3 ab··························408
21.3.1 压力配置····················409
21.3.2 结果查看····················409
21.4 VisualVM···················410
21.4.1 查看 CPU···················410
21.4.2 查看线程···················412
21.4.3 监控远程服务················413
21.5 JD-GUI·····················414

参考文献··························416

第一篇 基 础 篇

对于一名研发人员，研发语言是最基础的。语言是研发者和计算机打交道的基本工具，通过编写不同语言的代码，然后编译生成可执行文件，这样就完成了编程的最基本工作。一名研发人员每天都通过语言来实现业务逻辑，对语言的了解和精通可以让工作更加得心应手。由于 Java 的兼容性、覆盖度和热度都是语言中较为出色的，所以本书采用 Java 语言作为全书使用的语言。

通过语言的学习，可以操作依赖库进行功能逻辑的研发，但是众多依赖库之间的关系和管理也是一个问题，第 2 章介绍的 Maven 将使依赖工作变得简单，并且可以让工程更加方便地整合起来。

在日常工作中，编码工作通常不是由一个人独立完成的，大部分情况下都是一个研发小组同时编写同一工程内不同模块的代码，特殊情况下会出现不同的人编写同一段代码，这样就对工程的代码管理提出了挑战。每个人编写同一工程内不同的代码，那么代码如何合并？不同的人编写同一段代码，这种冲突如何处理？第 3 章将介绍 Svn 和 Git 这两种代码管理工具，它们可以很好地解决工作中的代码管理问题。

Java 服务通常运行在 Linux 环境下，在工作中使用的功能环境和生产环境的服务器都是基于 Linux 的，所以对 Linux 命令的了解是 Java 研发人员必备的技能。第 4 章将介绍 Linux 的常用命令，并且启动一个 Java 服务。

本篇的内容虽然看似彼此之间关系并不紧密，但确实是一名研发人员每天都在接触的东西。希望读者通过对本篇的阅读，能够有所得，并且应用在日常工作中。

第1章 Java 概要

Java 是一门面向对象的编程语言，它选择性地吸取了 C++语言的优点，并在其基础上丰富了自己的体系。Java 在健壮性、可移植性、安全性等多个方面均有所突破。同时 Java 的单一继承性和引用的概念（无指针）也使语言更易理解。本章通过讲解 Java 中常用的能力使读者能够快速地了解和使用 Java。如果您已经对 Java 有非常深的理解，那么此章也可以作为 Java 核心内容的提炼笔记，需要时可以随手翻阅。

1.1 Java 环境搭建

开始学习 Java，往往需要配置 Java 的基础环境，环境配置好后，就可以编写 Java 代码并编译运行。可以选择用记事本编写代码、用 JDK 编译运行，但是本书推荐直接采用 Eclipse[①]作为 Java 的 IDE 工具编译并运行代码。

1.1.1 Java 基础环境搭建

Java 基础环境搭建简单来说就是下载对应的 JDK，安装后配置对应的环境变量。下面是基础环境搭建的详细流程。

1）首先通过官网下载对应的 JDK 版本，官网地址为 http://www.oracle.com/technetwork/java/javase/downloads/index.html，本书采用统一的 JDK 版本 1.8。下载时注意选择 JDK 的版本和对应的操作系统。

2）下载完成后进行"傻瓜式"安装，安装完成后进行环境变量的配置。在计算机中找到环境变量设置的地方，添加对应的变量名和值，见表 1-1。

表 1-1 Java 环境变量配置

变 量	值
JAVA_HOME	C:\Program Files\Java\jdk1.8.0_144
CLASSPATH	.;%JAVA_HOME%\lib\dt.jar;%JAVA_HOME%\lib\tools.jar;
Path	%JAVA_HOME%\bin;%JAVA_HOME%\jre\bin;

3）JDK 基础环境安装完毕，下面通过命令行检测是否正确安装。在命令行输入 java-version，若输出结果类似图 1-1，则说明基础环境搭建正确。

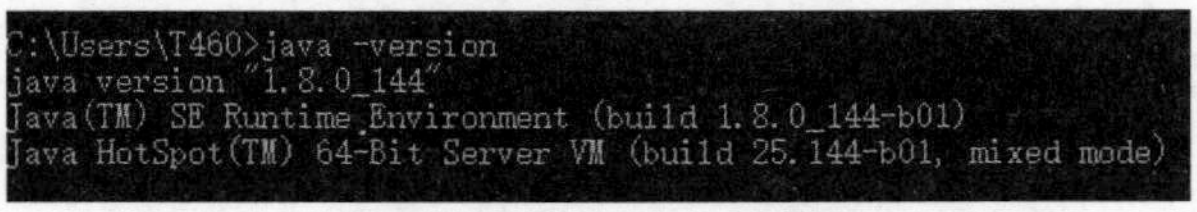

图 1-1 命令行输出结果

[①] Eclipse 不仅仅具备 IDE 能力，但是对于新手来说可以简单地把它理解为 IDE。

1.1.2　Eclipse 的安装

IDE（提供程序开发环境）是每个从事编程工作的人必须接触的工具，一个好的 IDE 能够大大地提高研发效率，Eclipse 就是这样一款开源的工具。本节介绍 Eclipse 工具的安装，安装完成后使用 Eclipse 编写简单的代码，通过控制台输出这段代码的运行结果"Hello World!"。

Eclipse 的安装非常简单，通过如下几步即可完成：

1）通过搜索找到下载地址，或者直接去 Eclipse 的官网 https://www.eclipse.org/downloads/ 进行下载。下载时选择 eclipse-packages，然后选择 Eclipse IDE for Java EE Developers 的正确系统版本即可。

2）下载后的文件为 eclipse-jee-oxygen-2-win32-x86_64.zip，解压此文件到你想安装的目录下，在解压后的文件中找到 eclipse.exe 可执行文件并运行。

3）运行后程序会让你选择工作空间，设定好工作空间文件夹后即可进入程序。

1.1.3　第一个 Java 程序

Java 基础环境和编译运行环境已经准备妥当，下面运行第一个程序"Hello World!"。

1）在 Eclipse 的菜单中选择"File->New->Project->Java Project"。Project name 设置为 HelloWorld，可以设置工程存放路径或者默认，然后单击 finish 按钮。

2）此时会生成一个 HelloWorld 的工程，鼠标右键单击此工程，在快捷菜单中选择"New->Class"，添加 Name 为 HelloWorld，可以设置 Package 为自定义名字（一般为域名的反转）或者直接使用默认名称，然后单击 finish 按钮。

3）现在已经创建了第一个 Java 的类 HelloWorld。在此类中填写如下内容：

```java
public class HelloWorld {
    public static void main(String[] args) {
        System.out.println("Hello World!");
    }
}
```

4）在此文件上用鼠标右键单击，选择"Run AS->Java Application"，可在 Console 窗格中看到输出为"Hello World!"。

以上是一个最简单的 Java 程序。如果在编写代码时经常犯拼写错误，可以设置 Eclipse 代码提示来解决这个问题。把"Window->Preferences->Java->Editor->Content Assist"中的 Auto activation triggers for Java 设置改为 abcdefghijklmnopqrstuvwxyz.即可。这也是直接使用 IDE 的好处之一。

1.2　基本类型与运算

Java 是一种面向对象的编程语言，并且 Java 的单根继承结构导致所有的对象都是由 Object 派生而来，但 Java 中也存在一些特例，即 Java 的基本类型。

1.2.1　基本类型概述

Java 的基本类型包含表示真假的 boolean，表示字符的 char，表示数值的 byte、short、int、long、float、double，表示空的 void。Java 基本类型的取值范围及默认值见表 1-2。

表 1-2 Java 基本类型

基本类型	大小	最小值	最大值	包装器类型	默认值
boolean	–	–	–	Boolean	false
char	16-bit	Unicode 0	Unicode $2^{16}-1$	Character	'\u0000'(null)
byte	8-bit	−128	+127	Byte	byte(0)
short	16-bit	-2^{15}	$+2^{15}-1$	Short	short(0)
int	32-bit	-2^{31}	$+2^{31}-1$	Integer	0
long	64-bit	-2^{63}	$+2^{63}-1$	Long	0L
float	32-bit	IEEE$_{754}$	IEEE$_{754}$	FLoat	0.0f
double	64-bit	IEEE$_{754}$	IEEE$_{754}$	Double	0.0d
void	–	–	–	Void	–

下面针对各个类型，通过程序验证其具体情况。创建一个 JavaBasicTypes 类，编写如下代码：

```java
public class JavaBasicTypes {
    static char charval;
    static byte byteval;
    static short shortval;
    static int intval;
    static long longval;
    static float floatval;
    static double doubleval;

    public static void basicTypesRange() {
        //char 便于展示范围，所以做了类型转换
        System.out.println("char size = " + Character.SIZE);
        System.out.println("char min = " + (int)Character.MIN_VALUE);
        System.out.println("char max = " + (int)Character.MAX_VALUE);
        System.out.println("char default = " + (int)charval);
        //byte
        System.out.println("byte size = " + Byte.SIZE);
        System.out.println("byte min = " + Byte.MIN_VALUE);
        System.out.println("byte max = " + Byte.MAX_VALUE);
        System.out.println("byte default = " + byteval);
        //short
        System.out.println("short size = " + Short.SIZE);
        System.out.println("short min = " + Short.MIN_VALUE);
        System.out.println("short max = " + Short.MAX_VALUE);
        System.out.println("short default = " + shortval);
        //int
        System.out.println("int size = " + Integer.SIZE);
        System.out.println("int min = " + Integer.MIN_VALUE);
        System.out.println("int max = " + Integer.MAX_VALUE);
        System.out.println("int default = " + intval);
        //long
        System.out.println("long size = " + Long.SIZE);
        System.out.println("long min = " + Long.MIN_VALUE);
        System.out.println("long max = " + Long.MAX_VALUE);
        System.out.println("long default = " + longval);
        //float
        System.out.println("float size = " + Float.SIZE);
```

```
            System.out.println("float min = " + Float.MIN_VALUE);
            System.out.println("float max = " + Float.MAX_VALUE);
            System.out.println("float default = " + floatval);
            //double
            System.out.println("double size = " + Double.SIZE);
            System.out.println("double min = " + Double.MIN_VALUE);
            System.out.println("double max = " + Double.MAX_VALUE);
            System.out.println("double default = " + doubleval);
        }

        public static void main(String[] args) {
            basicTypesRange();①
        }
    }
```

运行结果如下：

```
        char size = 16
        char min = 0
        char max = 65535
        char default = 0
        byte size = 8
        byte min = -128
        byte max = 127
        byte default = 0
        short size = 16
        short min = -32768
        short max = 32767
        short default = 0
        int size = 32
        int min = -2147483648
        int max = 2147483647
        int default = 0
        long size = 64
        long min = -9223372036854775808
        long max = 9223372036854775807
        long default = 0
        float size = 32
        float min = 1.4E-45
        float max = 3.4028235E38
        float default = 0.0
        double size = 64
        double min = 4.9E-324
        double max = 1.7976931348623157E308
        double default = 0.0
```

  以上代码使用它们的包装类取出对应的值，此方法是获取基本类型边界值的最快方法。程序的输出结果和预期的一样，7 个基本类型都正确输出了它们的范围值和默认值。这段代码提前使用了静态类和静态成员变量，具体讲解在后面章节会有涉及。另外本例仅用于演示，具体工作中应尽量保证所有值都已经被正确初始化。

---

① 静态方法 basicTypesRange 可以直接被 main 方法调用，具体原理后面会有讲述，本章如非必要不再展示 main 方法的代码。

1.2.2 操作符

1.2.1 节介绍了 Java 基本类型的概念，那么基本类型如何使用呢？基本类型的使用与操作符（运算符）是分不开的。操作符用于进行变量或者对象之间的计算或者关系判断，没有操作符就无法做任何运算、比较或者赋值。操作符主要分为以下几类，分别是算术操作符、赋值操作符、关系操作符、逻辑操作符、位操作符和其他操作符。

（1）算术操作符：包括加号（+）、减号（-）、乘号（*）、除号（/）以及取模操作符（%，除法的余数）、自增和自减运算符（++和--）。二元㊀算术操作符与等号连接使用可以达到简化书写的目的，例如 a+=b㊁表示 a=a+b。代码如下：

```java
public static void testArithmeticOperator() {
    int i = 123;
    int j = 5;
    System.out.println("i + j = " + (i+j));
    System.out.println("i - j = " + (i-j));
    System.out.println("i * j = " + (i*j));
    System.out.println("i / j = " + (i/j));
    System.out.println("i % j = " + (i%j));
    System.out.println("i++ = " + (i++));
    System.out.println("i = " + i);
    System.out.println("++i = " + (++i));
    System.out.println("i = " + i);
    System.out.println("i-- = " + (i--));
    System.out.println("i = " + i);
    System.out.println("--i = " + (--i));
    System.out.println("i = " + i);

    int sum = i + j;
    System.out.println("sum = " + sum);
    i += j;
    System.out.println("i += j =" + i);
}
```

运行结果如下：

```
i + j = 128
i - j = 118
i * j = 615
i / j = 24
i % j = 3
i++ = 123
i = 124
++i = 125
i = 125
i-- = 125
i = 124
--i = 123
i = 123
sum = 128
```

㊀ 算式中出现的数据个数，二元表示操作符处理两个操作数的关系。自增和自减为一元。

㊁ 此种方法虽然书写简单，但是新手没搞清楚之前不建议使用。

i += j =128

由输出结果可见，运用算术操作符，可以进行对应的数学运算。请注意除法对于 int 类型来讲是直接去掉小数点后面数字的，而不是四舍五入；自增自减操作符写在不同的位置得到的结果是不同的；简化的运算符赋值写法会改变左侧变量的值。

（2）赋值操作符：从前面的例子可以看到一个常用的符号"="，它的目的就是把右边的值赋值给左边。有些书把"+="操作符也归入赋值操作符，但是作者认为这仅仅是算术操作符与赋值操作符的一种简化合并写法，列入算术操作符或赋值操作符均可，这里就不再过多介绍。

（3）关系操作符：主要包含 6 种操作符，具体含义见表 1-3。

表 1-3 关系操作符

操 作 符	描 述
==	检查左右两侧操作数是否相等，相等为真
!=	检查左右两侧操作数是否不等，不等为真
>	检查左侧操作数是否大于右侧操作数，大于为真
<	检查左侧操作数是否小于右侧操作数，小于为真
>=	检查左侧操作数是否大于或等于右侧操作数，大于或等于为真
<=	检查左侧操作数是否小于或等于右侧操作数，小于或等于为真

下面通过代码演示关系操作符的使用方法及判断结果。

```
public static void testRelationalOperator() {
    int value = 10;
    System.out.println("value == 10 is " + (value == 10));
    System.out.println("value != 10 is " + (value != 10));
    System.out.println("value != 11 is " + (value != 11));
    System.out.println("value > 9 is " + (value > 9));
    System.out.println("value < 9 is " + (value < 9));
    System.out.println("value >= 10 is " + (value >= 10));
    System.out.println("value <= 8 is " + (value <= 8));
}
```

运行结果如下：

```
value == 10 is true
value != 10 is false
value != 11 is true
value > 9 is true
value < 9 is false
value >= 10 is true
value <= 8 is false
```

可以把关系操作符用于变量之间的比较，本例为了直观直接使用数值进行比较。

（4）逻辑操作符：包含逻辑与操作符"&&"，逻辑或操作符"||"，逻辑非操作符"!"。逻辑与操作符当两侧都为真时为真，逻辑或操作符当两侧有一个为真时为真，逻辑非操作符表示取反。

下面所示代码演示了逻辑操作符的使用方法。逻辑或操作符稍有特殊：当第一个表达式为真时，不再执行第二个表达式，这种情况称为短路。

```
public static void testLogicalOperator() {
```

```
        System.out.println("logical true && true is " + (true && true));
        System.out.println("logical true && false is " + (true && false));
        System.out.println("logical true || true is " + (true || true));
        System.out.println("logical true || false is " + (true || false));
        System.out.println("logical false || false is " + (false || false));
        System.out.println("logical !true is " + (!true));
        System.out.println("logical !false is " + (!false));

        int i = 10, j = 20;
        System.out.println("logical || short circuit is " + ((i++) > 5 || ((j++) > 2)));
        System.out.println("i = " + i + " j = " + j);
    }
```

运行结果如下：

```
logical true && true is true
logical true && false is false
logical true || true is true
logical true || false is true
logical false || false is false
logical !true is false
logical !false is true
logical || short circuit is true
i = 11 j = 20 // j 没有进行自增运算
```

（5）位操作符：用来操作整数基本数据类型中的二进制位。这种用法在实际使用中比较少用。下面以整数类型 int 为例，讲解位操作符的用法。代码如下：

```
    public static void testBitwiseOperator() {
        int i = 15;
        int j = 11;
        System.out.println("i binary is " + Integer.toBinaryString(i));
        System.out.println("j binary is " + Integer.toBinaryString(j));
        System.out.println("i & j = " + (i & j) + ",Binary = " + Integer.toBinaryString(i & j));
        System.out.println("i | j = " + (i | j) + ",Binary = " + Integer.toBinaryString(i | j));
        System.out.println("i ^ j = " + (i ^ j) + ",Binary = " + Integer.toBinaryString(i ^ j));
        System.out.println("~i = " + (~i) + ",Binary = " + Integer.toBinaryString(~i));
        System.out.println("i << 2 = " + (i << 2) + ",Binary = " + Integer.toBinaryString(i << 2));
        System.out.println("i >> 2 = " + (i >> 2) + ",Binary = " + Integer.toBinaryString(i >> 2));
        int k = -1;
        System.out.println("-1 Bianry = " + Integer.toBinaryString(k));
        System.out.println("k << 2 = " + (k << 2) + ",Binary = " + Integer.toBinaryString(k << 2));
        System.out.println("k >> 2 = " + (k >> 2) + ",Binary = " + Integer.toBinaryString(k >> 2));
        System.out.println("k >>> 2 = " + (k >>> 2) + ",Binary = " + Integer.toBinaryString(k >>> 2));
    }
```

运行结果如下：

```
i binary is 1111
j binary is 1011
i & j = 11,Binary = 1011
i | j = 15,Binary = 1111
i ^ j = 4,Binary = 100
~i = -16,Binary = 11111111111111111111111111110000
i << 2 = 60,Binary = 111100
```

```
i >> 2 = 3,Binary = 11
-1 Bianry = 11111111111111111111111111111111
k << 2 = -4,Binary = 11111111111111111111111111111100
k >> 2 = -1,Binary = 11111111111111111111111111111111
k >>> 2 = 1073741823,Binary = 111111111111111111111111111111
```

上面例子中使用的 Integer.toBinaryString()方法，是转化整型数为二进制数的展示。代码中先打印了两个整数的二进制的展示形式，然后通过位操作符对数字进行操作，获取结果。位操作符的含义见表 1-4。

表 1-4 位操作符

位 操 作 符	作 用
&	左右两个操作数按位进行与操作，即都为 1 才为 1
|	左右两个操作数按位进行或操作，即有一个为 1 则为 1
^	左右两个操作数按位进行异或操作，不同则为 1
~	对操作数按位取反
<<	按位向左移动
>>	按位向右移动（左侧用符号位的数字补齐）
>>>	不带符号的向右移动

表中难以理解的地方就是负数的位操作，但是这种操作使用较少，待使用时再详细了解即可。

（6）三元操作符：此操作符较为特殊，因为它有 3 个操作数。简单来讲，此操作符通过第一个操作数的判断条件是否为真，在后面两个操作数中选择一个。

```
Condition？value0：value1；
```

如果 Condition 为真，选择 value0，否则选择 value1。代码如下：

```java
public static void testConditionOperator() {
    System.out.println("condition operator trueCondition = " +
                (true?"conditionTrue":"conditionFalse"));
    System.out.println("condition operator falseCondition = " +
                (false?"conditionTrue":"conditionFalse"));
}
```

运行结果如下：

```
condition operator trueCondition = conditionTrue
condition operator falseCondition = conditionFalse
```

（7）字符串操作符：前面的例子中已经大量使用此操作符，字符串的连接通过操作符"+"或者"+="实现，其他类型和字符串进行"+"操作时会转化为字符串。代码如下：

```java
public static void testStringOperator() {
    String stringValue = "string value ";
    int i = 1,j = 2;
    System.out.println(stringValue + i + j);
    System.out.println(stringValue + (i + j));
    System.out.println(i + j + stringValue);
    stringValue += "add other string ";
    System.out.println(stringValue);
}
```

运行结果如下：

```
string value 12
string value 3
3string value
string value add other string
```

可见，字符串连续与整型进行"+"操作时，由于操作符的结合性，是自左向右连接了数字，而不是进行了数字的加法操作。当用（）把数字括起来后，数字才可以正确相加，这又涉及了操作符的优先级，见表1-5。

表1-5 操作符的优先级和结合性

优先级	操作符	类型	结合性
1	()	括号操作符	由左至右
1	[]	方括号操作符	由左至右
2	!、+（正号）、-（负号）	一元操作符	由右至左
2	~	位操作符	由右至左
2	++、--	自增自减操作符	由右至左
3	*、/、%	算术操作符	由左至右
4	+、-	算术操作符	由左至右
5	<<、>>	位操作符	由左至右
6	>、>=、<、<=	关系操作符	由左至右
7	==、!=	关系操作符	由左至右
8	&	位操作符	由左至右
9	^	位操作符	由左至右
10	|	位操作符	由左至右
11	&&	逻辑操作符	由左至右
12	||	逻辑操作符	由左至右
13	?:	条件操作符	由右至左
14	=	赋值操作符	由右至左

1.2.3 类型转换与越界

表示数值的基本类型之间是可以进行相互转换的，当取值范围比较小的类型向较大类型转换时可以获得更高的精度或者更大的存储空间；当取值范围比较大的类型向较小的类型转换时，往往意味着丢失或者其他问题。boolean 是不能和其他类型进行转换的。下面用几个例子来说明这些问题。

（1）类型转换：以 int 为初始数据类型，赋整数的最大值，然后向其他类型转换。代码如下：

```
public static void testConversion() {
    int i = Integer.MAX_VALUE;
    System.out.println("max int = " + i);
    short j = (short)i;
    System.out.println("max int to short = " + j);
    long k = i;
    System.out.println("max int to long = " + k);
    float x = i;
    System.out.println("max int to float = " + x);
    double y = i;
```

㊀ 本节对操作符优先级的使用不过多介绍，新手如果还不理解优先级的情况请一律使用括号进行明确的操作区域分隔。

```
        System.out.println("max int to double + " + y);
    }
```

运行结果如下：

```
max int = 2147483647
max int to short = -1
max int to long = 2147483647
max int to float = 2.14748365E9
max int to double + 2.147483647E9
```

可见，当 int 转为 short 的时候，存在风险；当 int 转为 long 时一切正常；当 int 转为 float 时数据会丢失一部分；当 int 转为 double 时一切正常。

（2）越界：观察几种类型已经为可表示的最大值时，再进行加运算会发生什么；或者几种类型已经为可表示的最小值时，再进行减运算会发生什么。代码如下：

```
public static void testOutRange() {
    int i = Integer.MAX_VALUE;
    System.out.println("max int = " + i);
    i = i + 1;
    System.out.println("max int + 1 = " + i);

    int j = Integer.MIN_VALUE;
    System.out.println("min int = " + j);
    j = j - 1;
    System.out.println("min int - 1 = " + j);

    double x = Double.MAX_VALUE;
    System.out.println("max double = " + x);
    x = x + Double.MAX_VALUE;
    System.out.println("max double + max double = " + x);

    double y = -Double.MAX_VALUE;⊖
    System.out.println("- max double = " + y);
    y = y - Double.MAX_VALUE;
    System.out.println("- max double - max double = " + y);
}
```

运行结果如下：

```
max int = 2147483647
max int + 1 = -2147483648
min int = -2147483648
min int - 1 = 2147483647
max double = 1.7976931348623157E308
max double + max double = Infinity
- max double = -1.7976931348623157E308
- max double - max double = -Infinity
```

越界的运算往往出现意想不到的结果，虽然实际使用这些类型时，只要正确地分配了对应的类型（例如不要给手机号分配 byte 类型），一般都不会出现这种情况。但是如果代码编写中出现了错误，还是会遇到越界的情况。

⊖ 感兴趣的读者可以了解一下为什么使用-Double.MAX_VALUE 而不使用 Double.MIN_VALUE。

（3）boolean 的使用：boolean 不可以通过其他类型转换而来（与 C++不同），boolean 的值只能是 true 或者 false。代码如下：

```
public static void testBoolean() {
    boolean bool = false;
    int i = 1;
    //bool = (boolean)i;
    bool = (i>=1)?true:false;
    System.out.println("boolean value = " + bool);
}
```

运行结果如下：

```
boolean value = true
```

在上述代码中，注释掉的部分是错误的使用方法，一定要注意。

（4）运算中的转换与赋值：当不同类型同时参与一组运算时，往往伴随着类型转换，而类型的转换都是向上（更大）转换。代码如下：

```
public static void testOperation() {
    int i = 6,j = 5;
    int k = i/j;
    System.out.println("int i/j = " + k);
    double x = i/j;
    System.out.println("double i/j = " + x);
    double y = (double)i/j;
    System.out.println("double (double)i/j = " + y);
    double z = i*1.0/j;
    System.out.println("double i*1.0/j = " + z);
}
```

运行结果如下：

```
int i/j = 1
double i/j = 1.0
double (double)i/j = 1.2
double i*1.0/j = 1.2
```

- 当 int 相除的结果赋给 int 类型时，会去掉小数点后面的数。
- 当 int 相除的结果赋给 double 类型时，其实是先得出 int 的整数值，然后用这个得出的整数值赋给 double 类型，所以还是会丢失数据，但是精度提高了。
- 当 int 相除时进行显式的类型转换，则结果为 double 类型。
- 当以 int 先乘以一个 double 类型的值，此 int 值已经升级为 double 类型，计算结果赋给 double 类型可以得到正确的值。

1.3 流程控制

程序在执行时会出现各种情况，例如上一节通过关系操作符和逻辑操作符得出的结果，会走向不同的程序分支，如何实现分支的选择就属于流程控制。另外程序还会出现不停执行某语句，直到执行条件不成立为止的情况，这也属于流程控制。Java 处理流程控制的关键字和语句包含 if-else、while、do-while、for、return、break、continue、switch。本节讲解以上主要流程控制语句的使用方法。

1.3.1 If-else

if-else 语句主要是根据 if 语句的判断结果，选择不同的分支路径。此语句有几种不同的写法：if 后面可以没有 else 语句；if-else 语句一起使用；或者 else 后面可以再连接一个 if 的判断语句，继续进行条件判断。代码如下：

```java
public static void testIfElse(int num) {
    System.out.println("num = " + num);
    if(num < 10) {
        System.out.println("num < 10");
    }
    
    if(num < 100) {
        System.out.println("num < 100");
    }else {
        System.out.println("num >= 100");
    }
    
    if(num < 50) {
        System.out.println("num < 50");
    }else if(num>=50 && num <100) {
        System.out.println("num>=50 && num<100");
    }else {
        System.out.println("num > 100");
    }
}
```

运行结果如下：

```
num = 51
num < 100
num>=50 && num <100
```

在上面的例子中，传入的参数为 51，可见第一个 if 条件判断不成功，所以对应的代码段没有执行；第二个 if 语句判断成功，所以显示了 num<100；最后，在 else 后面的 if 语句判断成功，所以显示 num>=50 && num <100。

1.3.2 Switch

当使用 if-else 语句时，如果需要判断的条件过多，那么会出现很多个 if-else 语句，这样的代码可读性是很差的，当出现这种情况时推荐使用 switch 语句。switch 语句列出了所有待选条件，当符合条件判断时则执行相应的代码。例如：

```java
public enum⊖ Color {
    RED, GREEN, BLACK, YELLOW
}
public static void testSwitch(Color color) {
    switch (color) {
```

⊖ 枚举类型可以先理解为对几个同类常量值的封装。

```
            case RED:
                System.out.println("color is " + Color.RED);
                break;
            case GREEN:
                System.out.println("color is " + Color.GREEN);
                break;
            case BLACK:
                System.out.println("color is " + Color.BLACK);
                break;
            case YELLOW:
                System.out.println("color is " + Color.YELLOW);
                break;
            default:
                break;
        }
    }
```

当传入参数为 Color.RED 时，输出为：

```
color is RED
```

swtich 语句的主要写法如上所示，用 case 列举各种情况进行匹配，当匹配成功时执行相应的代码段，代码段的后面用 break 结束执行。break 的主要作用就是结束当前的选择语句或者循环语句。如果去掉 case 后面对应的 break 语句，那么代码将继续执行下一个 case 的内容。连续执行的特性在实际工作中会有用处，但是在没有彻底搞清楚之前不建议使用。

1.3.3 For

for 循环其实是依靠三个字段来达成循环的目的，三个字段分别是初始值、结束条件、游标移动。设置一个游标的初始值，每次循环移动游标，达到结束条件时结束循环。例如：

```java
public static void testFor() {
    int[] array = new int[10];
    for(int i=0;i<10;i++) {
        array[i] = i;
    }

    for(int j:array) {
        System.out.print(j+" ");
    }
}
```

运行结果如下：

```
0 1 2 3 4 5 6 7 8 9
```

上例中使用了两种 for 循环的用法，第一种是基本的 for 循环使用方法，用 for 循环实现了数组的赋值。第二种方法是对已有的数据进行遍历，是 for 循环的简单写法。

1.3.4 While

while 也是一种循环控制的方法，while 后面跟随一个判断条件，当条件成立时则执行后面程序段的语句。do-while 方法则是先执行语句，再进行条件判断。例如：

```java
public static void testWhile() {
    int[] array = new int[10];
    int i = 0;
    while(i<array.length) {
        array[i] = i;
        i++;
    }
    
    int j = 0;
    do {
        System.out.print(array[j]+" ");
        j++;
    } while (j<array.length);
}
```

运行结果如下：

```
0 1 2 3 4 5 6 7 8 9
```

for 和 while 都是 Java 进行循环操作的方法，但是写循环时一定要谨慎，除了有目的的无穷循环[○]以外一定要确定循环可以退出，即有结束条件并且可以结束。还有就是循环的嵌套，例如 for 语句中又嵌套了一层 for 语句，一定要确定这种写法不会对程序的执行造成很大的影响，嵌套循环的时间复杂度是两个循环执行次数相乘，应尽量优化这种嵌套写法，例如使用便于查找的容器来替代其中的一层循环等。除非确实必要[○]，尽量不要写 3 层以上嵌套的循环，这种循环会让程序完全失控。

1.3.5 break 与 continue

break 与 continue 在循环中起着重要的作用。break 可以直接退出整个循环，当循环嵌套时，退出 break 所属的循环；continue 可以结束本次循环，进行下次循环。例如：

```java
public static void testBreakAndContinue() {
    int[] array = new int[10];
    for(int i=0;i<10;i++) {
        array[i] = i;
    }
    
    for(int j:array) {
        if(j == 3) {
            continue;
        }
        if(j == 6) {
            break;
        }
        System.out.print(j+" ");
    }
}
```

[○] 无穷循环：for(;;)和 while(true)这两种写法。
[○] 除非任何方式都不能实现目标逻辑，否则都不要写 3 层及以上的嵌套。

运行结果如下：

```
0 1 2 4 5
```

上面的代码对前面的 for 循环的例子进行修改，在打印时设置了条件判断，当 j 为 3 时，直接进行下次循环，所以 3 没有打印出来；当 j 为 6 时，直接退出整个循环，所以 6 以后的数字没有打印。

1.3.6 Return

return 语句可以退出当前的方法，并且可以带出返回值；如果一个 void 返回值的方法没有写 return，那么在方法的结尾有一个隐式的 return。return 语句后面的代码段都不会执行，但是有一个例外——finally。例如：

```java
public static void testReturn(int num) {
    System.out.println("testReturn start*******");
    if(num == 1) {
        return;
    }else if(num == 2) {
        try {
            System.out.println("testReturn try *******");
            return;
        } finally {
            System.out.println("testReturn finally*******");
        }
    }
    System.out.println("testReturn end*******");
}

public static void main(String[] args) {
    testReturn(2);
}
```

在 main 方法中传不同参数的输出分别为：

```
0:
testReturn start*******
testReturn end*******
1:
testReturn start*******
2:
testReturn start*******
testReturn try *******
testReturn finally*******
```

return 语句的执行就代表一个方法的结束。return 语句后面可以跟随一个变量用于返回一个值，例如 return 0；至于 finally 这个特例会在 1.8 节讲述，这里仅作为演示。

1.4 对象

Java 是一种面向对象的语言，什么是面向对象以及如何使用对象是本节要介绍的内容。

1.4.1 什么是对象

什么是对象？试想身边常用的任何物品，拿正在使用的手机举例。把手机比喻成对象，那么手机的硬件例如 CPU、显示屏、电池就是对象里的字段；打电话、使用 app、上网等就是对象里的方法。面向对象的核心其实就是把任何事物抽象为类，这个事物具备的能力就是抽象出来的方法，这个事物具备的各个实际物品就是抽象出来的字段。下面以学生为例，编写一个学生类并创建它的实例[一]。

```java
public class Student {
    private int age;
    private String name;

    public int getAge() {
        return age;
    }
    public void setAge(int age) {
        this.age = age;
    }
    public String getName() {
        return name;
    }
    public void setName(String name) {
        this.name = name;
    }
}
```

观察上面的代码，这个类名叫 Student（Java 的 public 的类名必须和文件名相同）。这个类从学生这个群体中抽象出来两个字段，一个是 age（年龄），一个是 name（名字）。可以通过 get 或者 set 方法对字段进行获取和设置操作，例如 getAge()方法得到学生的年龄。下面根据这个抽象出来的类，创建第一个实体（实例）。

```java
public static void main(String[] args) {
    Student student = new Student();
}
```

通过 new 关键字，可以创建某个类的实例。这样就完成了 Java 面向对象最基本的抽象和实例创建的过程。其中类是抽象，new 是创建此类型单个实例个体。

1.4.2 方法

前面代码中已经大量使用了方法，读者对方法的使用应该也有一个初步的了解。方法主要包含 4 个内容，按照顺序分别是：返回值、方法名、参数、方法体。也可以用其他关键字来修饰一个方法，以达到其他能力，例如方法的可见范围和静态[二]。

普通方法的调用格式是 Object.fun(arg);。下面编写代码对上一节创建的实例进行方法的调用。

```java
public static void main(String[] args) {
    Student student = new Student();
```

㊀ 可以叫作创建一个类的实例，或者创建一个类的对象。
㊁ public 等关键字可以修饰方法的可见范围，static 可以设置方法为静态。

```
        student.setAge(12);
        student.setName("xiaoming");
        System.out.println("student age = " + student.getAge());
        System.out.println("student name = " + student.getName());;
    }
```

运行结果如下：

```
student age = 12
student name = xiaoming
```

在代码中已经演示了创建对象以及方法的调用，为了揭示对象更多的特性，需要再创建一个类 School。具体代码如下：

```
import java.util.ArrayList;
import java.util.List;

public class School {
    private String address;
    private String name;
    List<Student> stList = new ArrayList<Student>();

    public String getAddress() {
        return address;
    }
    public void setAddress(String address) {
        this.address = address;
    }
    public String getName() {
        return name;
    }
    public void setName(String name) {
        this.name = name;
    }
    public List<Student> getStList() {
        return stList;
    }
    public void setStList(List<Student> stList) {
        this.stList = stList;
    }
}
```

上面的代码中用到了 import 关键字，它的作用是引用其他类，本例中它引用了 List 容器类[一]。以后在代码中使用其他的类时，也需要用此关键字引入。

代码中的 School 类是对学校的抽象，包含的字段有地址、名字以及学生列表。可以实现新的方法，用于把学生添加到学校的学生列表中。具体代码如下：

```
    public void addStudent(int age,String name) {
        Student student = new Student();
        student.setAge(age);
        student.setName(name);
        addStudent(student);
```

[一] List 类是一种容器类，用于存放列表。

```
    }
    public void addStudent(Student student) {
        stList.add(student);
```

以上两个方法负责把学生添加到学生列表,下面使用这两个方法向学校中添加学生。

```
    public static void main(String[] args) {
        School school = new School();
        school.setAddress("beijing");
        school.setName("qinghua");
        school.addStudent(18, "xiaoming");
        Student student = new Student();
        student.setAge(19);
        student.setName("daming");
        school.addStudent(student);
        System.out.println(school.getStList().size());
    }
```

通过上面的代码可见,两个相同名字的方法都可以用于把学生放入学生列表,这就是 Java 的方法重载机制。方法重载就是同名方法,但是方法的参数数量或者类型不同⊖。

在代码中会发现一个问题,每次创建一个学生对象的时候,总是要调用两次 set 方法用于设置学生的字段属性,有没有什么办法能够方便地创建对象呢?

1.4.3 初始化

对象的初始化是通过构造器实现的,构造器就是与类名相同并且没有返回值的那个方法。如果一个类没有明确地编写构造器,那么编译器会默认生成一个构造器。构造器可以有多个,每个构造器的参数列表不同。下面针对 Student 类编写它的构造器。

```
    public Student(int age,String name) {
        this.age = age;
        this.name = name;
    }

    public Student() {
        this.age = 0;
        this.name = "todo";
    }
```

如果只写上面带参数的构造器,那么之前编写的代码无法正确编译,因为默认的构造器没有自动生成,而之前的代码都是通过默认构造器创建的对象,所以必须添加无参数的构造器。添加 Student 的构造器后,可以修改 School 类的 addStudent 方法为:

```
    public void addStudent(int age,String name) {
        Student student = new Student(age,name);
        addStudent(student);
        /*Student student = new Student();
```

⊖ 方法返回值不同不能形成重载。

```
            student.setAge(age);
            student.setName(name);*/
        //addStudent(student);
    }
```

这种调用构造器创建对象的方法会使代码更简洁，同时也保证程序的健壮，否则对象中的字段没有正确初始化，会在不可知的地方发生问题。同时也注意到，之前的代码用/*、*/和//框起来了，/*和*/可以使其中间的代码失效，//可以使它后面的代码失效。通常用它们注释无用代码或者添加说明性文字。

除了构造器的方式，也有其他的方法初始化字段的值，例如直接在声明字段的时候赋值，或者通过初始化代码块来进行赋值。下面演示这两种写法。

```
public class Student {
    private int age = 18;
    private String name = "todo";

    public Student() {
        //this.age = 0;
        //this.name = "todo";
    }
    …⊖
    public static void main(String[] args) {
        Student student = new Student();
        //student.setAge(12);
        //student.setName("xiaoming");
        System.out.println("student age = " + student.getAge());
        System.out.println("student name = " + student.getName());;
    }
}
```

运行结果如下：

```
student age = 18
student name = todo
```

以上截取了部分 Student 类修改后的写法，由输出可见字段被设置了初始值。下面添加初始化块，再执行 main 方法观察字段值最后的输出结果。

```
public class Student {
    private int age = 18;
    private String name = "todo";

    {
        age = 20;
        name = "Construct";
    }
    …
}
```

运行结果如下：

⊖ 本书在代码段中可能会包含"…"这样的内容，这是表示在书中省略了部分已介绍代码或者 get、set 方法，如果要查看"…"表示的实际内容，可以查看本书附带的工程源码。

```
student age = 20
student name = Construct
```

最后的输出结果是初始化块中赋值的结果，这里涉及了一个初始化顺序的问题，最后赋值的结果才会被打印出来，所以可知初始化块的执行时间晚于变量声明时赋值的时间。

在目前已经介绍的内容中，初始化顺序先是变量声明时的赋值，然后是初始化块，最后是构造器。对于一个普通的类，这个顺序是一定的，但是当引入静态和继承之后，会在这个顺序的基础上插入其他步骤。

1.4.4 This 与 Static

在上面的例子中，使用 set 方法设置实例某个字段的值，方法里的语句是 this.name = name;。这里 this 的作用就是指代调用这个方法的实例，这句话的意思就是把调用的实例的 name 字段的值设为传入的参数的值。普通成员方法都是默认有 this 的，其意义就是指代调用的实例。

static 修饰的方法称为静态方法。静态方法与普通成员方法不同，静态方法里没有 this，所以它不能指代调用的实例，或者说静态方法不关心是哪个实例调用它，它只对所属的类负责。下面对 Student 类进行改造，统计创建的学生的人数。

```java
public class Student {
    private int age = 18;
    private String name = "todo";

    private static int count = 0;

    public static int getCount() {
        return count;
    }
    …
    public Student(int age,String name) {
        count++;
        this.age = age;
        this.name = name;
    }

    public Student() {
        count++;
        this.age = 0;
        this.name = "todo";
    }

    public static void main(String[] args) {
        for(int i=0;i<10;i++) {
            Student student = new Student();
        }
        System.out.println("student count = " + Student.getCount());
    }
}
```

运行结果如下：

```
student count = 10
```

这里仅展示和静态相关的内容，添加了一个静态的变量 count 和静态方法 getCount()。

每次调用构造器时把计数变量自增。这样在 main 方法中创建了 10 个对象后，计数器会正确记录创建实例的个数。注意静态方法的调用方式，是通过类名调用的，而不是通过实例调用的。

对于静态和非静态的区别，只要记住，静态变量是属于类的，一个类仅有一份[⊖]；非静态变量是属于实例的，每个实例一份；静态方法是没有 this 的，所以不能像普通方法那样调用，静态方法不能直接调用类里的非静态变量和方法，因为非静态的变量和方法需要 this。

静态的引入会对类的构造顺序造成影响，当第一次使用某个类时，会先初始化类的静态变量然后执行静态初始化块，之后才会按照上面所讲的类的构造顺序进行构造。当创建此类的第二个实例时则不会执行静态的构造，因为静态的数据只构造一次。

1.4.5 访问权限

Java 的访问权限分为 4 种，分别是公开访问权限、保护访问权限、包访问权限、私有权限。这四种权限的写法和使用范围见表 1-6。

表 1-6 访问权限

权限名称	关键字	权限范围	用 法
公开访问权限	public	所有都可访问	一些希望别人使用的方法或者公开的 API
保护访问权限	protected	派生子类可用	不希望所有人都可以使用，但是希望此方法子类可以使用或者更改
包访问权限	（default）	默认访问权限，没有关键字，同一包内可以访问	限于同一包内，仅希望同一个包里的其他类可以使用它
私有访问权限	private	仅自己类内部可以使用	类里的完全私有方法，包含类的对象都不可以调用，一般用于实现类的私有能力，并不对外开放

关于权限的使用不再举例，前面的例子中已经大量使用了权限设置。一般来讲注意以下几点即可：不要把字段设为 public 权限，要分配正确的 get 和 set 方法；不要把所有方法都设为 public，要适当地对外暴露方法；除非明确包访问权限的用意，否则一定要合理分配一个显式的权限设置；protected 权限需要明确类里的继承关系时再使用，否则这个权限等同于 private。

1.4.6 垃圾回收

前面用大量的篇幅讲解了如何创建一个实例以及如何使用这个实例。但是这些实例使用完之后去了哪里呢？这个问题对于 C++语言来说是必须解决的，如果用 C++语言创建完对象后置之不理，整个程序肯定会崩溃。但是这个问题对于 Java 来讲就没有那么重要了，Java 有一套自动回收的机制用于处理创建出来的实例，而本书所涉及的内容基本都不需要手动清理垃圾，实际使用中真正需要手动清理的地方也很少，所以就不再赘述，有兴趣的读者可以查阅关于 Java 清理的相关资料。

⊖ 设计模式中的单例模式就是通过静态实现的。

1.5 继承和多态

继承是指派生类继承基类的属性和某些行为，多态是指派生类在基类的基础上进行重写从而表现出来的不同性状。本节从基础的 Object 和类似的组合讲起，一点点了解继承和多态的原理和用法。

1.5.1 Object

前面编写了好多代码用来创建类对象以及调用相应的方法，但是类不仅仅是这些内容。观察下面代码的输出，了解类和对象的其他特性。

```java
public class Person {
    public long id;
    public String name;

    public Person(long id,String name) {
        this.id = id;
        this.name = name;
    }

    public static void main(String[] args) {
        Person person = new Person(1, "xiaoming");
        System.out.println(person.toString());
    }
}
```

运行结果如下：

```
com.javadevmap.Person@7852e922
```

这里使用了 public 的字段，是为了使代码看起来简单一些，实际项目中不这样使用。在这个例子中，创建了一个 person 对象，然后调用了 toString()方法，但是类里并没有这个方法，这个方法是从哪里来的呢？

这就涉及 Java 的单根继承结构。在 Java 中，所有的类都继承自 Object 类。也就是说除了基本类型，其他类都是一种 Object。而 toString 方法就是 Object 里的方法，通过类的继承而来。这种单根继承结构也为 Java 的内存回收提供了很大的便利。

继承听起来很费解，举个例子。例如常用的手机是一种物质，看不见的原子也是一种物质，那么把这些东西的通用性全部抽离出来，用物质这个统称来代替它们是可以的。对于 Java 语言，这个统称就是 Object，所有物质包含的属性，例如大小，重量就相当于 Object 里的字段或方法。而继承就是在这个统称之上再进行细分，从而凸显自己的特性。

再回到代码中，toString 方法其实是把类的内容转化为 String 进行输出，但好像并没有得到期望输出的内容⊖。是否可以通过某种办法输出期望的数据？代码如下：

```java
@Override
public String toString() {
    return "id = " + this.id + " name = " + this.name;
}
```

⊖ toString 的默认输出这里不再介绍，请查阅相关资料。

运行结果如下：

```
id = 1 name = xiaoming
```

可以在 Person 类中重写这个方法，用于替换 Object 的默认 toString 实现，从而达到正确输出的目的。

Object 还有一个 equals()方法，用于对象的比较。通过下面的代码演示这个方法的使用。首先不重写此方法，观察输出的结果。

```
public static void main(String[] args) {
    Person person1 = new Person(1, "xiaoming");
    Person person2 = new Person(1, "xiaoming");
    System.out.println("person 1 == person 2 = " + (person1 == person2));
    System.out.println("person 1 equals person 2 = " + (person1.equals(person2)));
}
```

运行结果如下：

```
person 1 == person 2 = false
person 1 equals person 2 = false
```

虽然代码中给创建的两个对象赋的值是相同的，但是无论用"=="比较还是用 equals 比较，比较的结果都是不同的。下面重写 equals 方法，再执行程序观察输出的结果。

```
@Override
public boolean equals(Object obj) {
    if(obj == null) {
        return false;
    }
    Person person = (Person)obj;
    if((this.id == person.id) && (this.name==person.name)) {
        return true;
    }
    return false;
}
```

运行结果如下：

```
person 1 == person 2 = false
person 1 equals person 2 = true
```

通过重写 equals 方法，对对象里的字段值进行比较，字段值相同即两个对象相同，最后两个对象比较的结果是 true。那么为什么"=="比较的结果还是 false 呢？其实"=="比较的是对象的地址，两个对象地址不同所以不同，equals 默认的方法比较的也是对象的地址，需要覆盖 equals 的默认实现才能正确进行比较。

1.5.2 组合

在了解继承之前，先弄清楚什么是组合。正如前面章节所讲，手机相对于物质来讲，属于继承关系，它继承了物质这个大概念下的一些属性；手机相对于触摸屏来讲，属于包含关系，这个包含叫作组合。在之前的 Person 类中写一个内部类⊖Eyes，添加一个 Eyes 的实例为 person

---

⊖ 内部类是指在一个类中、方法中或者表达式中等非文件最外层范围定义的类，普通内部类由于定义在其他类的内部作用域中，所以其含有一些特性，例如可以访问外层类的字段等，大部分内部类是为了简化代码实现等目的才使用的；静态内部类比较特殊，它的能力没有被外部类所限制，类似于外部类的能力。内部类的内容也比较多，建议读者自行了解。

的字段。

```java
public class Person {
    public long id;
    public String name;
    public Eyes eyes = new Eyes();

    public static class Eyes {
        public String left = "zuoyan";
        public String right = "youyan";
    }
    …
}
```

在这个例子中,用 Person 包含了一个静态内部类的实例,当然 Person 也可以包含非内部类的实例,这里这样写是为了展示方便。Person 是对一个事物的抽象,但是这个事物是由很多部分组成的,每个部分也可以抽象出来,最后在 Person 中组合在一起。

在实际项目中组合的应用是非常多的,当组装业务逻辑的时候,常常会把数据库的操作类实例、对外服务调用的操作类实例等和业务相关的类实例组合在一起以实现业务逻辑。

1.5.3 继承

如前面所讲,继承是在同一种共性基础上的细分和丰富。在基类中定义此类事物的共性,在派生类中对基类中定义的共性进行具体的实现或者修改,并且添加自己的特性。下面通过动物的例子来说明这个问题。

```java
public class Animal {
    public int weight;

    public Animal(int weight) {
        this.weight = weight;
    }

    public void move() {
        System.out.println("animal can move!");
    }
}

public class Tiger extends Animal{
    public String roar = "ao";

    public Tiger(int weight,String roar) {
        super(weight);
        this.roar = roar;
    }

    @Override
    public void move() {
        System.out.println("tiger can run!");
    }
}
```

代码中先定义了动物的基类 Animal，在基类中定义了动物共有的属性 weight 和方法 move()。然后在派生类 Tiger 中通过 extends 关键字声明此类继承了 Animal 基类，添加 Tiger 自己的特性 roar，并且重写⊖了基类的 move()方法。在派生类的构造器中，通过 super 传递需要基类构造的属性。

```
    public static void main(String[] args) {
        Tiger tiger = new Tiger(500, "ao!");
        tiger.move();
        System.out.println("tiger weight = " + tiger.weight + " tiger roar = " + tiger.roar);

        Animal animal = new Animal(1000);
        animal.move();
        System.out.println("animal weight = " + animal.weight);
    }
```

运行结果如下：

```
tiger can run!
tiger weight = 500 tiger roar = ao!
animal can move!
animal weight = 1000
```

上面代码中创建了 Tiger 类的实际对象，其不仅包含了新定义的字段，还包含了基类的字段，并且 move()方法具有不同的表现。

引入了继承后，一个对象的构造顺序又变得更加复杂了，当创建一个派生类对象的时候，原则是先构造此派生类的基类部分，再构造派生类新定义的部分。那么设想一个问题，当派生类和基类中都包含静态字段时，整个的构造顺序是什么样的呢？希望读者自己动手，设计一个方法，验证这种情况的构造顺序。毕竟编程不是背概念，自己动手才能丰衣足食。

1.5.4 多态

说到多态就不得不说动态绑定，动态绑定是指在执行时判断所作用对象的实际类型。多态的实现基于动态绑定，是指用基类的引用指向派生类的实例，当调用方法时再确定是应该调用基类的方法还是派生类的方法。基于上面的例子，再添加一个派生类 Fish 来说明这个问题。

```
    public class Fish extends Animal{
        public String livein = "water";

        public Fish(int weight,String livein) {
            super(weight);
            this.livein = livein;
        }

        @Override
        public void move() {
            System.out.println("fish can swim!");
        }
    }
```

⊖ 父类定义的方法仅仅定义了一类事物的通性，但子类的特性没有很好地表现，这种情况下子类一般重写这个方法以突显子类的特性，此方法会覆盖父类的方法。

派生类 Fish 继承自基类 Animal，并且重写了 move()方法。下面分别创建 Animal、Tiger、Fish 这 3 个类的实例，然后把它们加入同一个数组，最后用 Animal 的类型分别引用这 3 个实例进行操作并观察其结果。

```
public static void main(String[] args) {
    Animal animal = new Animal(1000);
    Tiger tiger = new Tiger(500, "ao!");
    Fish fish = new Fish(10, "water");

    Animal[] animals = new Animal[3];
    animals[0] = animal;
    animals[1] = tiger;
    animals[2] = fish;

    for(Animal temp:animals) {
        System.out.println("animal weight = " + temp.weight);
        temp.move();
    }
}
```

运行结果如下：

```
animal weight = 1000
animal can move!
animal weight = 500
tiger can run!
animal weight = 10
fish can swim!
```

在这段代码的 for 循环中，都是用 Animal 类型的引用指代数组中的实例，但是在调用 move()方法时却有不同的表现，这就是多态。多态是用基类指代派生类，在实际调用时调用派生类的实现。通过基类的引用可以调用在基类中定义的字段，例如 weight，但是不能使用在派生类中添加的字段。

1.5.5 接口

设想一种情况，当把一组 Tiger 类的实例放入一个容器（List）中，希望按照每个实例的重量从小到大进行排序，应该怎么办？一个标准的容器是有 sort()排序方法的，排序需要基于大小的比较，这个比较方法需要通过 Tiger 类来提供。但是容器都是通用的，在比较的时候容器需要一个通用的引用来指代具体的对象，而这种引用又不适合于使用 Object，因为如果那样的话，Object 将会非常庞大并且需要对非常多的这种需求提供基础方法。因此需要有一种机制对能力进行说明，并且实现对象通用能力的引用，这就是接口[○]。

由于 Java 的单根继承结构，所以没法像 C++一样通过多重继承来引入更多的能力；而一些能力又不能全部封装进 Object 基类；一些通用的容器却又需要一种通用的引用来指代不同的类；在这种情况下，通过接口来封装一些通用的能力，具体的类继承接口并且实现这种能力，通用的容器就可以用这个接口引用具体的类实例进行比较了。下面提前透露容器的内容，来看看容器的排序是怎么实现的。

○ 接口类不可以实例化，并且它不会影响 Java 的单根继承结构。

```java
public class Tiger extends Animal implements Comparable<Tiger>{
    public String roar = "ao";

    public Tiger(int weight,String roar) {
        super(weight);
        this.roar = roar;
    }

    @Override
    public void move() {
        System.out.println("tiger can run!");
    }

    @Override
    public String toString() {
        return "the tiger weight is " + weight;
    }

    @Override
    public int compareTo(Tiger o) {
        if(this.weight < o.weight) {
            return -1;
        }else if(this.weight == o.weight) {
            return 0;
        }else {
            return 1;
        }
    }

    public static void main(String[] args) {
        Tiger tiger1 = new Tiger(498, "ao!");
        Tiger tiger2 = new Tiger(430, "ao!");
        Tiger tiger3 = new Tiger(500, "ao!");
        Tiger tiger4 = new Tiger(488, "ao!");
        Tiger tiger5 = new Tiger(590, "ao!");
        List<Tiger> list = new ArrayList<Tiger>();
        list.add(tiger1);
        list.add(tiger2);
        list.add(tiger3);
        list.add(tiger4);
        list.add(tiger5);
        System.out.println("sort before " + list);
        Collections.sort(list);
        System.out.println("sort after " + list);
    }
}
```

运行结果如下：

```
sort before [the tiger weight is 498, the tiger weight is 430, the tiger weight is 500, the tiger weight is 488, the tiger weight is 590]
sort after [the tiger weight is 430, the tiger weight is 488, the tiger weight is 498, the tiger weight is 500, the tiger weight is 590]
```

Tiger 类实现了接口 Comparable 中的 compareTo()方法，此方法会允许继承此接口的类定义实例间的比较方式，这样当把一组实例放入容器后，通过 Collections.sort(list)即可快速排序。

1.5.6 抽象类

抽象类简单来讲，就是不可创建实例的基类。在之前的例子中，如果在 Animal 类的 class 前面添加一个 abstract 关键字，会发现之前创建 Animal 实例的地方都报错了。抽象类的定义取决于程序的设计，设计者希望 Animal 类作为基类可以指代派生出来的所有动物，同时不希望创建一个毫无意义的仅仅叫作动物的实例（而不知道具体的类型），这种情况下，会把基类定义为抽象类。

在抽象类中，可以定义抽象方法，就是在方法的前面添加 abstract，并且没有方法的实现。抽象方法要求派生类必须实现此方法。例如修改之前的 Animal 类，观察它的派生类的变化。

```java
public abstract class Animal {
    public int weight;

    public Animal(int weight) {
        this.weight = weight;
    }

    public void move() {
        System.out.println("animal can move!");
    }

    public abstract void eat();
}
```

在基类中添加抽象方法 eat()，如果不修改它的派生类，派生类会报错，要求实现此方法。是否使用抽象类还需要在实际项目中根据具体情况进行选择。

1.6 容器

容器是存放对象的地方，当大量的对象需要在内存中存在，并且单个对象分别使用很不方便的时候，就是容器应用的场景了。Java 存放数据的方式有很多种，例如固定大小的数据以及可以自动调整大小的容器类。而容器类经过 Java 多个版本的迭代，继承关系较为复杂，有些容器已经建议废弃，目前比较常用的就是 List、Set、Map，本节将一一使用它们并且了解它们的基础用法。在介绍容器类之前先看看数组。

1.6.1 数组

数组相对于容器类，效率更高[⊖]，但缺点也很明显，在生命周期内不可改变数组大小。数组有 length 字段，用于访问数组的大小。"[]"语法可以访问组数成员。数组的创建也有多种方式，例如用 new 创建或者直接填写数组元素。数组还有多维的能力，可以创建二维以上的数组。下面分别演示数组的这些用法。

⊖ 如果追求极限速度，数组确实效率更高，如果只是一般的应用，其实效率的体现并没有特别重要的意义。

（1）一维数组

```java
public class JavaArray {
    public static void testDirectConArray() {
        String[] strings = {"lilei","hanmeimei","lucy"};
        System.out.println("array length = " + strings.length);
        System.out.println(Arrays.asList(strings));
    }

    public static void testNewConArray() {
        String[] strings = new String[5];
        strings[0] = "A";
        strings[1] = "B";
        strings[2] = "C";
        strings[3] = "D";
        System.out.println("array length = " + strings.length);
        System.out.println(Arrays.asList(strings));
    }

    public static void main(String[] args) {
        testDirectConArray();
        testNewConArray();
    }
}
```

运行结果如下：

```
array length = 3
[lilei, hanmeimei, lucy]
array length = 5
[A, B, C, D, null]
```

代码中使用了两个方法，每个方法使用不同的方式创建一维数组。在第一种方法中，采用直接赋值的方式初始化数组，并且在打印时使用 Arrays.asList()方法将数组转化为 List 进行打印。如果不使用此方法，可以采用直接打印的方式打印数组，看看结果是否如期望的那样，并考虑一下为什么。在第二种方法中采用 new 来创建数组空间，并且逐个赋值。方法中创建的数组空间是 5 个，而实际只赋值了 4 个，但是打印时还是打印出了第五个空元素。

（2）二维数组

```java
public static void testTdimArray() {
    String[][] strings = {{"one","two","three"},{"four","five","six"},{"seven","eight","nine"}};
    for(String[] tempStrings:strings) {
        System.out.println(Arrays.asList(tempStrings));
    }
}

public static void testNewTidmArray() {
    String[][] strings = new String[2][];
    strings[0] = new String[2];
    strings[1] = new String[4];
    strings[0][0] = "up";
    strings[0][1] = "down";
    strings[1][0] = "east";
```

```
            strings[1][1] = "south";
            strings[1][2] = "west";
            strings[1][3] = "north";
            for(String[] tempStrings:strings) {
                System.out.println(Arrays.asList(tempStrings));
            }
        }

        public static void main(String[] args) {
//            testDirectConArray();
//            testNewConArray();
            testTdimArray();
            testNewTidmArray();
        }
```

运行结果如下：

```
[one, two, three]
[four, five, six]
[seven, eight, nine]
[up, down]
[east, south, west, north]
```

代码中使用两种方式创建二维数组，二维数组和一维数组的使用没有太大分别，只是多加了一个维度；嵌套的数组大小可以保持统一或者自定义不同的大小。

数组的使用和功能简单，虽然有效率高的优点，但是一般的业务逻辑很难体现其优势，通常情况下一般使用容器类来代替数组的使用。

1.6.2 List

容器 List 其实就是一个列表，但是 Java 对列表的实现分为两种，一种是类似数组的实现 ArrayList，一种是链表的实现 LinkedList。这两种 List 都可以通过 List 类进行引用并且调用方法，只是由于内部实现的不同存在性能上的差异，ArrayList 在插入方面不如 LinkedList，LinkedList 在获取列表中的值方面性能不如 ArrayList。可以设计实验方法来检验这两种 List 的性能差别，这里就不过多介绍了，仅介绍 List 的基本用法。

（1）ArrayList

```
public static void testArrayList() {
    List<String> list = new ArrayList<String>();
    list.add("one");
    list.add("two");
    list.add("three");
    list.add("four");
    System.out.println(list);
    System.out.println(list.get(3));
    list.remove("four");
    System.out.println(list);
    System.out.println("list contains one is " + list.contains("one"));
    list.add("five");
    list.set(3, "four");
    System.out.println(list);
    System.out.println("list index of two is " + list.indexOf("two"));
```

```
        System.out.println("sub list is " + list.subList(1, 3));
        if(!list.isEmpty()) {
            Object[] strings = list.toArray();
            System.out.println("array length is " + strings.length);
        }
    }
```

运行结果如下：

```
[one, two, three, four]
four
[one, two, three]
list contains one is true
[one, two, three, four]
list index of two is 1
sub list is [two, three]
array length is 4
```

在上面的例子中，使用了常用的 List 方法，添加、删除、包含判断、设置值、查询索引、生成子 List、转为数组等。List 的使用还有很多其他方法，大家可以查看类文档进行了解。List 的遍历可以用 foreach 的形式，也可以用迭代器的形式，下面代码演示 LinkedList 和迭代器如何配合使用。

（2）LinkedList

```
public static void testLinkedList() {
    String[] strings = {"one","two","three","four","five","six","seven","eight","nine","ten"};
    List<String> list = new LinkedList<String>();
    list.addAll(Arrays.asList(strings));
    Iterator<String> it = list.iterator();
    while(it.hasNext()) {
        String string = it.next();
        if(string=="three" || string=="six" || string=="nine") {
            it.remove();
        }
    }
    System.out.println(list);
}
```

运行结果如下：

```
[one, two, four, five, seven, eight, ten]
```

这个例子运用迭代器对 List 进行遍历，在遍历的过程中根据业务需要对数据进行处理，删除 List 中不需要的内容，很好地利用了 LinkedList 方便增删数据的特性。

1.6.3 Set

Set 是一个集合，它不保证存取的顺序[⊖]，它的主要特性就是存储值的唯一性，重复的添加操作对 Set 无用，集合中只会存储一份数据。要判断存储的对象是否相等，可使用 equals 和 hashCode 方法。本节主要介绍两种 Set，分别是 HashSet 和 TreeSet。

⊖ 如果必须保证存储的顺序，则有额外的开销，例如使用 LinkedHashSet，它用一个链表来维护顺序。

（1）HashSet

```java
public class Person {
    public long id;
    public String name;
    …
    @Override
    public boolean equals(Object obj) {
        if(obj == null) {
            return false;
        }
        Person person = (Person)obj;
        if((this.id == person.id) && (this.name==person.name)) {
            return true;
        }
        return false;
    }

    @Override
    public int hashCode() {
        return (int) this.id;
    }
}

public class JavaSet {
    public static void testSet() {
        Set set = new HashSet<>();
        set.add(new Person(1, "lilei"));
        set.add(new Person(2, "hanmeimei"));
        set.add(new Person(3, "lucy"));
        set.add(new Person(3, "lucy"));
        System.out.println(set);
        set.remove(new Person(2, "hanmeimei"));
        System.out.println(set);
        Iterator iterator = set.iterator();
        while (iterator.hasNext()) {
            Object object = iterator.next();
            if(object instanceof Person) {
                System.out.println("it is a person and " + object);
            }
        }
    }

    public static void main(String[] args) {
        testSet();
    }
}
```

运行结果如下：

```
[id = 1 name = lilei, id = 2 name = hanmeimei, id = 3 name = lucy]
[id = 1 name = lilei, id = 3 name = lucy]
it is a person and id = 1 name = lilei
it is a person and id = 3 name = lucy
```

对之前实现的 Person 类添加 hashCode 方法。使用 HashSet 作为集合容器，向集合中添加 Person 的实例。HashSet 通过 Hash 算法保证对象的快速读取，通过 hashCode 和 equals 方法保证对象不会重复，所以当重复添加对象的时候 Set 中并没有出现重复的内容。Set 包含的方法很多，例如 contains()就是一个常用方法，用于判断一个集合是否包含某个对象。读者可以在编写代码时了解 Set 其他方法的具体用法。

这个例子中使用 Set 的写法和之前的 List 写法并不相同，Set 后面的<>符号中没有添加具体的类型，这种情况下当遍历集合中的对象时，并不知道集合中的具体对象类型，所以需要使用 instanceof 动态判断对象的类型（当然也可以使用强制转换类型）；而如果使用在<>中添加类型的写法则不用进行这种判断，这种写法称为泛型，后面的章节会有介绍。

（2）TreeSet

```java
public static void testTreeSet() {
    TreeSet<Person> set = new TreeSet<>(new Comparator<Person>() {
        @Override
        public int compare(Person o1, Person o2) {
            if(o1.id < o2.id) {
                return -1;
            }else if(o1.id == o2.id) {
                return 0;
            }else {
                return 1;
            }
        }
    });
    set.add(new Person(40, "xiaoming"));
    set.add(new Person(29, "xiaoming"));
    set.add(new Person(41, "xiaoming"));
    set.add(new Person(32, "xiaoming"));
    set.add(new Person(50, "xiaoming"));
    set.add(new Person(37, "xiaoming"));
    System.out.println(set);
}
```

运行结果如下：

> [id = 29 name = xiaoming, id = 32 name = xiaoming, id = 37 name = xiaoming, id = 40 name = xiaoming, id = 41 name = xiaoming, id = 50 name = xiaoming]

TreeSet 是一种可排序的 Set，代码中没有采用之前的让对象实现接口从而排序的方法，而直接采用对 Set 设置比较方法来进行排序，这两种方法都可以实现排序的能力。在这里用 Person 作为泛型的类型，所以在内部类中可以直接进行对象的比较而不用进行类型转换。方法最后输出了一个有序的集合数据。

1.6.4 Map

Map 是通过键值对存储的，可以通过键来获取值。HashMap 是最常用的 Map，本节以它为例讲解 Map 的原理和实现。HashMap 通过散列的形式，以达到快速存取和空间控制的目的。以手机号为例，用手机号对 10000 取余，那么所有手机号就散列了 10000 个分组，分别是从 0 到 9999，这种散列的基础就是 hashCode 方法。散列后手机号映射到的分组值会重复，要把这些散列后重复的数据保存到某一分组中就用到了链表，在链表中要正确地取值就需要 equals 方法作为对象的比较依据。这就是作为 HashMap 的 Key 值的类为什么必须实现 hashCode 和 equals 这两个方法的原因，见表 1-7。

表 1-7 散列情况

0	138xxxx0000->139xxxx0000->137xxxx0000
1	138xxxx0001->139xxxx0001->137xxxx0001
2	138xxxx0002->139xxxx0002->137xxxx0002
…	…
9997	…
9998	…
9999	138xxxx9999->139xxxx9999->137xxxx9999

代码如下：

```java
public static void testHashMap() {
    HashMap<Person, String> map = new HashMap<>();
    map.put(new Person(1,"xiaoming"), "Musician");
    map.put(new Person(1,"xiaoming"), "Musician");
    map.put(new Person(2,"daming"), "Scientist");
    map.put(new Person(3,"xiaobai"), "Astronaut");
    System.out.println(map);

    for(Map.Entry<Person, String> entry : map.entrySet()) {
        System.out.println("key = " + entry.getKey() + " value = " + entry.getValue());
    }

    for(Person person : map.keySet()) {
        System.out.println(person);
    }

    for(String string : map.values()) {
        System.out.println(string);
    }

    Iterator<Map.Entry<Person, String>> its = map.entrySet().iterator();
    while (its.hasNext()) {
        Map.Entry<Person, String> entry = its.next();
        System.out.println("key = " + entry.getKey() + " value = " + entry.getValue());
    }
}
```

运行结果如下：

```
{id = 1 name = xiaoming=Musician, id = 2 name = daming=Scientist, id = 3 name = xiaobai=Astronaut}
key = id = 1 name = xiaoming value = Musician
key = id = 2 name = daming value = Scientist
key = id = 3 name = xiaobai value = Astronaut
id = 1 name = xiaoming
id = 2 name = daming
```

```
id = 3 name = xiaobai
Musician
Scientist
Astronaut
key = id = 1 name = xiaoming value = Musician
key = id = 2 name = daming value = Scientist
key = id = 3 name = xiaobai value = Astronaut
```

在上面的例子中，构建了一个 HashMap，key 值是之前经常使用的 Person 类对象，当然可以构建 key 值是基本类型包装器类的对象，具体使用什么作为 Key 需要在实际项目中进行判断，这里仅作为演示。代码中提供了几种遍历 HashMap 的方法，包含全量遍历 HashMap、只遍历 Key 和只遍历 Value。一些其他方法这里就不过多介绍，读者可查看相关文档进行了解。

本节已经演示了常用容器的使用方法并介绍了容器的不同特性，在实际的项目中需要根据业务的要求和容器的特性选择合适的容器。容器性能问题一般不会对业务造成太多困扰，除非特殊的业务逻辑，一般都不会遇到容器性能瓶颈。

1.7 泛型

正如前面例子所写，其实泛型最常见的使用场景就是在容器内，容器提供了存储对象的通用能力，其他所有类型的对象都可以放入容器之内，声明容器时，用具体的类型标明容器中使用的类型即可，这就是泛型的基本使用。

1.7.1 泛型的基本使用

泛型的基本使用前面已经有所涉及，在下面的例子中将创建一个继承结构，然后用基类的类型来声明容器，看看容器是否表现正常。并且创建一个泛型方法，观察其对类型的处理。

```java
class Fruit{
    public void print() {
        System.out.println("It is a Fruit!");
    }
}

class Apple extends Fruit{
    public void print() {
        System.out.println("It is an Apple!");
    }
}

class Orange extends Fruit{
    public void print() {
        System.out.println("It is an Orange!");
    }
}

public class TempleteTypeErase {

    public static <T> void print(List<T> list) {
        for(T t:list) {
            if( t instanceof Apple) {
```

```java
                    System.out.println("It is an Apple!");
                }else if( t instanceof Orange ) {
                    System.out.println("It is an Orange!");
                }else if( t instanceof Fruit ) {
                    System.out.println("It is an Fruit!");
                }
            }
        }

        public static void main(String[] args) {
            List<Fruit> fruits = new ArrayList<Fruit>();
            fruits.add(new Fruit());
            fruits.add(new Apple());
            fruits.add(new Orange());
            for(Fruit fruit : fruits) {
                fruit.print();
            }

            print(fruits);
        }
    }
```

运行结果如下：

```
It is a Fruit!
It is an Apple!
It is an Orange!
It is an Fruit!
It is an Apple!
It is an Orange!
```

使用泛型的容器，很好地保存了对象，并且取出对象后仍具备多态性，可见泛型容器的使用非常简单。代码中使用了一个静态方法 print，这个方法也是通过泛型定义的，在方法内不知道具体的数据类型㊀，所以通过动态类型检查来确定类型。

1.7.2 通配符

Fruit 是 Apple 类的基类，那么 List<Fruit>和 List<Apple>之间是什么关系呢？看看下面的例子。

```java
    public static void testExtendWildcard() {
        List<Apple> apples = new ArrayList<>();
        apples.add(new Apple());
        //List<Fruit> fruits = apples;
        List<? extends Fruit> fruits2 = apples;
        //fruits2.add(new Apple());
        fruits2.get(0).print();
    }

    public static void testSuperWildcard() {
        List<Fruit> fruits = new ArrayList<>();
        fruits.add(new Fruit());
        List<? super Apple> apples = fruits;
```

㊀ 在方法内部类型被擦除了，所以无法知道 T 到底是什么类型。

```
            apples.add(new Apple());
            ((Fruit)apples.get(0)).print();
        }
```

　　通过代码发现，List<Fruit>和 List<Apple>之间并没有关系，尤其需要注意的是，不要以为它们是继承关系。如果想用另一个容器指代 List<Apple>，就用到了通配符，可以用 List<? extends Fruit>来指代 List<Apple>。此用法大多作为方法的参数判断，例如某方法参数需要一个 Fruit 子类的容器。另外还需注意，无法通过 List<? extends Fruit>向容器中添加数据，只能获取数据。原因就是这个容器指代了一切继承自 Fruit 的类的容器，所以无法确定是否正确地向容器中添加了适当的类型。例如当? Extend Fruit 指代一个 Apple 容器时，如果向容器中添加 Orange 对象就会出现错误。

　　List<? super Apple>指代了 Fruit 至 Apple 及派生类继承关系链的对象，可以向其添加对象，因为添加的对象类型更加明确，可以通过类型转换成基类 Fruit 来获取和使用对象。

1.7.3　泛型接口

　　Java 的泛型区别于 C++的泛型实现主要在类型擦除。类型擦除就是说 Java 泛型只在编译期间进行静态类型检查，编译器生成的代码会擦除相应的类型信息，这样到了运行期间实际上 JVM 根本就不知道泛型所代表的具体类型。也就是说在运行期间无法识别 List<String>和 List<Integer>，因为它们都是 List<Object>。但是也不是毫无办法，可以定义一个泛型接口，然后让泛型类声明传进的具体类型必须实现此接口，这样还能保留部分接口能力。

```java
public interface PrintInterface<T> {
    public void print();
}
```

　　以上是声明的泛型接口，这个接口希望保留打印的能力。其实最常用的泛型接口是 Comparable<T>，它保留了对象比较的能力。

1.7.4　自定义泛型

　　声明了泛型接口，下面的代码将演示泛型接口在泛型类中是如何使用的。

```java
class Tomato implements PrintInterface<Tomato>{
    @Override
    public void print() {
        System.out.println("It is Tomato!");
    }
}

public class CustomTemplete<T extends PrintInterface<T>> {
    public T data;

    public void print() {
        data.print();
    }

    public static void main(String[] args) {
        CustomTemplete<Tomato> customTemplete = new CustomTemplete<>();
        customTemplete.data = new Tomato();
```

```
            customTemplete.print();
        }
    }
```

运行结果如下：

```
It is Tomato!
```

创建了一个继承自泛型接口的 Tomato 类，并且重写了接口方法用来打印数据。泛型类 CustomTemplete 通过<T extends PrintInterface<T>>表示，传入的类型必须继承此泛型接口。所以可以在 main 方法中调用泛型类的打印，如果不继承此接口，那么 data 对象在泛型内部并不知道自己具备什么能力（仅具备 Object 对象能力）。

以上介绍了泛型的常用方法和一些与其他语言不同的地方，同时还在 Java 泛型擦除的现实下保留了部分接口能力。

1.8 异常

对于使用 Java 编写的程序，编译器在编译的时候会进行语法检查等工作。但是有一些程序中存在的问题是编译阶段无法识别的，例如用 Java 实现了一个计算器，当用户输入除数为 0 的情况下，怎么办？这就是异常处理存在的原因。这些错误会导致程序无法继续进行，而异常处理就是处理这些错误的。

1.8.1 运行时异常

运行时异常是程序在执行过程中出现错误的调用而抛出的异常，这种异常都可以在编写时避免，而编译器也不要求对可能抛出运行时异常的代码段强制加上 try 语句。下面看几种常见的运行时异常。

```
public class JavaRuntimeException {
    public static void testDivisor() {
        int i = 6/0;
    }

    public static void main(String[] args) {
        testDivisor();
    }
}
```

运行结果如下：

```
Exception in thread "main" java.lang.ArithmeticException: / by zero
    at com.javadevmap.exception.JavaRuntimeException.testDivisor(JavaRuntimeException.java:5)
    at com.javadevmap.exception.JavaRuntimeException.main(JavaRuntimeException.java:9)
```

用一个整数除以 0 在算术上是明显错误的，但是这个问题编译器目前是不会报错的，只会在执行时抛出一个异常，异常包含错误的类型和代码的位置，可以很容易地找到出问题的地方并且优化。下面用异常捕获来处理这段代码。

```
public static void testDivisor() {
    try {
        int i = 6/0;
```

```
            System.out.println("i = " + i);①
        } catch (Exception e) {
            System.out.println("divisor can not be 0");
        }
    }
```

运行结果如下:

```
divisor can not be 0
```

例子中用 try/catch 进行了代码运行和异常捕获,try 语句块的意思是执行代码,catch 语句块的意思是对 try 中的代码异常进行处理。当然这里只作为演示,实际项目中一般还是先用 if 语句判断除数是否为 0,为 0 则直接进行提示或者其他的处理,而不用异常捕获来进行处理,这样就避免了运行时异常。

下面代码演示空引用异常:

```
public static void testNullPoint() {
    Person person = null;
    System.out.println(person.id);
}
```

运行结果如下:

```
Exception in thread "main" java.lang.NullPointerException
    at com.javadevmap.exception.JavaRuntimeException.testNullPoint(JavaRuntimeException.java:17)
    at com.javadevmap.exception.JavaRuntimeException.main(JavaRuntimeException.java:22)
```

这种空异常的避免办法一般也是在调用前对不确定是否已经初始化的对象进行非空判断,从而避免这种异常。

下面代码演示常见的 List 异常。

```
public static void testArrayRemove() {
    List<String> list = new ArrayList<>();
    list.add("one");
    list.add("two");
    list.add("three");
    int index = 0;
    for(String string : list) {
        System.out.println(string);
        index++;
        if(index==1) {
            list.remove(index);
        }
    }
}
```

运行结果如下:

```
one
Exception in thread "main" java.util.ConcurrentModificationException
    at java.util.ArrayList$Itr.checkForComodification(Unknown Source)
    at java.util.ArrayList$Itr.next(Unknown Source)
```

---

① 异常发生后,后面的语句不会执行。

```
    at com.javadevmap.exception.JavaRuntimeException.testArrayRemove(JavaRuntimeException.java:48)
    at com.javadevmap.exception.JavaRuntimeException.main(JavaRuntimeException.java:63)
```

这种情况，在 for 循环遍历过程中移除 List 元素是非常危险的，如何避免请参看 1.6 节。Java 的这种运行时异常有很多种，例如数组越界异常、类型转换异常等。异常的处理不是背出来的，在实际的代码中去解决异常才是最快的学习方法，本书附带的代码中包含了其他异常的几种情况，读者可以尝试模拟、处理和避免运行时异常。

1.8.2　检查性异常

运行时异常基本都可以避免，只要代码足够严谨就不会出现运行时异常。所以真正的代码中要处理的是检查性异常。这就涉及异常的抛出和捕获，在抛出异常的地方使用 throw 关键字抛出；在抛出异常的方法后面添加 "throws 异常类名"。异常捕获的地方使用 try-catch-finally。下面看一个读取文件的例子。

```java
public static String readFile(){
    boolean bool = true;
    StringBuilder builder = new StringBuilder();
    try {
        FileReader fReader = new FileReader("d:\\test.txt");
        char[] cs = new char[10];
        while (fReader.read(cs)!=-1) {
            builder.append(cs);
            cs = new char[10];
        }
        fReader.close();
    } catch (Exception e) {
        bool = false;
        e.printStackTrace();
    } finally {
        if(bool) {
            System.out.println("read file ok!");
        }else {
            System.out.println("read file fail!");
            builder.replace(0, builder.length(), "fail");
        }
    }
    return builder.toString();
}

public static void main(String[] args) {
    System.out.println(readFile());
}
```

运行结果如下：

```
read file ok!
1234567890
23456789
end
```

在这个例子中，使用文件读写类 FileReader 读取一个文件，在创建这个类的时候，编译器强制要求程序员必须把这个方法放入一段 try 语句中，或者使整个方法向外抛出异常（添加 throws 语句）由外层进行处理，否则编译不通过；这是区别于运行时异常的地方，运行时异常允许编译通过。

如果在文件的路径下没有找到对应的文件，则会抛出一个文件不存在的异常，这个异常会被 catch 捕获，并由 e.printStackTrace()方法打印到控制台。finally 语句是一定要执行的，它根据读取文件过程中判断是否发生异常来识别文件是否读取成功，如果发生异常则会把返回的字符串置为 fail，用于标明失败。在这个例子中连接字符串使用了 StringBuilder 类，如果读者感兴趣可以研究一下它的特性。

1.8.3 自定义异常

在实际编程中，例如数据库中的数据出现了业务逻辑上的错误等情况，希望通过抛出一个异常把问题暴露出来，而已有的异常类型不能说明问题的原意，所以需要自定义一个异常。自定义异常非常简单，只要继承相关的异常类就可以了。代码如下。

```java
public class CustomRuntimeException extends RuntimeException{
}
public class CustomException extends Exception{
}
public class CustomExceptionDemo {
    public static void testRuntimeException (){
        throw new CustomRuntimeException();
    }

    public static void testException() throws CustomException{
        throw new CustomException();
    }

    public static void main(String[] args) {
        try {
            testException();
        } catch (CustomException e) {
            System.out.println("catch CustomException");
        } catch (Exception e) {
            System.out.println("catch Exception");
        }
    }
}
```

运行结果如下：

```
catch CustomException
```

在上面的例子中，自定义了两个异常，但是两个异常的继承关系不一样。一个继承了 RuntimeException，另一个继承了 Exception，这两种继承关系会导致在实际使用中存在区别。继承自 RuntimeException 的异常，当用 throw 抛出的时候，包含它的方法不需要用 throws 声明要抛出异常，继承自 Exception 的异常则需要在方法上明确声明抛出异常类型。用 catch 异常捕获时，是按照顺序从第一个匹配的异常类型进行捕获的，一般都会把异常的基类 Exception 放到顺序的最后，防止它拦截了其他的捕获。

异常还包含其他一些方法可供使用，但是对于简单情况，只需要继承一个异常，并且用类的名字来区分异常类型就可以了。

1.9 I/O

使用之前讲解的 Java 内容，已经可以实现在程序内的很多业务处理能力了，但是在实际的业务中存在大量的交互和通信的需求，这就需要对 I/O 有相应的了解。I/O 使 Java 程序可以和控制台、文件、其他 Java 服务、数据库、缓存等各个组件进行信息的互通，有了 I/O 才能把程序组成一个庞大的系统，否则只能是一个个程序计算的孤岛。虽然很多组件都封装了简单易用的 I/O 操作，但是了解一些 Java 的基本 I/O 还是对研发者的工作有好处的。

Java 的 I/O 主要包含两种流，分别是字节流和字符流。字节流分为读入（InputStream）和输出（OutputStream）；字符流分为读入（Reader）和输出（Writer）。两者选择的根据是针对 Unicode⊖字符处理能力的，如果不需要 Unicode 基本都可以选择字节流。

1.9.1 控制台 I/O

在 IDE 中负责控制台输入输出的就是 Console 窗口，下面通过此窗口输入数据，然后通过两种不同的 I/O 流分别读取此数据，观察两种不同 I/O 的读取结果。代码如下：

```java
public static void testConsoleStreamIO() {
    try {
        char c;
        InputStream in = System.in;
        do {
            c = (char) in.read();
            System.out.println(c);
        } while (c!='q');
    } catch (Exception e) {
        System.err.println("catch Exception");
    }
}

public static void testConsoleBufferIO() {
    try {
        char c;
        BufferedReader in = new BufferedReader(new InputStreamReader(System.in));⊖
        do {
            c = (char) in.read();
            System.out.println(c);
        } while (c!='q');
    } catch (Exception e) {
        System.err.println("catch Exception");
    }
}
```

第一种方法直接获取系统的字节输入流，读取流数据分别显示到控制台；第二种方法用字

⊖ Unicode 是为了解决传统的字符编码方案的局限而产生的，它为每种语言中的每个字符设定了统一并且唯一的二进制编码，以满足跨语言、跨平台进行文本转换、处理的要求。

⊖ Java 5 之后，可以使用 Scanner 来读取输入。

符流封装了系统输入流，然后读取数据显示至控制台；使用这两种方法，当读取全英文时没有分别，当读取汉字时，使用字符流的方式能够准确输出汉字，而字节流则不能。这就是字节流和字符流最明显的区别。所以当读取二进制文件时，例如音频、图片等使用字节流是合适的方式，当读取汉字时使用字符流是合适的方式。

1.9.2 查看文件列表

File 类是 Java 对文件和目录进行操作的类，可以用它对文件进行创建、改名、删除等操作。File 类的使用相对简单，而且对应的 API 也很健全，只要正确使用即可。下面以遍历目录下的文件列表为例，简单介绍 File 类的使用。代码如下：

```java
public class FileListDemo {
    public static List<String> getFileListByDir(File dir){
        List<String> list = new ArrayList<>();
        for(File item : dir.listFiles()) {
            if(item.isDirectory()) {
                list.addAll(getFileListByDir(item));
            }else {
                list.add(item.getName());
            }
        }
        return list;
    }

    public static List<String> getFileList(String uri) {
        List<String> list = new ArrayList<>();
        File file = new File(uri);
        if(file.isDirectory()) {
            list.addAll(getFileListByDir(file));
        }else if(file.exists()&&file.isFile()){
            list.add(uri);
        }else if(!file.exists()) {
            System.out.println("file not found");
        }
        return list;
    }

    public static void main(String[] args) {
        System.out.println(getFileList("D:\\projects\\JavaDeveloperMap\\JavaBasic\\src"));
    }
}
```

运行结果如下：

> [JavaBasicTypes.java, JavaOperator.java, JavaProcessControl.java, JavaArray.java, JavaList.java, JavaMap.java, JavaSet.java, Animal.java, Fish.java, Tiger.java, CustomException.java, CustomExceptionDemo.java, CustomRuntimeException.java, FileExceptionDemo.java, JavaRuntimeException.java, ConsoleIO.java, FileListDemo.java, Person.java, School.java, Student.java]

main 方法中传进一个路径到 getFileList 方法，getFileList 方法会识别这个路径是文件夹还是文件，或者根本不存在。如果是文件夹，则进入 getFileListByDir 递归方法，搜索出路径下所有文件夹下的文件，加入到 List 中，递归完成后，得到一个文件清单。

1.9.3 文件 I/O

文件的读取在 1.8 节已经涉及，所以这里换一种文件的读取方法来演示文件的读写。Java 的 I/O 类比较有意思的地方是你可以通过流之间的包装，在最外层类对象中使用较为方便的功能。

```java
public class FileIO {
    public static void write(String uri) {
        try {
            File file = new File(uri);
            FileOutputStream outputStream = new FileOutputStream(file);
            OutputStreamWriter outputStreamWriter = new OutputStreamWriter(outputStream,"UTF-8");
            outputStreamWriter.write("你好，世界！\nhello world!");
            outputStreamWriter.close();
            outputStream.close();
        } catch (Exception e) {
            e.printStackTrace();
        }
    }

    public static String Read(String uri) {
        StringBuilder builder = new StringBuilder();
        try {
            File file = new File(uri);
            FileInputStream inputStream = new FileInputStream(file);
            InputStreamReader inputStreamReader =
                        new InputStreamReader(inputStream, "UTF-8");
            while (inputStreamReader.ready()) {
                char[] c = new char[128];
                inputStreamReader.read(c);
                builder.append(c);
            }
            inputStreamReader.close();
            inputStream.close();
        } catch (Exception e) {
            e.printStackTrace();
        }
        return builder.toString();
    }

    public static void main(String[] args) {
        write("d:\\test2.txt");
        System.out.println(Read("d:\\test2.txt"));
    }
}
```

运行结果如下：

```
你好，世界！
hello world!
```

在 write 方法中，通过 File 创建了 FileOutputStream 流，但是 FileOutputStream 无法简单地写入字符串，所以用 OutputStreamWriter 进行了包装，这样就可以把 String 类型写入文件。在 Read 方法中，使用 InputStreamReader 来包装 FileInputStream，从而实现文件内容的读取。

1.9.4 序列化

当把一个 Java 的对象存入文件或者进行网络通信时,需要把一个对象转为一串数据,并可以再反转回一个对象,这就是序列化的需求。Java 序列化有几种方式,例如对象的类继承 Serializable 接口;或者对象的类继承 Externalizable 接口,实现接口的两个方法;或者转换为其他的通用数据交换格式,例如 Json。

(1) Serializable 方法实现序列化

采用此种 Java 序列化方式较为简单,只要使需要序列化的对象类继承此接口,并且保证对象内的字段也是可序列化的,如果存在不可序列化或者无需序列化的字段,可以用 transient 关键字在字段前标注。代码如下:

```java
public class JavaSerialize {
    static public class Address implements Serializable{
        public double longitude;
        public double latitude;
        public String name;
        public transient Person person;
    }

    public static void write(String uri,Address address) {
        try {
            File file = new File(uri);
            FileOutputStream outputStream = new FileOutputStream(file);
            ObjectOutputStream objectOutputStream =
                            new ObjectOutputStream(outputStream);
            objectOutputStream.writeObject(address);
            objectOutputStream.close();
            outputStream.close();
        } catch (Exception e) {
            e.printStackTrace();
        }
    }

    public static Address Read(String uri) {
        Address address = null;
        try {
            File file = new File(uri);
            FileInputStream inputStream = new FileInputStream(file);
            ObjectInputStream objectInputStream = new ObjectInputStream(inputStream);
            address = (Address)objectInputStream.readObject();
            objectInputStream.close();
            inputStream.close();
        } catch (Exception e) {
            e.printStackTrace();
        }
        return address;
    }

    public static void main(String[] args) {
        Address address = new Address();
```

```
            address.latitude = 39.54;
            address.longitude = 116.23;
            address.name = "beijing";
            address.person = new Person(7, "xiaoming");
            write("d:\\test3.txt", address);
            Address result = Read("d:\\test3.txt");
            System.out.println("address name = " + result.name + " longitude = " + result.longitude + " latitude = " + result.latitude + " person = " + result.person);
        }
    }
```

运行结果如下:

```
address name = beijing longitude = 116.23 latitude = 39.54 person = null
```

在此例中，Address 类继承了 Serializable 接口，并且对 person 字段标注了 transient，表示无需序列化。创建 Address 类实例后，用 write 方法把序列化结果存入一个文件，然后用 read 方法从文件中读取数据并且反序列化回一个 Address 实例。

（2）Externalizable 方法实现序列化

此接口包含两个方法，分别是 readExternal 和 writeExternal，可以通过这两个方法完成序列化的定制。重写 Adddress 类，继承自 Externalizable 接口。代码如下：

```
…
static public class Address implements Externalizable{
    public double longitude;
    public double latitude;
    public String name;
    public transient Person person;
    @Override
    public void readExternal(ObjectInput arg0) throws IOException, ClassNotFoundException {
        longitude = arg0.readDouble();
        latitude = arg0.readDouble();
        name = (String)arg0.readObject();
    }
    @Override
    public void writeExternal(ObjectOutput arg0) throws IOException {
        arg0.writeDouble(longitude);
        arg0.writeDouble(latitude);
        arg0.writeObject(name);
    }
}
…
```

通过重写此 Address，最后达成的效果和上一个例子是相同的，但是此种写法给了编写者更大的灵活度，可以在两个方法中修改字段数据或者做其他的事情。在实际业务中具体采用哪种方法还需要根据实际需求来定。

（3）Json

Json 是一种轻量级的数据交换格式，可以把对象序列化为 Json 格式，序列化后会生成一个 Json 格式的字符串。上例中的数据序列化后变为：

```
{
    "latitude": 39.54,
```

```
            "longitude": 116.23,
            "name": "beijing",
            "person": {
                "id": 7,
                "name": "xiaoming"
            }
        }
```

这种格式非常简单易懂，而且方便研发人员直接阅读序列化后的数据，具体如何转换为 Json 格式会在后续章节讲解。

1.9.5 网络 I/O

Java 服务之间可以通过网络进行通信，从而可以实现程序间数据的互通，网络 I/O 是 Java 服务进行微服务化[一]的基础。网络通信一般较为复杂，但本书所涉及的内容一般不用考虑过多的网络部分，网络问题一般都由使用的服务框架解决。所以这里仅作为演示，了解 Java 基本的通信方式。在下面的例子中，用 Socket 套接字使用 TCP[二]协议进行通信，创建两个 Java 程序，分别是客户端程序和服务端程序。

（1）服务端程序

```java
public class NetIOServer extends Thread{
    private ServerSocket serverSocket;

    public NetIOServer(int port) throws Exception{
        serverSocket = new ServerSocket(port);
        //serverSocket.setSoTimeout(20000);
    }

    @Override
    public void run() {
        while (true) {
            try {
                Socket socket = serverSocket.accept();
                socketDialogue(socket);
            } catch (Exception e) {
                e.printStackTrace();
                break;
            }
        }
    }

    public void socketDialogue(Socket socket) {
        Thread thread = new Thread() {
            int count = 0;
            @Override
            public void run() {
                try {
```

[一] 微服务化其实主要是把一个单一的服务，通过能力的区别进行拆分，从而形成一个个模块更小、能力更集中的服务，服务之间通过网络进行通信，后续章节会有详细介绍。

[二] TCP 是网络七层协议中的面向连接的、可靠的、基于字节流的传输层通信协议。

```java
                    while (socket.isConnected()) {
                        DataInputStream inputStream =
                                    new DataInputStream(socket.getInputStream());
                        System.out.println("socket server receive: " + inputStream.readUTF());
                        Thread.sleep(5000);
                        DataOutputStream outputStream =
                                    new DataOutputStream(socket.getOutputStream());
                        outputStream.writeUTF("server say nihao" + count++);
                    }
                    socket.close();
                } catch (Exception e) {
                    e.printStackTrace();
                }
            }
        };
        thread.start();
    }

    public static void main(String[] args) {
        try {
            NetIOServer server = new NetIOServer(18088);
            server.start();
        } catch (Exception e) {
            e.printStackTrace();
        }
    }
}
```

在这段代码中，提前使用了线程，线程的使用在 1.10 节中讲解。在 main 方法中创建了一个 Server 监听线程，通过 ServerSocket 监听某端口号，当有链接请求时通过 accept 方法返回和客户端的连接。通过 Socket 得到客户端传过来的数据，并且延迟 5s 回复客户端一条数据。这个逻辑是写在无限循环中的，会一直监听连接的数据情况并且回复。

（2）客户端程序

```java
public class NetIOClient {
    public static void main(String[] args) {
        int count = 0;
        try {
            Socket socket = new Socket("127.0.0.1", 18088);
            DataOutputStream outputStream =
                        new DataOutputStream(socket.getOutputStream());
            outputStream.writeUTF("client say nihao");
            while (socket.isConnected()) {
                DataInputStream inputStream =
                            new DataInputStream(socket.getInputStream());
                System.out.println("socket client receive: " + inputStream.readUTF());
                Thread.sleep(5000);
                outputStream.writeUTF("client say nihao" + count++);
            }
            socket.close();
        } catch (Exception e) {
            e.printStackTrace();
```

```
            }
        }
    }
```

客户端的逻辑是请求与服务端的连接，连接成功后向服务端发送一条数据；发送完第一条数据后，会一直监听服务器的应答，并且延迟 5s 回复给服务端。所以在这个例子中，客户端和服务端之间会一直通信下去，直到手动结束它们。它们的输出情况为：

服务端输出：

```
socket server receive: client say nihao
socket server receive: client say nihao0
socket server receive: client say nihao1
socket server receive: client say nihao2
...
```

客户端输出：

```
socket client receive: server say nihao0
socket client receive: server say nihao1
socket client receive: server say nihao2
socket client receive: server say nihao3
...
```

以上仅演示了基本的服务间通信，具体项目中的通信情况会更加复杂，好在使用框架可以解决大部分网络问题，让研发人员能够专心完成业务逻辑。如果读者对网络通信很感兴趣，可以研究网络通信的 NIO 框架 Netty，相信会有不少收获。

1.10 并发

一个 Java 程序运行在一个进程[○]中，但是如上面例子所演示，有时希望一个程序可以同时做好多事情，例如监听端口、接收数据、逻辑计算等等，那么只有一个运算单元就明显不够了，所以这时需要启动好多个运算单元，这就是多线程。多个线程的执行其实是抢占 CPU 的时间，但是在感觉上好像在同时进行一样。本节介绍多线程的写法和一些重点。多线程其实是一个比较困难的知识点，尤其对于初次接触的新人来讲有些时候较为费解，对于多线程的学习一定要在实际问题中多多摸索才能真正理解。

1.10.1 多线程的实现

（1）Runnable 任务

```java
public class ThreadRunnable implements Runnable{
    private int start;
    private int end;
    public ThreadRunnable(int start,int end) {
        this.start = start;
        this.end = end;
    }

    @Override
```

○ 进程是程序运行的实体，计算机系统进行资源分配和调度的基本单位。

```java
        public void run() {
            int sum = 0;
            for(int i = start;i<=end;i++) {
                sum+=i;
            }
            System.out.println("thread is " + Thread.currentThread().getName()+ " start = " + start + " end = " + end + " sum = " + sum);
        }

        public static void main(String[] args) {
            ThreadRunnable runnable = new ThreadRunnable(100,1000);
            runnable.run();

            Thread thread = new Thread(new ThreadRunnable(200,2000));
            thread.start();
        }
    }
```

运行结果如下：

```
thread is main start = 100 end = 1000 sum = 495550
thread is Thread-0 start = 200 end = 2000 sum = 1981100
```

ThreadRunnable 是一个继承自 Runnable 的类，如果在主线程中创建这个类，并且调用 run 方法，其实它并没有什么特殊，只是正常执行求和的逻辑。Runnable 对多线程的作用就是可以把它传入一个 Thread 中，作为新建线程的执行任务，这样就实现了多线程。

（2）自定义 Thread

```java
    public class CustomThread extends Thread{
        @Override
        public void run() {
            try {
                Thread.sleep(1000);
            } catch (Exception e) {
                e.printStackTrace();
            }
            System.out.println(this);
        }
        public static void main(String[] args) {
            CustorThread thread1 = new CustorThread();
            thread1.setPriority(Thread.MAX_PRIORITY);
            CustorThread thread2 = new CustorThread();
            thread1.setPriority(Thread.MIN_PRIORITY);
            thread1.start();
            //thread2.setDaemon(true);
            thread2.start();
        }
    }
```

运行结果如下：

```
Thread[Thread-1,5,main]
Thread[Thread-0,1,main]
```

可以让一个类继承自 Thread 类，重写 run 方法来实现线程的任务单元，这样就可以不用把任务单元传给 Thread，只要创建此线程并且调用 start 即可实现多线程。在这个例子中用到了线程的几个方法，sleep 方法用来使线程休眠，setPriority 方法用来设置线程的优先级，setDaemon 方法用来设置后台线程⊖。

（3）线程池

以上两种方法都需要手动创建线程、启动线程，如果使用线程池进行托管，那么就省去了直接操作线程的麻烦，并且线程池中的线程还可以复用，也省去了重复创建线程的开销。代码如下：

```
public static void testCachedPool() {
    ExecutorService eService = Executors.newCachedThreadPool();
    for(int i=0;i<10;i++) {
        eService.execute(new ThreadRunnable(i*100,i*1000));
    }
    eService.shutdown();
}

public static void testFixedPool() {
    ExecutorService eService = Executors.newFixedThreadPool(5);
    for(int i=0;i<10;i++) {
        eService.execute(new ThreadRunnable(i*100,i*1000));
    }
    eService.shutdown();
}
```

以上是两种创建线程池的写法，只要把任务单元传入线程池即可执行多线程运算，而不用手动创建线程。这两种方法的区别就是 newFixedThreadPool 会规定最大线程数。

1.10.2 线程冲突

在上面的几个例子中，都会为每个线程创建独立的任务单元，目前看来执行的情况良好。设想一种情况，如果传入多个线程中的任务单元是相同的，并且使用了同一份数据，那么会发生什么？代码如下：

```
public class ThreadConflict{
    private int sum;

    public int getSum(int start,int end) {
        sum = 0;
        for(int i=start;i<end;i++) {
            sum +=i;
        }
        return sum;
    }

    public static void main(String[] args) {
        ThreadConflict threadConflict = new ThreadConflict();
        System.out.println("main thread sum = " + threadConflict.getSum(0, 1000));

        ExecutorService eService = Executors.newCachedThreadPool();
```

⊖ 当一个程序中所有线程都为后台线程，那么程序会退出。

```
                for(int i=0;i<10;i++) {
                    eService.execute(new Runnable() {
                        @Override
                        public void run() {
                            System.out.println(Thread.currentThread().getName() + " sum = " + threadConflict.getSum(0, 1000));
                        }
                    });
                }
                eService.shutdown();
            }
        }
```

运行结果如下：

```
main thread sum = 499500
pool-1-thread-1 sum = 499500
pool-1-thread-2 sum = 499500
pool-1-thread-3 sum = 499500
pool-1-thread-1 sum = 499500
pool-1-thread-4 sum = 499500
pool-1-thread-2 sum = 499500
pool-1-thread-6 sum = 499500
pool-1-thread-7 sum = 499500
pool-1-thread-5 sum = 327769
pool-1-thread-8 sum = 743833
```

代码中创建了一个 ThreadConflict 的对象，这个对象包含一个 sum 字段和一个求和的方法。把这个对象的执行任务传入线程池进行计算，结果有的线程执行结果是错误的。

Java 的多线程执行是抢占式的，当多个线程同时抢占同一资源进行运算时，有可能线程 A 运算到一半时线程 B 抢占了线程 A 重新开始计算，线程 B 计算完毕线程 A 抢占回资源继续运算，这时就会发生了错误，因为线程 A 抢占回的资源数据已经不是它离开时的数据了。在这种情况下，可以使用锁解决并发导致的资源抢占问题。

1.10.3 锁

在现实世界中锁住某些东西表示独占或者使用中，程序中的锁也是同样的意义。对于多个线程同时访问的单一资源，当前获得执行权限的线程可以把这个资源锁住，执行完毕再把锁打开。下面介绍几种简单的程序加锁方式。

（1）Synchronized 关键字

对上面的任务单元进行修改，改为如下内容，则程序运算可以输出正确的答案。

```
public synchronized int getSumBySyn(int start,int end) {
    sum = 0;
    for(int i=start;i<end;i++) {
        sum +=i;
    }
    return sum;
}
```

Synchronized 关键字把这个方法设置为同步方法，当有多个线程希望使用此方法时，此关键字只允许一个线程占用此方法，其他线程处于等待状态，先入线程执行完毕其他线程再分别单独抢占此方法。

（2）Lock

可以在这个对象中创建一个 ReentrantLock 的实例，对需要加锁的代码段的前面调用 lock 方法上锁，执行完毕调用 unlock 方法解锁。这样也可以避免其他线程抢占此公共资源。此实例还包含其他几种加锁的方式，例如使用 tryLock 方法。lock 方法和 tryLock 方法的区别是 tryLock 可以设置立刻返回或者等待一段时间再返回。

```java
public class ThreadConflict{
    private Lock lock = new ReentrantLock();

    public int getSumByLock(int start,int end) {
        lock.lock();
        try {
            sum = 0;
            for(int i=start;i<end;i++) {
                sum +=i;
            }
            return sum;
        } finally {
            lock.unlock();
        }
    }
    public int getSumByTryLock(int start,int end) {
        try {
            if(lock.tryLock(1, TimeUnit.SECONDS)) {
                sum = 0;
                for(int i=start;i<end;i++) {
                    sum +=i;
                }
                return sum;
            }
        } catch (InterruptedException e) {
            e.printStackTrace();
            return -1;
        } finally {
            lock.unlock();
        }
    }
}
```

常用的锁还有读写锁 ReentrantReadWriteLock，这里不再介绍，希望大家自己完成读写锁的学习。对于同步代码块或者锁的使用一定要精简，在确定会发生异步问题的地方才加入同步的逻辑，否则乱上锁会造成很大的性能问题，多线程的优势也得不到发挥。另外，被锁住的代码也要精简，不要把冗余的可以异步执行的代码放到同步代码块中。本节对于并发的意义以及并发会导致的问题通过几个例子都已经讲到了，希望读者能够很好地理解并且使用多线程。

1.11 反射与注解

对于一种语言来讲，前面的内容好像够全面了。那么 Java 的反射和注解为 Java 做出了什么贡献呢？

　　Java 的反射机制是指在运行状态中，对于任意一个类，都能够知道这个类的所有属性和方法；对于任意一个对象，都能够调用它的任意方法和属性。这种动态获取信息以及动态调用对象方法的功能称为 Java 语言的反射机制。

　　Java 反射机制的意义在于与框架结合，各个框架正是应用了 Java 的反射机制才能对业务代码进行加载和整合⊖。那么注解的意义是什么？可以把注解理解为 Java 对类、字段或方法的补充说明，Java 通过反射读到注解，通过注解的说明对被注解内容进行相应的操作，例如生成数据库表、字段、执行测试等。

1.11.1　反射

　　虽然平时使用反射较少，但是理解反射却非常重要，反射就像粘合剂一样，把 Java 框架和程序粘合在一起，下面通过代码了解反射的基本能力。

```java
public class JavaReflect {
    public long num = 123456;
    public String name = "xiaoming";

    @Override
    public String toString() {
        return "num=" + num +"/name=" + name;
    }

    public void print() {
        System.out.println(toString());
    }

    public void set(long num,String name) {
        this.num = num;
        this.name = name;
    }

    private void privatefun() {
        System.out.println("private fun");
    }

    @SuppressWarnings({ "rawtypes", "unchecked" })
    public static void main(String[] args) {
        try {
            Class reClass = JavaReflect.class;
            Class reClass2 = Class.forName("com.javadevmap.Reflect.JavaReflect");

            if(reClass==reClass2) {
                System.out.println("class is single");
```

⊖ 后续讲到 Spring 时相信大家会有更深的理解。

```java
            }

            Field[] fields = reClass.getFields();
            System.out.println("fields : " + Arrays.asList(fields));

            Method[] methods = reClass.getMethods();
            System.out.println("methods : " + Arrays.asList(methods));

            Constructor[] constructors = reClass.getConstructors();
            System.out.println("constructors : " + Arrays.asList(constructors));

            Object reObject = reClass.newInstance();
            Method method = reClass.getMethod("print");
            method.invoke(reObject, null);

            method = reClass.getMethod("set",long.class,String.class);
            method.invoke(reObject, 234567,"daming");

            method = reClass.getMethod("print");
            method.invoke(reObject, null);

            method = reClass.getDeclaredMethod("privatefun");
            method.setAccessible(true);
            method.invoke(reObject, null);
        } catch (Exception e) {
            e.printStackTrace();
        }
    }
}
```

运行结果如下：

```
class is single
fields : [public long com.javadevmap.Reflect.JavaReflect.num, public java.lang.String com.javadevmap.Reflect.JavaReflect.name]
methods : [public static void com.javadevmap.Reflect.JavaReflect.main(java.lang.String[]), public java.lang.String com.javadevmap.Reflect.JavaReflect.toString(), public void com.javadevmap.Reflect.JavaReflect.print(), public void com.javadevmap.Reflect.JavaReflect.set(long,java.lang.String), public final void java.lang.Object.wait() throws java.lang.InterruptedException, public final void java.lang.Object.wait(long,int) throws java.lang.InterruptedException, public final native void java.lang.Object.wait(long) throws java.lang.InterruptedException, public boolean java.lang.Object.equals(java.lang.Object), public native int java.lang.Object.hashCode(), public final native java.lang.Class java.lang.Object.getClass(), public final native void java.lang.Object.notify(), public final native void java.lang.Object.notifyAll()]
constructors : [public com.javadevmap.Reflect.JavaReflect()]
num=123456/name=xiaoming
num=234567/name=daming
private fun
```

从代码中可以看出，反射有两种获取 Class 对象的方式，但是常用的还是第二种方式。这两种方法生成的 Class 对象是同一个，每个类都有一个 Class 对象，并且唯一。可以通过反射获取字段、方法、构造器①。并且可以通过反射直接获取方法，调用带参数或者不带参数的方法，甚至可以调用私有方法。

如果没有特殊需求，一般的业务逻辑中不会带有反射，了解反射的用处和能力就好，如果有更高的要求，那么反射还是要仔细研究的。

1.11.2 注解

注解是丰富代码信息的一种方法，通过注解可以更加了解代码，并且通过注解解析和使用，能够方便管理代码和让编程更加简单。

（1）Java 内置了三种注解，分别是：
- @Override：表示当前方法覆盖基类的方法，如果基类不存在此方法，则编译器会报错（如果不使用注解，则编译器不会检查到）。
- @Deprecated：表示弃用的方法，如果其他类使用了这个注解标注的方法，编译器会生成警告。
- @SuppressWarnings：关闭警告的注解，使用此注解后编译器会关闭相应类型的警告。

（2）元注解

注解的定义是基于元注解的，元注解可以理解为注解的注解。通过元注解来修饰自定义注解，例如圈定使用范围或应用阶段等，见表 1-8。

表 1-8 元注解

@Target	表示注解的可用范围，ElementType 包含以下几种：CONSTRUCTOR（构造函数）、FIELD（字段）、LOCAL_VARIABLE（局部变量）、METHOD（方法）、PACKAGE（包）、PARAMETER（参数）、TYPE（类、接口、枚举）
@Retention	表示注解的应用级别，分为：SOURCE、CLASS、RUNTIME（最高）
@Documented	可以被 Javadoc 文档化
@Inherited	注解类型被自动继承

（3）自定义注解

自定义注解的写法和接口很像，只是在名字前面加上@即可。注解中用元注解修饰自定义注解的应用范围和应用级别；注解可以用的数据类型包含基本类型、String、Class、enum、Annotation 以及这些类型的数组。注解中的方法不能有参数，但是可以有默认值。

下面定义一个自定义注解，这个注解是用来辅助说明方法是做什么的，以及它对外暴露的接口是什么。

```
@Target(ElementType.METHOD)
@Retention(RetentionPolicy.RUNTIME)
@Inherited
public @interface MethodUrl {
    public int ID() default -1;
    public String Describe();
    public String URL();
}
```

---

① 反射调用不同的获取方法可以获取不同可见类型的字段、方法、构造器。

自定义注解通过@Target 设定，此注解用于方法；通过@Retention 设定注解有效期至运行期。注解的字段包含方法的描述和对外提供的 URL 路径。

（4）注解解析

如上例自定义注解，这个注解写完了有什么用呢？用于给方法做补充说明，难道仅仅在文档上有用吗？下面给一个类加上自定义注解，然后再通过一个解析的方法把注解解析出来，看看它的用处有多大。

```java
public class JavaAnnotation {
    public String name;

    @MethodUrl(ID=1,Describe="获取名字",URL="/JavaAn/getName")
    public String getName() {
        return name;
    }

    @MethodUrl(ID=2,Describe="设置名字",URL="/JavaAn/setName")
    public void setName(String name) {
        this.name = name;
    }
}
```

上面代码建立了一个类，这个类有两个方法，对每个方法都写了注解，标明了这个方法是做什么的，以及一个路径。下面的代码可以获取 JavaAnnotation 类的注解情况并输出。

```java
public static void main(String[] args) {
    try {
        Class clazz = Class.forName("com.javadevmap.Reflect.JavaAnnotation");
        Method[] methods = clazz.getMethods();
        for (Method method : methods) {
            MethodUrl methodUrl = method.getDeclaredAnnotation(MethodUrl.class);
            if(methodUrl!=null) {
                System.out.println(method.getName() + " function ID is " + methodUrl.ID() + " and url is localhost:8080" + methodUrl.URL() + " for " + methodUrl.Describe());
            }
        }
    } catch (Exception e) {
        e.printStackTrace();
    }
}
```

运行结果如下：

```
getName function ID is 1 and url is localhost:8080/JavaAn/getName for 获取名字
setName function ID is 2 and url is localhost:8080/JavaAn/setName for 设置名字
```

main 方法通过解析，输出了带自定义注解的两个方法的名字，还有一个 URL 地址，并且说明了用途。某些 Web 框架其实就是用这种方法把类中方法和访问地址关联起来，这样来自网络的调用就可以找到对应的方法，从而实现程序对外提供的 HTTP 服务。

注解的应用场景很多，例如下一节要讲的 JUnit 和本书后面的很多框架都用到了注解的内容。本节讲的是基本的注解使用原理，了解原理之后对以后章节的理解会更加轻松。

1.12 JUnit

当编写完代码，需要对自己写的功能进行测试时，可以直接写一个 main 方法来测试自己的代码，还有一种方式测试业务代码，即本节介绍的 JUnit 框架。JUnit 是一个针对 Java 语言的单元测试框架，通过使用 JUnit 可以保证程序的稳定性，并且减少花费在排错上的时间。

1.12.1 JUnit 的集成

JUnit 的基础环境简单来说就是下载对应的 JUnit 包，解压到本机，配置到项目的 Build Path 中。下面介绍两种集成 JUnit 的方法。

（1）如果当前项目为非 Maven 管理，可以使用下面两种方式集成 JUnit。
- 从 https://junit.org/junit5/ 下载 JUnit 最新版本的压缩文件，解压后，将需要的以 Junit 开头的 jar 包放到 Eclipse 项目里面的 libs①文件夹中，然后在引入的 jar 包上执行"鼠标右键单击->Build->Add Build Path"即可。
- 在项目上执行"鼠标右键单击->Build->Add Library"，在弹出来的界面中选中 JUnit，点击 next，选中 JUnit 的版本，一般选用 4.0 以上的版本，点击 Finish。

（2）当前项目为 maven 管理的项目。
在项目的 pom 文件中的 dependencies 元素下面添加如下代码：

```xml
<!-- junit test -->
<dependency>
    <groupId>junit</groupId>
    <artifactId>junit</artifactId>
    <version>4.12</version>
    <scope>test</scope>
</dependency>
```

添加完后，在项目的 pom.xml 文件上执行"鼠标右键单击->Run as->Maven install"即可使用。

1.12.2 JUnit 的基本使用

下面演示 JUnit 的基本使用。编写一个业务类，里面有两个方法如下：

```java
public class ServiceOfBusiness {
    private String serverCode;
    private boolean isTrueFlag;
    public ServiceOfBusiness() {
        super();
    }
    // Constructor
    public ServiceOfBusiness(String code, boolean isTrue) {
        this.serverCode = code;
        this.isTrueFlag = isTrue;
    }
    // prints the Flag
    public boolean printFlag() {
```

---

① libs 文件夹，此文件夹通常用来存放第三方依赖。如果当前项目没有 libs 文件夹，新建一个名为 libs 的目录即可。

```
            System.out.println(isTrueFlag);
            return isTrueFlag;
        }
        // prints the serverCode
        public String printServerCode() {
            System.out.println(serverCode);
            return serverCode;
        }
    }
```

对上面的两个业务方法进行测试。编写一个 JUnit 的测试类步骤如下：
（1）创建一个名为 TestJunit 的测试类。
（2）向测试类中添加名为 testPrintMethods() 的方法。
（3）在方法上添加注解@Test。
（4）执行 JUnit 的 assertEquals 方法来检查测试是否通过。
具体代码如下：

```
public class TestJunit {
    String code = "Hello World";
    ServiceOfBusiness service = new ServiceOfBusiness(code,false);
    @Test
    public void testPrintMethods() {
        assertEquals(code, service.printServerCode());
        assertEquals(false, service.printFlag());
    }
}
```

在 testPrintMethods 上执行"鼠标右键单击->run as->Junit test"，运行完后，会出现一个 JUnit 的测试结果窗口，如图 1-2 所示。

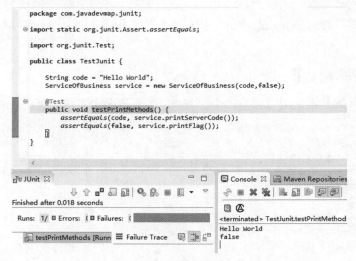

图 1-2 测试结果

其中的状态栏显示测试用例通过和未通过的比例，绿色表示通过，红色表示未通过，点击下边的未通过的测试方法，还可以看到未通过的原因信息。

测试类中注解的含义见表 1-9。

表 1-9 JUnit 的常用注解

注 解 名 称	含 义
@Before	运行前调用,一般用于初始化方法
@After	运行后调用,一般用于释放资源
@Test	测试方法,可以测试方法的执行情况
@BeforeClass	所有用例运行之前只执行一次,且方法必须为 static void
@AfterClass	所有用例运行之后只执行一次,且方法必须为 static void
@Ignore	忽略的测试方法

这样,使用 JUnit 就能编写一个测试用例来检验自己的代码。

第 2 章 Maven

在 Java 的世界中，依赖管理是不得不面对的问题。无论是外部的开源类库依赖，还是项目内部的模块间依赖，都需要进行依赖管理。可以说依赖管理是持续集成的核心内容之一。Maven 抽象定义了一个软件的完整生命周期，遵循这个模型，可轻松地管理自己的软件项目，避免不必要的学习成本，并促进软件项目管理的标准化、流程化。

2.1 Maven 安装和配置

从本节开始实际操作 Maven。首先介绍如何在 Windows 系统中安装 Maven 以及 Maven 的基本配置。

2.1.1 Maven 环境的搭建

Maven 的基础环境简单来说就是下载对应的 Maven 包，解压到本机，配置对应的环境变量。

（1）检查 JDK 环境是否配置成功，在安装 Maven 之前，需要检查当前 JDK 基础环境是否配置正确，Maven3.3+需要 JDK 7 及以上版本。在 Windows 的命令行输入 java –version，运行命令来检查 JDK 的版本，如图 2–1 所示。

```
C:\Users\T460>java -version
java version "1.8.0_144"
Java(TM) SE Runtime Environment (build 1.8.0_144-b01)
Java HotSpot(TM) 64-Bit Server VM (build 25.144-b01, mixed mode)
```

图 2–1 Java 版本

（2）首先通过官网https://maven.apache.org/download.cgi下载对应的 Maven 版本，在编写本书的时候，Maven 的最新版本为 3.5.2，这里下载 apache–maven–3.5.2，解压到常用软件安装目录，然后进行环境变量的配置。在计算机中找到设置环境变量的地方，添加对应的变量名和值，见表 2–1。

表 2–1 环境变量配置

变　量	值
M2_HOME	C:\apache–maven–3.5.2（请配置本地实际路径）
Path	%M2_HOME%\bin;

通过以上两步就配置好了 Maven 基础环境。在命令行输入 mvn –v，观察输出结果，会看到 Maven 的路径等信息，表明 Maven 已正确安装，如图 2–2 所示。

```
C:\Users\Administrator>mvn -v
Apache Maven 3.5.2 (138edd61fd100ec658bfa2d307c43b76940a5d7d, 2017-10-18T15:58:13+08:00)
Maven home: C:\apache-maven-3.5.2\bin\..
Java version: 1.8.0_65, vendor: Oracle Corporation
Java home: C:\Java\jdk1.8.0_65\jre
Default locale: zh_CN, platform encoding: GBK
OS name: "windows 10", version: "10.0", arch: "amd64", family: "windows"

C:\Users\Administrator>
```

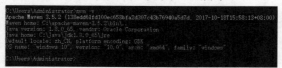

图 2–2 Maven 版本

2.1.2 在 Eclipse 中配置 Maven 的 settings 文件

打开 Eclipse 工具，在菜单栏选择"Window->Preferences"，在弹出的选项卡中左边选择 Maven 选项的"User Settings"，然后在右边的全局配置（Global Settings）中可以指定自己的 settings.xml 文件，如图 2-3 所示。

检测配置是否成功：打开 Eclipse 的"Window->Show View->Other"，然后选择"Maven->Maven Repositories"，打开 Maven 仓库，会出现 Local Repositories 仓库，在此仓库下可以看到在 setting.xml 中配置过的仓库，至此表示 Maven 配置成功，如图 2-4 所示。

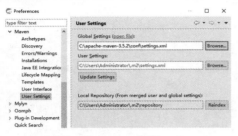

图 2-3　Maven 设置

图 2-4　Maven 仓库

2.2　Maven 使用

Maven 基础环境准备妥当后，创建一个简单的 Hello World 项目。本章会一步一步地编写代码并输出结果。

2.2.1　在 Eclipse 中创建第一个 Maven 项目

在 Eclipse 中选择"File->New->Maven Project"，在弹出的 tab 页面中直接点击 Next 按钮，在第二个 tab 页面中保持默认的 Archetype[⊖]（maven-archetype-quickstart）设置即可，点击 Next 按钮，在第三个 tab 页面中输入信息，见表 2-2，然后点击 Finish 按钮即可。

表 2-2　项目配置

变　量	值
Group Id	com.javadevmap.demo
Artifact Id	Hello-world-demo
Version	0.0.1-SNAPSHOT

常见的 Maven 项目代码结构如图 2-5 所示。

Maven 项目的目录属性一般如下。
- pom.xml：用于 Maven 的配置文件。
- /src：源代码目录。
- /src/main：工程源代码目录。
- /src/main/java：放置项目 Java 源代码目录。

图 2-5　项目结构

[⊖] Archetype 是 Maven 提供的快速构建项目骨架的工具。

- /src/main/resources[一]：放置项目的资源文件目录。
- /src/test：单元测试目录。
- /src/test/java：工程测试 Java 代码目录。
- /target：输出目录，项目输出存放在此目录中。

2.2.2 认识 pom 文件

Maven 项目的核心是 pom.xml 文件，此文件包含项目的基本信息、包依赖、项目构建等信息。下面为常用 pom 文件的基本内容。

```
<?xml version="1.0" encoding="UTF-8"?>
<project xmlns=http://maven.apache.org/POM/4.0.0
xmlns:xsi="http://www.w3.org/2001/XMLSchema-instance"
xsi:schemaLocation="http://maven.apache.org/POM/4.0.0 http://maven.apache.org/xsd/maven-4.0.0.xsd">
    <modelVersion>4.0.0</modelVersion>
    <groupId>com.javadevmap.demo</groupId>
    <artifactId>hello-world-demo</artifactId>
    <version>0.0.1-SNAPSHOT</version>
    <packaging>jar</packaging>
    <name>hello-world-demo</name>
    <url>http://maven.apache.org</url>
    <properties>
        <project.build.sourceEncoding>UTF-8</project.build.sourceEncoding>
    </properties>
    <dependencies>
        <dependency>
            <groupId>junit</groupId>
            <artifactId>junit</artifactId>
            <version>3.8.1</version>
            <scope>test</scope>
        </dependency>
    </dependencies>
    ...
</project>
```

第一行代码是 XML 头，然后是 project 元素，project 是 pom.xml 的根元素，同时还声明了 pom 命名空间以及 xsd 元素。

pom 文件里面的常见元素以及含义见表 2-3。

表 2-3 pom 元素含义

元素名称	元素作用
<project>	pom 的 xml 根元素
<parent>	声明继承
<modules>	声明聚合
<groupId>	声明项目属于哪个组织
<artifactId>	声明项目的唯一 ID
<version>	声明项目的版本

[一] resources 目录添加方法：执行"鼠标右键单击项目->new->source folder->Folder name"，填写/src/main/resources，点击确定，就生成了该目录。

(续)

元素名称	元素作用
\<packaging\>	指定当前项目构建类型，默认值 jar
\<build\>\<plugins\>\<plugin\>	插件
\<build\>\<pluginManagement\>	插件管理
\<repositories\>\<repository\>	仓库
\<pluginRepositories\>	插件仓库
\<pluginRepository\>\<dependencies\>\<dependency\>	依赖
\<dependencyManagement\>	依赖管理

这里着重说明一下 groupId、artifactId、version 这三个标签，它们定义了一个项目的基本坐标。任何 jar、pom 或 war 都是基于这三个标签进行区分的。

groupId 定义了项目属于哪个组，建议用公司名或组织名。一般来说，groupId 由三个部分组成，每个部分之间以 "." 分隔，第一部分是项目用途，例如用于商业的就是 com，用于非营利性组织的就是 org；第二部分是公司名，例如 tengxun、baidu、alibaba；第三部分是项目名。

artifactId 定义了当前 Maven 项目组中唯一的 ID，这里定义此项目的 artifactId 为 hello-world-demo。

version 指项目当前的版本，默认版本为 SNAPSHOT 版本，SNAPSHOT 意思为快照，说明当前项目处于开发迭代中，不是稳定版本。随着开发的推进，可以对版本依次递进修改，例如 xx.SNAPSHOT，xx.beta，1.0，2.0 等。

以上为 Maven 的常用标签，下面简要介绍 Maven 的其他标签，见表 2-4，读者了解即可。

表 2-4 pom 元素含义

元素名称	元素作用
\<properties\>	Maven 属性
\<reporting\>\<plugins\>	报告插件
\<name\>	名称
\<description\>	描述
\<organization\>	所属组织
\<licenses\>\<license\>	许可证
\<mailingLists\>\<mailingList\>	邮件列表
\<developers\>\<developer\>	开发者
\<contributors\>\<contributor\>	贡献者
\<issueManagement\>	问题追踪系统
\<ciManagement\>	持续集成系统
\<scm\>	版本控制系统
\<prerequisites\>\<maven\>	要求 Maven 最低版本，默认值为 2.0
\<build\>\<sourceDirectory\>	主源码目录
\<build\>\<scriptSourceDirectory\>	脚本源码目录
\<build\>\<testSourceDirectory\>	测试源码目录
\<build\>\<outputDirectory\>	主源码输出目录
\<build\>\<testOutputDirectory\>	测试源码输出目录
\<build\>\<resources\>\<resource\>	主资源目录
\<build\>\<testResources\>\<testResource\>	测试资源目录

（续）

元素名称	元素作用
\<build>\<finalName>	输出主构件的名称
\<build>\<directory>	输出目录
\<build>\<filters>\<filter>	通过 properties 文件定义资源过滤属性
\<build>\<extensions>\<extension>	扩展 Maven 的核心
\<profiles>\<profile>	POMProfile
\<distributionManagement>\<repository>	发布版本部署仓库
\<distributionManagement>\<snapshotRepository>	快照版本部署仓库
\<distributionManagement>\<site>	站点部署

2.2.3 运行 Maven 项目

当编写好业务代码后，需要构建运行项目。直接在项目的 pom.xml 文件上右击，选择"Run As"，就能看到常用的 Maven 命令，如图 2-6 所示。

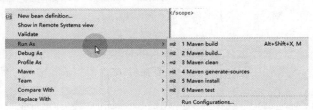

图 2-6　运行项目

选择要执行的 Maven 命令就能执行相关的构建操作，同时在 Eclipse 的 Console 中就能看到执行命令的结果输出。如果想执行自定义顺序的命令，只需要点击"Maven build ..."，在弹出的对话框的 Goals 输入中输入要执行的命令如 clean install 即可，如图 2-7 所示。

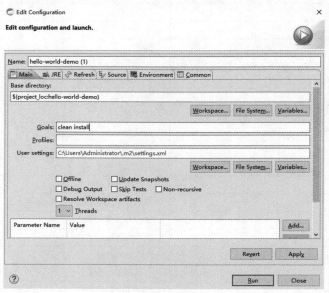

图 2-7　构建配置

点击 Run 按钮，可看到如图 2-8 所示的运行结果。

```
[INFO]
[INFO] --- maven-jar-plugin:2.4:jar (default-jar) @ hello-world-demo ---
[INFO] Building jar: D:\WorkSpace\hello-world-demo\target\hello-world-demo-0.0.1-SNAPSHOT.jar
[INFO]
[INFO] --- maven-install-plugin:2.4:install (default-install) @ hello-world-demo ---
[INFO] Installing D:\WorkSpace\hello-world-demo\target\hello-world-demo-0.0.1-SNAPSHOT.jar to C:\Us
[INFO] Installing D:\WorkSpace\hello-world-demo\pom.xml to C:\Users\Administrator\.m2\repository\cc
[INFO] ------------------------------------------------------------------------
[INFO] BUILD SUCCESS
[INFO] ------------------------------------------------------------------------
[INFO] Total time: 46.765 s
[INFO] Finished at: 2018-03-03T12:23:58+08:00
[INFO] Final Memory: 18M/151M
[INFO] ------------------------------------------------------------------------
```

图 2-8　运行结果

Goals 输入框常用的 Maven 命令见表 2-5。

表 2-5　Maven 命令

Maven 常用命令	命 令 作 用
mvn clean	清理（删除 target 目录下的编译内容）
mvn compile	编译项目
mvn test	编译并执行测试用例
mvn package	打包发布
mvn package -Dmaven.test.skip=ture	打包时跳过测试
mvn install:install-file -DgroupId=\<groupId\> -DartifactId=\<artifactId\> -Dversion=1.0.0 -Dpackaging=jar -Dfile=\<myfile.jar\>	安装指定版本到本地仓库

常用命令的含义如下：

- compile：编译当前项目，编译后的 class 文件会放在项目的 target/classes 文件夹中，这是 Maven 约定的存放位置。
- test：编译当前项目并执行测试用例，这里 Maven 可能会下载测试所依赖的构件，并且在测试之前，Maven 会编译主代码。
- package：打包当前项目，如果读者的 pom 文件里面的 packaging 元素设置的值为 jar，那么执行此命令会编译当前项目生成一个 jar 文件存放在 target 目录下面，默认生成的文件名称由 artifactId 和 version 拼接组成。
- install：安装到仓库命令，可以把生成的 jar 文件直接安装到本地的 Maven 仓库（默认仓库地址在当前用户目录下面的.m2/repository 文件夹）。

2.3　Maven 坐标和依赖

Maven 的一个重要功能是管理项目的依赖。本节讲解 Maven 坐标和它的作用、Maven 如何在实际的项目中进行应用以及常用的 Maven 使用技巧和实战经验。

2.3.1　什么是坐标

坐标是 Maven 中任何一个依赖包的唯一标识。任何一个构件都明确地定义了自己的坐标。Maven 的坐标包含以下几个元素：

- groupId 定义了当前 Maven 项目的归属组织。
- artifactId 定义了一个 Maven 项目或者模块的唯一名称。
- version 定义了当前 Maven 项目所处的版本。
- packaging 定义了当前 Maven 项目的打包方式。

- classifier 定义了用来帮助定义构建输出的一些附属构建。

上述 5 个元素中，groupId，artifactId，version 必须定义，packaging 可选（默认为 jar），classifier 是不能直接定义的，因为附属构件不是项目直接默认生成的，而是由附加的插件帮助生成的。

Maven 会根据 pom 文件里面配置的坐标元素，到 Maven 内置的中央仓库(https://repo1.maven.org/maven2/)里面寻找对应的依赖包，例如在 pom 文件中定义了 groupId=junit、artifactId=junit、version=4.12，Maven 就会检查本地仓库有没有对应文件，如果没有就会从中央仓库找到对应的文件并下载到本地的仓库，提供给工程使用。

2.3.2 什么是 Maven 依赖

在项目开发的时候，或多或少会依赖第三方组件或者其他工程模块，Maven 依赖即指引用的第三方组件或者模块。这体现在本工程的 pom 文件里面 project 下面的 dependencies 标签下，此标签可以包含一个或多个 dependency 元素，来声明当前项目所依赖的一个或多个依赖。类似如下形式：

```xml
<project>
...
    <dependencies>
        <dependency>
            <groupId>junit</groupId>
            <artifactId>junit</artifactId>
            <version>4.12</version>
            <scope>test</scope>
        </dependency>
    </dependencies>
...
</project>
```

每个依赖的 groupId、artifactId、version 这三个元素构成一个依赖的基本坐标，Maven 根据这个坐标才能找到对应的依赖。

下面讲解在 Maven 项目中如何引入本地的包，例如在其他项目中生成一个 jar 包，引入到现有的项目中来。

方法一：将待引入的包放在指定目录下（如 lib 目录下），修改项目的 pom 文件，加入依赖并且将 scope 设置为 system。pom 文件配置如下所示：

```xml
<dependency>
    <groupId>com.javadevmap.demo</groupId>
    <artifactId>hello-world-demo</artifactId>
    <version>0.0.1-SNAPSHOT</version>
    <scope>system</scope>
    <systemPath>${project.basedir}/lib/hello-world-demo-0.0.1.jar</systemPath>
</dependency>
```

方法二：将待引入的 jar 包安装到本地 repository 中，例如将 hello-world-demo-0.0.1.jar 安装到本地仓库。

1）先把待引入的 jar 包放在一个目录下，打开命令行，进入 jar 包所在的目录，执行 mvn install 命令，具体如下：

```
mvn install:install-file
 -Dfile=hello-world-demo-0.0.1.jar
 -DgroupId=com.javadevmap.demo
 -DartifactId= hello-world-demo
 -Dversion=0.0.1-SNAPSHOT
 -Dpackaging=jar
```

2）在项目的 pom 文件中加入包对应的依赖：

```
<dependency>
    <groupId>com.javadevmap.demo</groupId>
    <artifactId>hello-world-demo</artifactId>
    <version>0.0.1-SNAPSHOT</version>
</dependency>
```

这样就可以将本地的 jar 文件配置到自己的 Maven 项目中进行使用了。

2.3.3 Maven 依赖的 scope 范围

Maven 在不同的生命周期使用的 classpath 是不同的，例如执行项目的测试和 Maven 项目运行的时候，这两者之间的 classpath 是有差异的。常用的 JUnit 构件就是如此，在测试阶段会引入，但在运行 Maven 项目的时候是不需要的。

Maven 的依赖范围与 classpath 的关系见表 2-6。

表 2-6　scope 范围

scope	编译期	测试期	运行期	说　　明
compile	Y	Y	Y	编译依赖范围，scope 默认使用 compile
test	--	Y	--	只在测试期依赖，如 JUnit 包
provided	Y	Y	--	运行期由容器提供，如 servlet-api 包
runtime	--	Y	Y	编译期间不需要直接引用
system	Y	Y	--	编译和测试时由本机环境提供

根据上面的信息，可以轻松地引入其他构件，协助开发程序。在引用其他依赖时如果出现引用冲突，可以通过 Maven 的传递性依赖解决这一问题。

如果依赖没有声明依赖范围，那么其依赖范围就是默认的 compile。假如 A 依赖 B，B 依赖 C，那么 A 对 B 是第一直接依赖，B 对 C 是第二直接依赖，A 对 C 是传递性依赖。第一直接依赖范围和第二直接依赖范围决定了传递性依赖的范围。

在表 2-7 中，左边第一列表示第一直接依赖范围，最上面一行表示第二直接依赖的范围，中间的单元格表示传递性依赖范围，表格中的"-"表示依赖无法传递。

表 2-7　依赖传递

第一直接依赖 \ 第二直接依赖	compile	test	provided	runtime
compile	compile	-	-	runtime
test	test	-	-	test
provided	provided	-	provided	provided
runtime	runtime	-	-	runtime

根据上面的规则举例，例如项目 Proj 中，有一个直接依赖 A，其依赖范围为 compile，而 A 依赖里面又有一个 B 的直接依赖，其依赖范围为 runtime，那么显然 B 是 Proj 项目的传递性依

赖。参照表2-7,第一直接依赖为compile,第二直接依赖为runtime,因此B对项目Proj是一个范围为runtime的传递性依赖。

2.3.4 Maven的依赖调解原则

当项目里面依赖变多的时候,多个项目之间难免存在引用不同版本依赖的情况,这样容易出现依赖版本不一致,导致项目的构建出现问题。要解决此问题需要明白Maven的依赖调解原则。下面介绍这两个原则。

(1)路径最短者优先。

这里"->"符号代表依赖,"()"代表版本号。例如A->B(2.0)指的是A依赖版本号为2.0的B构件。下面是两条依赖链条。

A->B->C->X(1.0)
A->D->X(2.0)

可以发现两个依赖链条上都有版本X,而且X的版本是不一致的,根据路径最短者优先原则,X(1.0)的版本路径长度为3,而X(2.0)的版本路径长度为2,那么X(2.0)会被依赖使用。

(2)依赖路径长度相等的前提下,顺序最靠前的那个依赖优先。

A->B->X(1.0)
A->D->X(2.0)

上面两个不同版本的X的路径长度是一样的,依据Maven定义的依赖调解的第二原则:第一声明者优先。这里X(1.0)声明在前面,会被工程使用。

2.3.5 Maven仓库使用

Maven通过仓库来管理构件,仓库分为两种类型:本地仓库和远程仓库。当Maven根据坐标查找构件的时候,它首先会查看本地仓库,如果本地仓库存在此构件,直接使用。如果不存在,Maven就会去远程仓库查找,发现需要的构件后,下载到本地仓库再使用。如果在本地和远程仓库都没有找到,那么Maven就会显示找不到构件的错误提示信息。

本地仓库是在用户当前操作系统上存放构件的地方,默认在当前用户目录下面都有一个路径名称为.m2/repository/的仓库目录。

中央仓库是Maven核心自带的远程仓库,包含了大部分开源的构件。在默认配置下,当本地仓库没有Maven需要的构件时,它会尝试从中央仓库进行查找下载。

私服是另一种远程仓库,例如许多公司为了节省带宽和时间,会在内部搭建一个私服,也就是内部使用的Maven仓库,可以存放公司内部的构件或者其他开源构件。例如常见的Nexus服务。

众所周知,国内开发很头疼的一件事就是Maven仓库的下载速度太慢。所以一般使用国内公开仓库,常见的有阿里云仓库(http://maven.aliyun.com/nexus/content/groups/public/)。

下面介绍如何修改仓库地址。修改Maven根目录下的conf文件夹中的setting.xml文件,对应内容如下:

```
<mirrors>
    <mirror>
        <id>alimaven</id>
        <name>aliyun maven</name>
        <url>
            http://maven.aliyun.com/nexus/content/groups/public/
```

```
            </url>
            <mirrorOf>central</mirrorOf>
        </mirror>
    </mirrors>
```

这样就把 Maven 的中央仓库的地址修改成阿里云仓库地址了。

当然也可以定义本地仓库的目录地址，修改 settings.xml 文件，设置本地仓库的实际存储路径。例如：

```
<settings>
    <localRepository>E:\Repository</localRepository>
</settings>
```

2.4 Maven 生命周期和插件

Maven 的生命周期是对项目构建生命周期的一个抽象。在 Maven 出现之前，项目构建的生命周期早已存在。例如开发人员对项目的清理、编译、测试以及部署。

2.4.1 Maven 生命周期

Maven 有一套完善、易扩展的生命周期。包含了项目的清理、初始化、编译、打包、集成测试、验证、部署和站点生成等。可以映射到目前几乎所有软件的生命周期上。

每个生命周期包含了一系列阶段（phase），这些阶段有自己的顺序，并且前后阶段是有依赖关系的。Maven 的生命周期如图 2-9 所示。

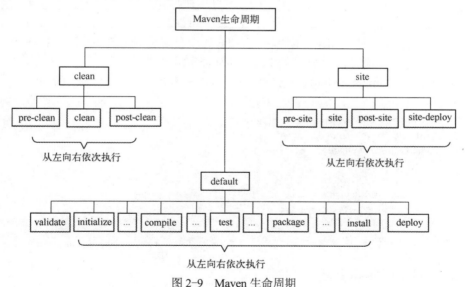

图 2-9 Maven 生命周期

下面以常用的 Maven 命令为例，讲解其执行的生命周期阶段：

（1）mvn clean：该命令调用 clean 生命周期的 clean 阶段。实际执行的阶段为 clean 生命周期的 pre-clean 和 clean 阶段。

（2）mvn test：该命令调用 default 生命周期的 test 阶段。实际执行的阶段为 default 生命周期的 validate、initialize 等直到 test 的所有阶段。这也解释了为什么在执行测试的时候，项目代码能够自动得到编译。

（3）mvn clean install:调用 clean 生命周期的 clean 阶段和 default 生命周期的 install 阶段。实际执行为 clean 生命周期的 pre-clean、clean 阶段以及 default 生命周期的从 validate 至 install 的所有阶段。

（4）mvn clean deploy site-deploy:调用 clean 生命周期的 clean 阶段、default 生命周期的 deploy 阶段以及 site 生命周期的 site-deploy 阶段。实际执行的阶段为 clean 生命周期的 pre-clean、clean 阶段，default 生命周期的所有阶段和 site 生命周期的所有阶段。

2.4.2 Maven 插件

Maven 常用的插件见表 2-8。

表 2-8 Maven 常用插件

插件名称	插件的 artifactId
自动定义打包	maven-assembly-plugin
SCP 文件传输	copy-maven-plugin
源码分析	maven-pmd-plugin
代码格式检查	maven-checkstyle-plugin
单元测试报告	maven-surefire-report-plugin
TODO 检查报告	taglist-maven-plugin
Java 代码的度量工具	javancss-maven-plugin
JavaDoc	maven-javadoc-plugin
FireBug 检查	findbugs-maven-plugin
查找重复依赖	duplicates-finder-plugin

Windows 系统默认使用 GBK 编码格式，Java 项目经常使用的编码为 UTF-8，需要在 compiler 插件中进行相应设置，否则中文乱码可能会导致编译错误。

使用插件 maven-compiler-plugin 设定编译的 JDK 版本和编码格式，如下所示：

```
<plugin>
    <groupId>org.apache.maven.plugins</groupId>
    <artifactId>maven-compiler-plugin</artifactId>
    <version>3.7.0</version>
    <configuration>
        <source>1.8</source>
        <target>1.8</target>
        <encoding>UTF-8</encoding>
    </configuration>
</plugin>
```

2.4.3 生命周期与插件的关系

Maven 的生命周期本身是不做任何实际工作的，实际的任务操作都交给插件来完成。生命周期抽象了构建的各个步骤，定义了步骤执行的顺序，但是没有具体实现，具体的实现由插件来完成。即 Maven 通过这种插件的机制，使得每个构建的步骤都可以绑定一个或者多个插件的行为，而且 Maven 内置了很多默认的插件。让使用者在大多数的时间里，感觉不到插件的存在。当然该机制提供了足够的扩展空间，使用者可以自己配置插件或者自定义插件来构建特定的行为。

2.5 Maven 聚合和继承

Maven 的聚合特性能够把一个项目的各个模块聚合在一起进行构建。Maven 的继承特性能够帮助抽取相同的依赖和插件等配置，在简化 pom 的同时，还能够促进各个模块配置的一致性。

2.5.1 聚合应用的场景

假设有两个 Maven 项目，分别为工程 child01 和工程 child02，如图 2-10 所示。两个项目并行开发时，需要分别到两个模块目录中执行 mvn 命令进行构建。如果并行的项目更多会造成命令执行操作非常烦琐。而 Maven 聚合可以实现执行一次命令，构建多个模块。下面对 child01 和 child02 两个项目进行聚合操作。

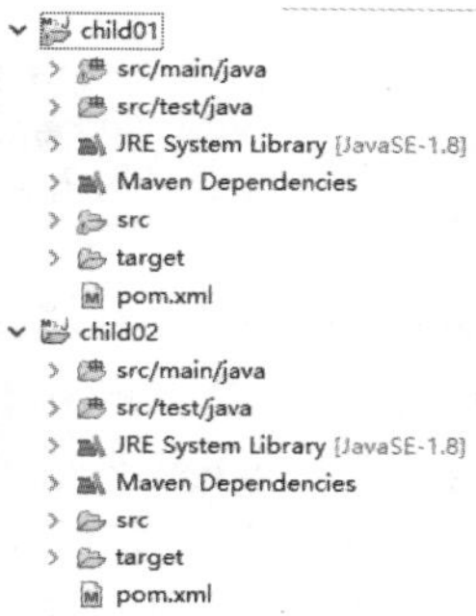

创建另外一个项目 parent，通过该项目构建项目组所有模块。parent 作为一个 Maven 项目，必须拥有自己的 pom 文件。在 Eclipse 中创建此项目时要选择 maven-archetype-site-simple。此 parent 项目不作为业务代码开发使用。

图 2-10 项目结构

parent 项目 pom 文件关键部分配置如下：

```xml
<project xmlns="http://maven.apache.org/POM/4.0.0"
  xmlns:xsi="http://www.w3.org/2001/XMLSchema-instance"
  xsi:schemaLocation="http://maven.apache.org/POM/4.0.0
  http://maven.apache.org/xsd/maven-4.0.0.xsd">
  <modelVersion>4.0.0</modelVersion>

  <groupId>com.javadevmap.demo</groupId>
  <artifactId>parent</artifactId>
  <version>0.0.1-SNAPSHOT</version>
  <packaging>pom</packaging>
  <modules>
      <module>../child01</module>
      <module>../child02</module>
  </modules>
  ...//省略
</project>
```

parent 项目中的 packaging 配置必须为 pom。在 parent 项目中运行 clean package 命令，就会分别在 child01/child02 下生成对应的 jar 包，如图 2-11 所示。

```
[INFO]
[INFO] --- maven-install-plugin:2.4:install (default-install) @ parent ---
[INFO] Installing D:\WorkSpace\parent\pom.xml to C:\Users\Administrator\.m2\repos
[INFO]
[INFO] Reactor Summary:
[INFO]
[INFO] child01 ............................................. SUCCESS [  3.175 s]
[INFO] child02 ............................................. SUCCESS [  0.769 s]
[INFO] parent .............................................. SUCCESS [  0.018 s]
[INFO] ------------------------------------------------------------------------
[INFO] BUILD SUCCESS
[INFO] ------------------------------------------------------------------------
[INFO] Total time: 4.147 s
[INFO] Finished at: 2018-03-03T21:48:19+08:00
[INFO] Final Memory: 11M/150M
[INFO] ------------------------------------------------------------------------
```

图 2-11 运行结果

2.5.2 Maven 的继承

如果子项目 child01，child02 需要继承项目 parent 中的 pom 配置，那么就需要使用 Maven 的继承。在子项目中添加配置：

```xml
<parent>
    <groupId>com.javadevmap.demo</groupId>
    <artifactId>parent</artifactId>
    <version>0.0.1-SNAPSHOT</version>
    <relativePath>../parent/pom.xml</relativePath>
</parent>
```

parent 元素中的属性对应的都是父项目中的内容。在 parent 元素中还有一个属性 relativePath，maven 会通过这个路径去查找父项目的 pom.xml 文件，如果找不到会从本地仓库中查找。relativePath 的默认值是 ../pom.xml，也就是默认父项目的 pom 在上一层目录。由于当前 parent 项目与已有项目平级，这里就需要指定 pom 文件的位置。可继承的 pom 元素见表 2-9。

表 2-9 可继承的 pom 元素

元素	说明
groupId	项目组 ID
version	项目版本
description	项目描述
organization	项目的组织信息
inceptionYear	项目的创建年份
url	项目的 url 地址
developers	项目的开发者信息
contributors	项目贡献值信息
distributionManagement	项目的部署配置
issueManagement	项目的缺陷跟踪系统信息
ciManagement	项目持续集成系统信息
scm	项目的版本控制系统消息
mailingLists	项目的邮件列表信息
properties	自定义的 maven 属性
dependencies	项目的依赖配置
dependencyManagement	项目的依赖管理配置
repositories	项目的仓库配置
build	项目的源码目录配置、输出目录配置、插件配置、插件管理配置等
reporting	项目的报告输出目录配置等

2.5.3 Maven 中 dependencyManagement 的使用

子项目都会继承父项目的依赖关系，如果子项目不需要父项目的依赖关系，Maven 提供的 dependencyManagement 元素能让子项目继承到父项目的依赖配置等属性（如版本信息），确保子模块的灵活性，同时 dependencyManagement 元素下的依赖声明不会引入实际的依赖。

父项目中使用该元素声明的依赖既不会给父项目引入依赖也不会给子项目引入依赖，但是该配置会被继承。如果子项目中不声明经过父项目 dependencyManagement 修饰的依赖，那么子项目就不会引入该依赖。子项目如果要使用父项目中经过 dependencyManagement 修饰的依赖，只需要定义 groupId 和 artifactId 即可。例如 parent 项目有以下配置：

```xml
<dependencyManagement>
    <dependencies>
        <dependency>
            <groupId>junit</groupId>
            <artifactId>junit</artifactId>
            <version>4.12</version>
            <scope>test</scope>
        </dependency>
        <dependency>
            <groupId>com.google.code.gson</groupId>
            <artifactId>gson</artifactId>
            <version>2.8.2</version>
        </dependency>
    </dependencies>
</dependencyManagement>
```

子项目要使用父项目的 JUnit 和 Gson 依赖，不需要添加版本号信息，只需向 dependencies 中加入如下依赖配置：

```xml
<dependencies>
    <dependency>
        <groupId>junit</groupId>
        <artifactId>junit</artifactId>
    </dependency>
    <dependency>
        <groupId>com.google.code.gson</groupId>
        <artifactId>gson</artifactId>
    </dependency>
</dependencies>
```

Maven 继承机制以及 dependencyManagement 元素能解决不同模块相同依赖构件版本不一致问题。注意，是 dependencyManagement 而非 dependencies。也许读者已经想到在父模块中配置 dependencies，那样所有子模块都自动继承，不仅达到了依赖一致的目的，还省掉了大段代码。这么做是有问题的，例如将模块 child01 的依赖 spring-aop 提取到了父模块中，但模块 child02 不需要 spring-aop，却也直接继承了。dependencyManagement 就没有这样的问题，dependencyManagement 只会影响现有依赖的配置，但不会引入依赖。

在多模块 Maven 项目中，dependencyManagement 几乎是必不可少的，用它能够有效地维护依赖一致性，消除多模块插件配置重复。

2.5.4　Maven 中的 pluginManagement 的使用

与 dependencyManagement 类似，也可以使用 pluginManagement 元素管理插件。一个常见的用法就是希望项目所有模块使用 Maven Compiler Plugin 的时候，都使用 Java 1.8，以及指定 Java 源文件编码为 UTF-8，这时可以在父模块的 pom 文件中对 pluginManagement 进行如下配置：

```xml
<build>
    <pluginManagement>
        <plugins>
```

```xml
            <plugin>
                <groupId>org.apache.maven.plugins</groupId>
                <artifactId>maven-compiler-plugin</artifactId>
                <version>3.7.0</version>
                <configuration>
                    <source>1.8</source>
                    <target>1.8</target>
                    <encoding>UTF-8</encoding>
                </configuration>
            </plugin>
        </plugins>
    </pluginManagement>
</build>
```

这段配置会被应用到所有子模块的 maven-compiler-plugin 中，由于 Maven 内置了 maven-compiler-plugin 与生命周期的绑定，因此子模块就不再需要任何 maven-compiler-plugin 的配置了。

通常所有项目对于任意一个依赖的配置都应该是统一的，但插件却不是这样，例如你希望模块 A 运行所有单元测试，模块 B 要跳过一些测试，这时就需要配置 maven-surefire-plugin 插件来实现，那样两个模块的插件配置就不一致了。也就是说，简单地把插件配置提取到父 pom 的 pluginManagement 中往往不适合所有情况，因此在使用的时候就需要注意了，只有那些普适的插件配置才应该使用 pluginManagement 提取到父 pom 中。

虽然 Maven 只是用来帮助构建项目和管理依赖的工具，pom 也并不是正式产品代码的一部分，但也应该认真对待 pom。随着敏捷开发和 TDD[一]等方式越来越被人接受，测试代码得到了开发人员越来越多的关注。因此不能仅满足于一个能用的 pom，而应该积极地修复 pom 中使用不当的地方。

[一] TDD 是测试驱动开发（Test-Driven Development）的英文简称，是敏捷开发中的一项核心实践和技术，也是一种设计方法论。

第 3 章 代 码 管 理

本章会介绍两个优秀的版本管理工具 Svn 和 Git。

Svn 是集中式版本控制系统，版本库存放在中央服务器，必须联网才能工作。

Git 是分布式版本控制系统，也就是每个研发人员从中心版本库的服务器上拉取代码后会在自己的计算机上克隆一个自己的版本库。工作的时候不需要联网，因为版本都在自己的计算机上。

3.1 Svn

Svn（Apache Subversion）是一个开放源代码的版本控制系统。文件存放在中心版本库，记录每一次文件和目录的修改。Svn 允许把数据恢复到早期版本，或是检查数据修改的历史。Svn 可以通过网络访问它的版本库，从而使用户在不同的计算机上进行操作，在编写代码的时候，会生成很多不同版本的代码，使用 Svn 可以有效管理不同版本的代码，从而把研发人员从烦琐的版本管理中解放出来。

3.1.1 Svn 客户端的安装

Svn 的使用需要客户端软件，本书选用 TortoiseSVN[⊖]客户端。TortoiseSVN 客户端的安装非常简单，通过如下几步即可完成软件的安装：

（1）通过搜索找到下载地址，或者直接去 TortoiseSVN 的官网 https://tortoisesvn.net/downloads.html 进行下载。下载时选择中文 TortoiseSVN 的正确版本即可。在编写本书时，最新版本为 LanguagePack_1.9.7.27907-x64-zh_CN.msi[⊖]。

（2）双击执行安装。运行后程序会让你选择安装路径，设定好文件夹后即可一步步操作直至安装完成。Svn 版本信息如图 3-1 所示。

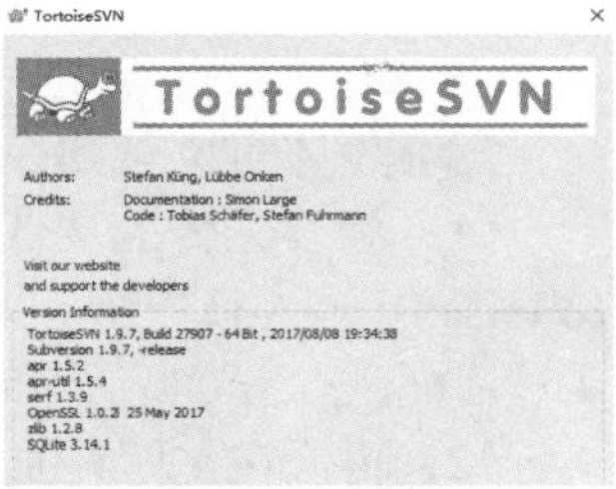

图 3-1 Svn 版本信息

3.1.2 Svn 基本使用

下面介绍使用 Svn 来管理项目。

1. 获取项目文件 SVN Checkout

可以根据 Svn 服务器的地址，例如 svn://39.106.10.196/javadevMapSvn[⊖]以及用户名和密码，将项目 Checkout 到本地。

（1）首先在本地创建一个空的文件夹。在文件夹内执行"鼠标右键单击->SVN checkout"。如图 3-2 所示。

⊖ TortoiseSVN 是一个免费的 Svn 客户端，并且很好地结合了 Windows 系统，可视化界面使它直观且易于使用。

⊖ msi 文件是 Windows Installer 的数据包，它实际上是一个数据库，包含安装一种产品所需要的信息和在很多安装情形下安装（和卸载）程序所需的指令和数据。

⊖ 此处及后面使用的 IP 地址、域名等，均为本书编写时临时使用的测试环境地址，不对读者提供接口或访问能力。

（2）在弹出的对话框中，输入 Svn 的服务器地址，如图 3-3 所示。

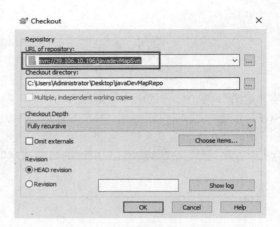

图 3-2　Svn 快捷命令　　　　　　　　　图 3-3　配置 Svn 服务器地址

（3）输入地址后，点击 OK，会弹出一个对话框，输入用户名和密码，记得勾选保存认证，否则每次操作都需要输入用户名和密码。

（4）进行如上操作后，可以看到 Svn 服务器的文档已经下载到本地，如图 3-4 所示。

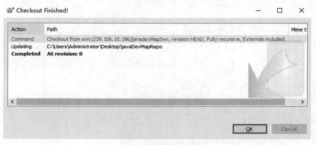

图 3-4　拉取文件成功

2. 提交本地文件到 Svn 服务器

（1）将要上传的文件放到 Svn 管理的文件夹内，然后在文件夹上执行"鼠标右键单击->TortoiseSvn->Add …"，在弹出的对话框中勾选需要上传的文件，然后点击 OK。会弹出一个确认对话框，点击 OK。此时会发现文件夹或者文件上面会有一个小加号。

（2）再次在文件夹的空白处执行"鼠标右键单击->SVN Commit…"，在弹出的输入框中填写本次提交的备注，然后点击 OK 即可。

3. Svn 更新（SVN Update）

更新本地代码至 Svn 服务器上最新版本，只要在需要更新的文件夹上或者在文件图标旁空白处执行"鼠标右键单击-> SVN Update"即可(如果要获取指定历史版本的内容，执行"鼠标右键单击->Update to revision…"）。

4. 删除 Svn 服务器上的文件

如果被删除的文件还未加入 Svn 版本库，直接删除文件即可；如果要删除的文件已加入版本库，则使用如下方法删除：

方法一，选择要删除的文件，执行"鼠标右键单击->TortoiseSVN-> Delete"，然后选择待删除文件的上级目录，执行"鼠标右键单击-> SVN Commit…"，并填写备注。

方法二，在计算机中直接删除该文件，然后选择被删除文件的父目录，执行"鼠标右键单击-> SVN Commit…"，在变更列表中选择被删除的文件，填写备注并提交。

5. SVN 还原（SVN Revert）

进入 Svn 文件夹中，执行"鼠标右键单击->Tortoise SVN->Update to revision…"，然后会弹出一个窗口，如图 3-5 所示。

例如回退到第 4 个版本只需要选择 Revision，并在输入框中填写相应的版本号，然后点击 OK 即可。

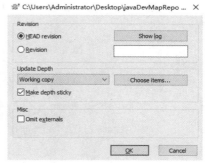

图 3-5　Svn 还原

6. 锁定和解锁（Get lock and Release lock）

Svn 具备文件锁定的能力，锁定后他人将无法修改此文件。

选中要锁定的文件，执行"鼠标右键单击->TortoiseSVN->Get lock..."进行锁定，系统会弹出锁定信息框。当文本文件锁定后，需要通过解锁，他人才能继续对文件进行修改。在被锁定的文件上执行"鼠标右键单击->TortoiseSVN-> Release lock…"进行解锁。

7. 重命名文件（Rename）

要修改文件名，可选中需要重命名的文件或文件夹，然后执行"鼠标右键单击->TortoiseSVN ->Rename…"，在弹出的对话框中输入新名称，点击 OK 按钮，并将修改文件名后的文件或文件夹通过执行"鼠标右键单击->SVN Commit…"提交到 Svn 服务器上。

8. 获取历史文件（Show log）

Show log 有显示日志的作用，主要是显示某文件或目录已经执行的操作，包含被谁修改了以及修改的时间和日期。执行"鼠标右键单击->TortoiseSVN-> Show log"，会显示某路径下的所有文件版本信息。

3.1.3　Svn 解决冲突

为什么会产生冲突代码呢？原因很简单，因为不同的人，同时修改了同一个文件的同一个地方，其中一个人提交了，另一个人就提交不了了。如果另一个人要提交代码，需要先进行更新，然后解决冲突后才能提交。

如果产生冲突，Svn 会生成如下 3 个文件，用于帮助解决冲突。如图 3-6 所示。

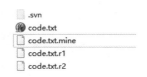

图 3-6　Svn 冲突文件

- code.txt.mine 是你修改后准备提交的版本，即没有提交成功的版本。
- code.txt.r1 是冲突前本地的版本。
- code.txt.r2 是别人赶在你之前提交的版本。

查看 code.txt 的内容，具体如下：

```
<<<<<<< .mine
lisi change it||||||| .r1
=======
Zhangsan change it>>>>>>> .r2
```

其中，<<<<<<<.mine 与=======之间的代码是你自己的，而=======与>>>>>>>.r2 之间的代码是别人与你冲突的部分。

解决方案如下。

（1）在发生冲突的文件上执行"鼠标右键单击-> TortoiseSVN->Edit conflicts"，弹出编辑对话框，如图 3-7 所示。

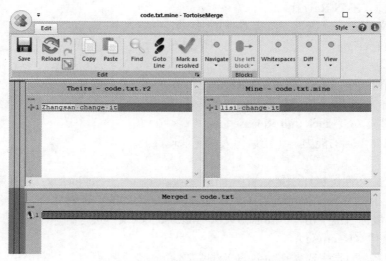

图 3-7　Svn 冲突编辑对话框

注意上面一共有三个窗口，三个窗口的含义见表 3-1。

表 3-1　冲突解决窗口含义

窗口名称	含义
Theirs	窗口为服务器上当前最新版本
Mine	窗口为本地修改后的版本
Merged	窗口为合并后的文件内容显示

对话框中有许多颜色，含义见表 3-2。

表 3-2　冲突解决窗口颜色含义

颜色	含义
浅黄色	新增或修改的内容
白色	表示没有发生任何变化的部分
红色	当前行出现了冲突

在红色块处单击鼠标右键，就会弹出如图 3-8 所示的菜单，根据菜单中的选项，进行文件的合并操作。

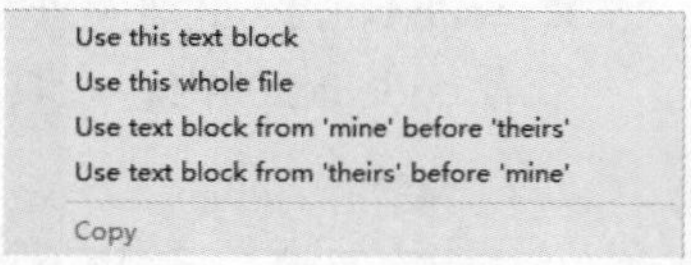

- use this text block：选取选中行的内容。
- use this whole file：选取选中行所在文件的全部内容。
- use text block from mine before theirs：先用自己的内容，接着再用别人的。
- use text block from theirs before mine：先用别人的内容，接着再用自己的。

图 3-8　Svn 解决冲突选项

（2）根据上面的操作步骤，解决完冲突后，选择 save，然后选择"Mark as resolved"，最后在 Svn 文件夹中执行"鼠标右键单击->SVN Commit…"即可。

3.1.4 Svn 分支

在建立项目版本库时，可首先建好项目文件夹，并在其中建立 trunk, branches, tags 三个空的子目录。这三个目录的作用如下：
- trunk 是主分支，是日常开发进行的地方。
- branches 是分支。一些阶段性的发布版本，是可以继续进行开发和维护的，则放在 branches 目录中。或者对于项目不同的开发版本也可以放在此分支中。
- tags 目录一般是只读的，这里存储阶段性的发布版本，只是作为一个里程碑的版本进行存档。

3.2 Git

Git 不同于 SVN，它可以在没有中央系统的情况下进行代码的提交和管理，本节详细介绍 Git 的用法。

3.2.1 Git 客户端安装

本书选用 msysGit 作为 Git 客户端。Git 客户端的安装非常简单，通过如下几步即可完成：

（1）通过搜索找到下载地址，或者直接去 Git 的官网 https://gitforwindows.org/进行下载。下载时选择中文 Git 的正确版本即可。在编写本书时，最新版本为 Git-2.16.2-64-bit.exe。

（2）双击执行安装。运行后程序会让你选择安装路径，设定好文件夹后即可一步步操作直至安装完成。安装完成后，在开始菜单里单击"Git Bash Here"，弹出一个类似命令行窗口的界面，就说明 Git 安装成功。

（3）设置全局用户名和邮箱。执行"鼠标右键单击->Git Bash Here"，在弹出的命令行中分别输入下面的命令。

```
$ git config --global user.name "Your Name"
$ git config --global user.email "Youremail@example.com"
```

3.2.2 Git 基本使用

在使用 Git 之前，有必要了解一下 Git 的几个重要概念：Git 工作区、暂存区和版本库。
- 工作区（Working directory）：简单地说是在计算机里能看到的目录。
- 暂存区（stage）：用来暂时存放工作区中修改的内容。
- 版本库（Repository）：工作区里有一个名为 .git 的隐藏目录，这个目录不算工作区，而是 Git 的版本库。

Git 基础环境准备妥当后，需要先从远程仓库克隆项目文件。准备好远程仓库，例如 git@gitee.com:hwhe/JavaDeveloperMap.git[①]，这里打开"Git Bash Here"命令行，用命令 git clone 克隆一个本地库。

```
$ git clone git@gitee.com:hwhe/JavaDeveloperMap.git
```

[①] Gitee：码云(gitee.com)是开源中国社区团队推出的基于 Git 的快速的、免费的、稳定的在线代码托管平台。

本节使用的远程仓库是通过 Oschina 创建的，Oschina 给出的地址不止一个，还可以用 https://gitee.com/hwhe/JavaDeveloperMap.git 这样的地址。实际上，Git 支持多种协议，默认使用 ssh，但也可以使用 https 等其他协议。区别为 https 方式每次 push 都必须输入口令。

Git 常用命令如下：

- git clone 复制一个仓库到本地

用 git clone 复制一个 Git 仓库到本地，能够查看该项目，或者进行修改。

```
$ git clone [url]
```

- git add 添加文件到缓存

```
$ git add test.java
```

- git status 查看当前 Git 状态，来确定当前是否有修改

```
$ git status
```

- git commit 保存到本地仓库中

git commit 是将修改推送到本地仓库中。使用-m 选项可以设置提交注释。执行此命令之前，需要先执行 git add 将修改放入暂存区中。

```
$ git commit -m '本地提交的备注'
```

- git push 推送到远程仓库

git push 命令用于将本地分支的更新，推送到远程主机。它的格式与 git pull 命令相仿。

```
$ git push <远程主机名> <本地分支名>:<远程分支名>
```

注意，分支推送顺序的写法是<来源地>:<目的地>，所以 git pull 后面跟<远程分支>:<本地分支>，而 git push 后面跟<本地分支>:<远程分支>。

如果省略远程分支名，则表示将本地分支推送到与之存在"追踪关系"的远程分支（通常两者同名），如果该远程分支不存在，则会被创建。

```
$ git push origin master
```

上面命令表示，将本地的 master 分支推送到 origin 主机的 master 分支。如果后者不存在，则会被创建。

- git pull 同步远程分支到本地

```
$ git pull
```

如果本地没有配置 SSH 公钥，则需要根据提示输入用户名和密码才能更新。配置 SSH 公钥，可以免去输入用户名和密码，具体配置方法在 3.2.5 节会有详细介绍。

- git reset 和 revert 代码回滚

第一种情况，还没有 push，只是在本地 commit。

1）找到之前提交的 git commit 的 id 信息

```
$ git log
```

2）找到想要撤销的 id，执行 git reset 命令，完成撤销，同时将代码恢复到 commit_id 对应的版本。注意：hard 参数的作用是使修改的代码也回滚到 commit_id 的版本。

```
$ git reset --hard <commit_id>
```

或者

```
$ git reset <commit_id>
```

不带参数 hard，则仅完成 commit 命令的撤销，不对代码修改进行撤销，可以直接通过 git commit 重新提交本地修改的代码。git reset 常用参数见表 3-3。

表 3-3　git reset 常用参数

窗 口 名 称	含　　义
—mixed	为默认方式。回退到某个版本，只保留源码，回退 commit 和 index 信息
—soft	回退到某个版本，只回退 commit 的信息
—hard	彻底回退到某个版本，本地的源码也会变为目标版本的内容

第二种情况，commit push 代码已经更新到远程仓库。

对于已经把代码 push 到线上仓库，如果回退本地代码也想同时回退线上代码，使线上、线下代码保持一致，需要用到下面的命令：

```
$ git revert <commit_id>
```

revert 之后本地代码会回滚到指定的历史版本，这时再 git push 即可把线上的代码更新。

git reset 是回退到某次提交，提交及之前的 commit 都会被保留，但是此次之后的修改都会被退回到暂存区；git revert 是生成一个新的提交来撤销某次提交，此次提交之前的 commit 都会被保留。

当前使用的开发工具 Eclipse 默认带有 Git 插件，操作更加便利。例如导入项目，执行"file->import->Git"，根据提示输入对应 Git 信息，即可完成 Git 项目导入。

当要提交代码、更新项目时，只需在项目上执行"鼠标右键单击->Team"操作对应的菜单即可，如图 3-9 所示。

图 3-9　Eclipse 中的 Git 菜单

3.2.3　Git 分支管理

在 Git 开发过程中，经常需要使用分支操作把当前的代码从开发主分支 master 上分离开来，在不影响主分支的情况下继续工作。分支也是 Git 的优秀特性之一。

- 查看已有的分支

```
$ git branch –a
```

- 创建一个分支

```
$ git branch <分支名>
```

- 切换分支

```
$ git checkout –b <分支名>
```

- 合并分支(注意是当前的分支合并其他分支):

```
$ git merge <其他分支名>
```

3.2.4　Git 标签

Git 可以针对某一个时间点的版本打一个标签（tag），例如当前开发了一个稳定版本，可以用"git tag …"命令打一个标签留作标记。

标签相关命令如下：
- 查看所有标签

    $ git tag

- 创建标签

    $ git tag <标签名>

- 删除标签

    $ git tag -d <标签名>

- 切换标签

    $ git checkout <标签名>

- 发布标签提交到 Git 服务器

    $ git push origin <标签名>

通常的 git push 不会将标签对象提交到 Git 服务器，所以需要进行显式的操作。

3.2.5　在 Git 中配置 SSH

Git 使用 https 协议，每次 pull、push 都会提示输入密码，如果使用 Git 协议，然后使用 SSH 密钥，这样可以免去每次都输入密码的麻烦。

初次使用 Git 的用户要使用 Git 协议一般需要三个步骤：

（1）生成密钥对。

Git 服务器都会选择使用 SSH 公钥来进行授权，每个用户必须提供一个公钥用于授权，没有的话就要生成一个。SSH 公钥默认储存在账户主目录下的 ~/.ssh 目录中。目录中有.pub 后缀的文件就是公钥，另一个文件则是密钥（例如 id_rsa 和 id_rsa.pub）。

假如没有这些文件，甚至连.ssh 目录都没有，可以用 ssh-keygen 来创建。该程序包含在 MsysGit 包里，打开"Git Bash Here"命令行执行以下命令：

    $ ssh-keygen -t rsa -C "your_email@youremail.com"

然后，系统会提示输入密码（建议输入密码具备一定的安全性，当然不输入也是可以的），按照提示设置完成后，本地的密钥对就生成了。

（2）设置远程仓库（以码云为例）上的公钥。

进入本机当前登录用户下面的"~/.ssh"目录，执行"鼠标右键单击-> Git Bash Here"，执行"cat ~/.ssh/id_rsa.pub"命令，查看生成的公钥。

    $ cat ~/.ssh/id_rsa.pub

复制生成的公钥的内容，登录码云账号，点击"头像->设置"，然后点击左边菜单的 SSH 公钥，复制上面的公钥内容，粘贴进"公钥 value"文本域内。在公钥标题域，起一个名字。

（3）把 Git 的 remote url 修改为 Git 协议。

如果之前的代码已经使用 Git 协议，则这一步可以略过，如果之前使用的是用 https 协议下载的代码，可以使用如下命令查看并修改协议类型。

    $ git remote -v
    origin https://gitee.com/hwhe/spring-mvc.git (fetch)
    origin https://gitee.com/hwhe/spring-mvc.git (push)

从码云项目上复制 Git 协议的相应的 url，执行 git remote set-url 来调整 url。

```
$ git remote set-url origin git@gitee.com:hwhe/JavaDeveloperMap.git。
```

3.2.6 用 Git stash 暂存代码

本节讲解 git stash，它可用来暂存当前正在进行的工作，例如想 pull 最新代码，又不想提交当前代码。常常遇到的情况就是，为了修改一个紧急的 bug，先暂存当前代码，然后迁出之前的代码，修改完 bug 后提交代码，最后从暂存区取出暂存的代码。

git stash 使用步骤如下：

（1）保存当前修改。

git stash 会把所有未提交的修改（包括暂存的和非暂存的）都保存起来，用于以后恢复当前工作内容。同时注意，git stash 是本地的，不会随 push 命令上传到服务器上。

```
$ git stash
$ git status
```

通过上面的命令，可以将当前代码恢复到最近提交的状态，这样就可以先去执行其他紧急的任务。例如切换到其他分支，修改 bug，修改完后，再切换到之前的分支，继续上次没修改完的内容。

（2）恢复之前临时缓存的内容。

```
$ git stash pop
```

上面的命令可以将缓存堆栈中的第一个 stash 删除，并将对应修改应用到当前的工作目录下。也就是恢复到上一个 stash 命令执行之前的状态。

（3）查看所有的 stash。

```
$ git stash list
```

（4）移除 stash，注意这里的 stash 的名称，可以从 stash list 输出里面找到。

```
$ git stash drop <stash 名称>
```

第 4 章 Linux 命令

在日常工作中，通常测试环境和生产环境的软件都是运行在 Linux 服务器上，了解并掌握基本的 Linux 命令对研发者来讲尤为重要。本章将讲解 Linux 系统的常用命令。

4.1 Linux 简介

Linux 是一套免费使用和自由传播的类 UNIX 操作系统，是一个基于 POSIX 和 UNIX 的多用户、多任务、支持多线程和多 CPU 的操作系统。它能运行主要的 UNIX 工具软件、应用程序和网络协议。Linux 是一个性能稳定的多用户操作系统。

4.2 Linux 常用命令

Linux 提供了大量的命令，可以有效地完成大量的工作，如磁盘操作、文件存取、目录操作、进程管理、文件权限设定等。Linux 发行版最少的命令也有 200 多个，这里只介绍比较重要和使用频率最高的命令。

Linux 命令格式如下，命令参数含义见表 4-1。

```
$ command [-options] [parameter1] …
```

表 4-1　Linux 命令参数

名　称	含　义
command	命令名，相应功能的英文单词或单词的缩写
[-options]	选项，可用来对命令进行控制，也可以省略，[]代表可选
parameter1 …	传给命令的参数：可以是零个、一个或者多个

（1）查看帮助文档

1）--help

在 Linux 命令后加上--help 参数可显示命令自带的帮助信息。例如：

```
$ ls --help
```

2）man(manual)

man 是 Linux 提供的一个手册，包含了绝大部分的命令、函数使用说明，该手册分成很多章节（section），使用 man 时可以指定不同的章节来浏览。例如：

```
$ man ls
```

（2）Linux 命令自动补全

用户敲出命令的前几个字母后，可按下 tab 键，系统会自动补全命令。

（3）查看历史输入的命令

```
$ history
```

当系统执行过一些命令后，执行 history 命令会将执行过的命令列举出来。也可按上下键翻看以前的命令。

（4）df -h 查看当前磁盘使用情况。

  $ df -h

（5）free 显示当前系统未使用的和已使用的内存大小，还可以显示内核使用的内存缓冲区大小。"-h"的作用是人性化输出内存单位。

  $ free -h

输出内容中每列的含义见表 4-2。

表 4-2　free 命令输出内容

列　名　称	含　　义
total	内存总数
used	已使用的内存数
free	空闲的内存数
shared	当前已经废弃不用的内存数
buffers/cache	缓存内存数
available	估计可提供内存数

（6）执行 shell 脚本文件。假设已经有了一个具有执行权限的 shell 脚本文件 start.sh，那么此文件有以下两种执行方式：

1）sh 命令执行 shell 脚本

  $ sh start.sh

2）"./"前缀启动 shell 脚本

  $ ./start.sh

（7）打包和解压文件。在 Linux 系统中经常需要打包和解压文件。下面介绍 zip 包和 tar 包的打包和解压命令。

1）zip 包的打包命令

  $ zip test.zip fileZip

将 fileZip 打包成一个 zip 格式的名为 test.zip 的压缩包。

2）解压 test.zip。

  $ unzip test.zip

3）tar 包的打包命令

  $ tar -czvf png.tar.gz test.png

将当前目录下面的 test.png 打包成一个名为 png.tar.gz 的包。

4）解压 png.tar.gz

  $ tar -xzvf png.tar.gz

tar 命令参数的含义见表 4-3。

表 4-3 tar 命令参数

参数	含义
-c	创建压缩文档
-x	解压
-z	调用 gzip 压缩（解压）
-v	显示命令运行全过程
-f	使用文件名。注意，这个参数只能是最后一个参数，后面只能接文件名，即 -f 文件名

4.3 Linux 文件管理

本节将介绍文件操作的基本命令以及文件权限的级别和使用。

4.3.1 Linux 文件操作命令

（1）查看文件信息：ls

ls 是英文单词 list 的简写，其功能为列出目录的内容，是 Linux 系统最常用的命令之一，它类似于 DOS 下的 dir 命令。

```
$ ls
```

常用参数见表 4-4。

表 4-4 ls 常用参数

参数	含义
-a	显示指定目录下所有的子目录与文件，包括隐藏文件
-l	以列表形式显示文件的详细信息
-h	配合-l 以友好的方式显示文件大小

（2）分屏显示：more

查看内容时，当信息过长无法在一屏上显示时，会出现快速滚屏，使得用户无法看清文件的内容，此时可以使用 more 命令，每次只显示一页，按下空格键可以显示下一页，按下 q 键退出显示，按下 h 键可以获取帮助。

```
$ more <文件名>
```

（3）管道命令：|

一个命令的输出可以通过管道作为另一个命令的输入。"|" 的左端输入东西（写），右端读取东西（读）。

```
$ ls -lh | more
```

（4）切换工作目录：cd

cd 命令可以切换工作目录，常用的 cd 命令见表 4-5。

表 4-5 cd 常用命令

参数	含义
cd ~	切换到当前用户的主目录(home/用户目录)
cd .	切换到当前目录
cd ..	切换到上级目录
cd -	切换到上次所在的目录

（5）显示当前路径：pwd

使用 pwd 命令可以显示当前的工作目录，该命令很简单，直接输入 pwd 即可，后面不带参数。

```
$ pwd
```

（6）创建目录：mkdir

```
$ mkdir mapDir
```

通过 mkdir 命令可以创建一个新的目录。参数-p 可递归创建目录。需要注意的是新建目录的名称不能与当前的目录中已有的目录或文件同名，并且目录创建者必须对当前目录具有写权限。

（7）删除文件：rm

可通过 rm 删除文件或目录。使用 rm 命令要小心，因为文件删除后不能恢复。为了防止文件误删，可以在 rm 后使用-i 参数以逐个确认要删除的文件。

```
$ rm test.txt
```

rm 命令常用参数以及含义见表 4-6。

表 4-6 rm 参数

参　数	含　义
-i	删除已有文件或目录之前先询问用户
-f	强制删除文件或目录
-r 或 -R	递归删除，指定目录下的所有文件与子目录一并处理
-v	显示指令详细执行过程

（8）删除目录：rmdir

可使用 rmdir 命令删除一个目录。执行此命令时必须离开待删除目录，并且目录必须为空目录，否则提示删除失败。

```
$ rmdir mapDir
```

（9）文件搜索：find

```
$ find / -name test.log
```

上面命令的含义是：从根文件系统开始搜索名称为 test.log 的文件。

4.3.2 Linux 文件权限

Linux 的文件权限分为读、写、执行三种，可以使用 ls -l 进行查看。见表 4-7。

表 4-7 文件权限

参　数	含　义
r	可读的权限，数值表示为 4
w	写文件的权限，对于目录，可以创建新的文件，数值表示为 2
x	可执行，对于目录，可以进入，数值表示为 1
-	无对应的权限

仔细观察，会发现与权限相关的标记总共有 9 个，每 3 个为一个小组。三个小组的含义见表 4-8。

表 4-8　分组权限

组　别	含　义
第一组	文件所有者的权限
第二组	文件所有组的权限
第三组	其他人权限，不是所有者，也不在所有组里面

例如，要修改文件权限，执行如下命令：

```
$ chmod 754 javaMap.txt
```

根据表 4-8，上面的命令含义如下。

（1）7=4+2+1，表示文件所有者：可读可写可执行权限。

（2）5=4+1，表示文件所属组：拥有可读可执行权限，但是没有写权限。

（3）4 代表其他：拥有可读权限。

4.4　Linux 启动服务

在 Linux 中常用的启动 Java 服务方式有两种。一种是放在容器里面的 war 包，一种是 jar 包形式的服务。接下来，分别演示这两种服务的启动方式。

（1）以 tomcat 容器为例，在 Linux 环境下将 war 包放到 tomcat 容器后，启动 tomcat，运行服务。常用的命令如下：

1）进入 tomcat 的 bin 目录

```
$ cd /usr/local/tomcat8.524/bin
```

2）执行 tomcat 启动命令

```
$ ./catalina.sh start
```

3）关闭 tomcat 服务

```
$ ./catalina.sh stop
```

4）重启 tomcat 服务

```
$ ./catalina.sh restart
```

5）查看服务启动日志

```
$ cd /usr/local/tomcat8.524/logs
$ tail -f catalina.out
```

6）查看 tomcat 服务是否存活

```
$ ps -ef | grep tomcat
```

（2）在 Linux 服务器上以后台服务的方式启动 jar 程序：

1）nohup 命令

在讲解如何后台启动 Java 程序之前，先讲解一下 nohup 命令。

nohup 是不挂断的运行命令，即如果你正在运行一个进程，而且要在退出账户时该进程不结束，那么可以使用 nohup 命令。该命令可以在退出账户或者关闭终端之后继续运行相应的进程。nohup 可以理解为不挂起的意思（no hang up）。此命令的语法为：

```
nohup Command [ Arg ... ] [&]
```

nohup 命令可以和数据流进行交互。Linux 操作系统中有三个常用的数据流，见表 4-9。

表 4-9 数据流

数据流数字	含 义
0	标准输入流 stdin
1	标准输出流 stdout
2	标准错误流 stderr

在 nohup 命令中，当使用"> console.txt"参数时，实际是"1 > console.txt"的省略用法，即标准输出重定向到 console.txt 文件中；而"<console.txt"参数实际是"0 < console.txt"的省略用法，是将 console.txt 输入到标准输入流中。

2）启动 Java 服务

启动一个 jar 包服务，要先定位到 jar 包所在的目录下面，然后执行下面的命令：

```
$ nohup java –jar xx.jar >/dev/null &
```

注意/dev/null 代表空设备文件。该命令表示上面的启动 Java 服务的输出是不需要的，直接导向空的设备。

jps 命令可以查看有权限访问的 Java 虚拟机进程。使用此命令可以查看运行的 Java 程序。

```
$ jps
```

可以使用不同的命令参数，对启动的 Java 程序进行配置。

```
$ nohup java –jar –Xmx512M –Xms128M test.jar > nohup.out 2>&1 &
```

以上命令指定 Java 程序启动时内存为 128MB，最大内存为 512MB。同时输出内容不打印到屏幕上，而是输出到 nohup.out 文件中。 2>&1 是将标准错误流重定向到标准输出，这里的标准输出已经指定为 nohup.out 文件，所以标准错误流也输出到 nohup.out 文件中。最后一个&，是让该命令在后台执行。

Java 程序启动时堆内存参数见表 4-10。

表 4-10 Java 程序启动时堆内存参数

数据流数字	含 义
Xms	程序启动时占用内存大小
Xmx	程序运行期间最大可占用的内存大小
Xss	每个线程的堆栈大小

第二篇　服务框架篇

第一篇分别讲解了 Java 语言、Maven 工程构建、代码管理和服务器命令，这些内容组合起来距离一个可以支撑业务运行的服务来说好像还很远。

如果一个业务请求通过 HTTP 协议发送到 Java 程序，程序如何承接请求、如何转化协议为对象、如何把请求映射到业务处理逻辑？如果业务请求希望得到一个可以展示的页面，那么如何绘制页面并且正确地返回给请求方？如果某个业务希望永久保存某些数据，那么就不能把数据存储在程序内存中，数据需要通过数据库进行持久化操作，如何通过程序操作数据库？如果这些问题要自己一一解决将会非常困难。好在现在有许多非常成熟的服务框架可以使用，例如常用的 SSH（Struts+Spring+Hibernate）框架或者 SSM（Spring+ SpringMVC+MyBatis）框架。

框架可以帮助解决上面列出的大部分通用的基础问题，研发人员只需要使用框架进行正确的配置并完成业务的特殊逻辑就可以实现业务需求。本篇会介绍这些框架的原理和核心的用法，通过 Spring 进行服务管理，通过 Spring MVC 进行业务流程控制，通过 MyBatis 进行持久化操作。然后会介绍 Spring Boot 工程，它会使框架对业务的影响更小，配置更简单。

通过 SSM 框架已经可以进行程序的业务研发工作了，但是此框架对于承载大型互联网业务可能存在一些弊端⊖。例如在同一工程中，代码间的耦合和依赖会是一个问题；部分修改尤其是基础模块的修改需要整个项目的编译及系统测试，会是一种人力成本的浪费；某些模块能力需要对外开放时，传统架构对单一模块的扩展无法支撑也是一个问题。诸如此类的问题非常多，其实汇聚为一点，就是大系统不灵活。

为了解决以上问题，微服务的理念应运而生。简单来讲，微服务就是把一个大系统拆分成很多个小系统，每个小系统独立运行，承担某一部分能力，这样可以让单独一个小系统承担的任务更加纯粹，负责这块系统的研发人员也可以更加专注。微服务的好处很多，总体来看它确实优化了很多传统框架无法解决的问题。但是把一个大程序拆分成很多小程序，也会带来很多问题。因为程序间是通过网络调用的，不像传统架构是程序内部调用，那么程序间的网络通信是微服务要解决的问题；还有程序间的调用关系链及依赖问题、服务间互相调用时的服务发现问题等。微服务的引入带来的问题主要通过微服务框架进行解决，所以本篇还会介绍 Spring Cloud 微服务框架。

希望读者通过本篇的学习，能够对服务整体的运行原理有所认识，能够亲手搭建一个传统框架的服务和一套微服务框架的服务。

⊖ 虽然作者认为错误都是人造成的，框架不应该承担这些责任，但是如果能够通过框架降低人为犯错的概率也是个不错的选择。

第 5 章　Spring

Spring[一]是一个轻量级 JavaSE/JavaEE 开发应用框架[二]，可以一站式地构建企业应用。Spring 是模块化的，几乎涵盖了开发所需要的所有组件，如果业务需求超出其能力，也可方便集成第三方组件。Spring 可以管理对象，还提供了适用于安全控制、异常处理、日志记录等场景的面向切面的能力，同时，Spring 提供与第三方框架无缝集成能力，进一步方便业务开发和拓展。

5.1 Spring 概述

Spring 框架由 7 个核心模块组成。Spring 模块构建在核心容器之上，核心容器定义了创建、配置和管理 Bean 的方式。

5.1.1 核心模块

下面简要描述每个模块的作用。
- Spring Core：Core 封装包是框架的最基础部分，提供 IoC 和依赖注入特性。
- Spring Context：构建于 Core 封装包基础上的 Context 封装包，提供了一种框架式的对象访问方法。
- Spring DAO：DAO (Data Access Object)提供了 JDBC 的抽象层，消除了冗长的 JDBC 编码，能解析数据库厂商特有的错误代码。JDBC 封装包还提供了一种比编程性更好的声明性事务管理方法，不仅实现了特定接口，而且对所有的 POJOs（plain old Java objects）都适用。
- Spring ORM：ORM 封装包提供了常用的"对象/关系"映射的集成层。
- Spring AOP：Spring 的 AOP 封装包提供了面向切面的编程实现，从而减弱了代码的功能耦合，使切面逻辑和功能逻辑清晰地分离开。
- Spring Web：Spring 中的 Web 包提供了基础的针对 Web 开发的集成特性。Spring 提供对常见框架如 Struts、webwork、JSF 的支持，能够管理这些框架，将资源注入给这些框架，也能在这些框架的前后插入拦截器。
- Spring Web MVC：Spring 中的 MVC 封装包提供了 Web 应用的 Model-View-Controller（MVC）实现。

5.1.2 预备知识

在学习 Spring 之前，需要理解以下几个概念。
- POJO：Plain Old Java Objects，简单的 Java 对象。
- 容器：在日常生活中容器就是一种盛放东西的器具，从程序设计角度看，容器是管理其他

[一] Spring 官网是 http://spring.io。
[二] 框架：提供了一些基础功能，简化开发，让研发人员更加专注业务逻辑的组件集合。

第 5 章 Spring

对象的对象。因为存在放入、拿出等操作，所以容器还要管理对象的生命周期。
- 控制反转：即 Inversion of Control，缩写为 IoC，控制反转还有一个名字叫作依赖注入（Dependency Injection），就是由容器控制程序之间的关系，而非传统实现中由程序代码直接操控。
- Bean：一般指被容器管理的对象，在 Spring 中指 Spring IoC 容器管理的对象。

5.2 构建第一个 Spring 工程

Spring 的核心是 IoC 容器，其他所有技术都是基于容器实现的。下面创建一个 Spring 项目，通过 Spring 创建一个 Product 类实例，来演示 IoC 功能。

（1）构建工程

在 Eclipse 中执行"File->New->Maven Project"命令，在弹出的对话框中选择"Select an Archetype"以及下面的"Maven-archetype-quickstart"选项，点击 Next 按钮，在弹出的对话框中输入 Maven 的坐标信息，具体输入信息如下：

```
<groupId>com.javadevmap</groupId>
<artifactId>SpringBasic</artifactId>
<version>0.0.1-SNAPSHOT</version>
<packaging>jar</packaging>
```

然后点击 Finish 按钮，完成基本的 Maven 项目构建。

（2）添加依赖

在 pom 文件中加入 Spring Context 依赖。具体如下：

```
<dependency>
    <groupId>org.springframework</groupId>
    <artifactId>spring-context</artifactId>
    <version>4.3.3.RELEASE</version>
</dependency>
```

（3）添加配置文件

Spring 配置文件是用于指导 Spring 工厂进行 Bean①生产、注入及 Bean 实例分发的核心组成部分。在项目的 resources 文件夹下，新建 Spring 配置文件 spring-bean.xml。即在 resources 文件夹上执行"鼠标右键单击->new->Spring Bean Configuration File"，文件内容具体如下：

```
<?xml version="1.0" encoding="UTF-8"?>
<beans xmlns="http://www.springframework.org/schema/beans"
    xmlns:xsi="http://www.w3.org/2001/XMLSchema-instance"
    xsi:schemaLocation="http://www.springframework.org/schema/beans
    http://www.springframework.org/schema/beans/spring-beans.xsd">
    <!-- 配置信息 -->
</beans>
```

（4）创建 Product 的 POJO 类，具体如下：

```
public class Product {
    private int id;
    private String name;
```

---

① 本章用 Bean 表示类实例，bean 表示配置文件中的配置。

```java
        public Product() {
        }
        public Product(int id, String name) {
            System.out.println("invoke method -- Product(int id, String name)");
            this.id = id;
            this.name = name;
        }
        public int getId() {
            return id;
        }
        public void setId(int id) {
            System.out.println("invoke method -- setId");
            this.id = id;
        }
        public String getName() {
            return name;
        }
        public void setName(String name) {
            System.out.println("invoke method -- setName");
            this.name = name;
        }
    }
```

（5）配置 bean 标签

需要在 Spring-bean.xml 文件中，配置相应的资源信息。这里 bean 标签的作用以及含义会在 5.3.3 节介绍。

```xml
<?xml version="1.0" encoding="UTF-8"?>
<beans xmlns="http://www.springframework.org/schema/beans"
    xmlns:xsi="http://www.w3.org/2001/XMLSchema-instance"
    xsi:schemaLocation="http://www.springframework.org/schema/beans
    http://www.springframework.org/schema/beans/spring-beans.xsd">
    <!-- 配置信息 -->
    <bean id="beanId" class="com.javadevmap.bean.Product"></bean>
</beans>
```

（6）构建测试

接下来，在 src/test/java 目录下面创建一个 JUnit 的测试类，具体如下：

```java
public class TestIocCaseStart {
    ApplicationContext ctx;
    @Test
    public void testCase() {
        ctx=new ClassPathXmlApplicationContext("spring-bean.xml");
        //从容器中获得 id 为 beanId 的 bean
        Product product=(Product)ctx.getBean("beanId");
        System.out.println("ApplicationContext.getBean() = "+product);
    }
}
```

运行结果如下：

```
ApplicationContext.getBean() = com.javadevmap.bean.Product@3d921e20
```

执行完上面几步，即可通过 Spring 实例化一个 Product 对象，可以直接使用这个 Product 对象，而不用关心如何创建以及销毁，这些都是由 Spring 进行管理的。至此，就搭建完成一个最简单的 Spring 的开发环境。接下来会在此环境上进行其他功能的演示。

5.3 IoC

上一节中的 Product 类实例化是由 Spring 的什么功能实现的呢？这要归功于 Spring 的 IoC 模块。IoC 是 Spring 框架的核心内容，既可使用 XML 配置，也可以使用注解配置。Spring 容器在初始化时先读取配置文件，根据配置文件创建与组织对象，并存入容器中，运行程序时再从 IoC 容器中取出需要的对象实例。

5.3.1 IoC 和 DI 基本原理

IoC（Inversion of Control），即"控制反转"，是一种设计思想。在 Java 开发中，IoC 意味着将定义好的对象交给容器控制。这与传统 Java SE 主动通过 new 创建对象的程序设计方式不同。

DI（Dependency Injection），即"依赖注入"。组件之间的依赖关系由容器在运行期决定，即由容器动态地将某个依赖关系注入到组件之中。依赖注入的目的是为了提升组件重用的频率，并为系统搭建一个灵活、可扩展的平台。通过依赖注入机制，只需要简单的配置，而无需任何代码便可指定目标需要的资源，完成其自身的业务功能，而不需要关心具体的资源来自何处，由谁实现。

在传统应用开发中，自己主动控制并创建依赖或注入对象的方式称为正转；而反转则是由容器来创建及注入依赖对象。

5.3.2 IoC 的配置使用

Spring 中 IoC 的配置有两种方式，一种是 XML 实现方式，一种是 IoC 注解实现，两种方式各有利弊，接下来一一讲解。

Spring 的 XML 配置文件的根元素是 beans，每个组件使用 bean 元素来定义，bean 元素可以有许多属性，其中有两个重要的属性：id 和 class。id 表示组件的默认名称，class 表示组件的类型。

通过 Spring 获取 Bean 对象有如下几种方式：
- 不指定 id，只配置必需的全路径类名即 class 属性，IoC 容器会为其生成一个标识，客户端必须通过接口"T getBean(Class<T> requiredType)"获取 Bean。
- 指定 id，必须在 IoC 容器中唯一。
- 指定 name，必须在 IoC 容器中唯一。
- 同时指定 id 和 name，id 就是标识符，而 name 就是别名，必须在 IoC 容器中唯一。

新建 spring-bean-ioc.xml 文件，使用上面介绍的几种方式配置之前定义的 product 类。核心内容如下：

```
<?xml version="1.0" encoding="UTF-8"?>
<beans xmlns="http://www.springframework.org/schema/beans"
    xmlns:xsi="http://www.w3.org/2001/XMLSchema-instance"
    xsi:schemaLocation="http://www.springframework.org/schema/beans
    http://www.springframework.org/schema/beans/spring-beans.xsd">
```

```xml
<!-- 配置信息 -->
<bean class="com.javadevmap.bean.Product"></bean>
<bean id="beanId" class="com.javadevmap.bean.Product"></bean>
<bean name="beanName" class="com.javadevmap.bean.Product"></bean>
<bean id="beanId01" name="beanName01" class="com.javadevmap.bean.Product"></bean>
</beans>
```

那么如何获取上面配置的 Bean 对象呢？这需要用到 ApplicationContext 容器。ApplicationContext 是 Spring 中较高级的容器，可以加载配置文件中定义的 bean，构建为 Bean 对象，对 Bean 集中管理，按需分配 Bean。

最常使用的 ApplicationContext 接口实现类有以下三种：

- FileSystemXmlApplicationContext：该容器从 XML 文件中加载已定义的 bean。需要提供给构造器 XML 文件的完整路径。
- ClassPathXmlApplicationContext：该容器从 CLASSPATH 中的 XML 文件加载已定义的 bean。不需要提供 XML 文件的完整路径，正确配置 CLASSPATH 环境变量即可，因为容器会从 CLASSPATH 中搜索 bean 配置文件。
- WebXmlApplicationContext：该容器会在一个 Web 应用程序的范围内加载已在 XML 文件中定义的 bean。

当获取 Application Context 的上下文后，就可以通过 getBean() 方法得到所需要的 Bean。这个方法通过配置文件中的 bean ID 来返回一个真正的对象。

根据 ApplicationContext 容器获取上面配置的 Bean，测试类如下：

```java
public class TestIocCaseIoC {
    ApplicationContext ctx;
    @Test
    public void testCase() {
        ctx=new ClassPathXmlApplicationContext("spring-bean-ioc.xml");
        //从容器中获得 id 为 Product 的 bean
        //Product proClass=(Product)ctx.getBean(Product.class);
        //System.out.println("不指定 id，只配置必须的全限定类名= "+proClass);
        Product pro=(Product)ctx.getBean("beanId");
        System.out.println("指定 id 获取  ="+pro);
        pro=(Product)ctx.getBean("beanName");
        System.out.println("指定 name 属性获取 = "+pro);
        pro=(Product)ctx.getBean("beanId01");
        System.out.println("指定 id 和 name 获取= "+pro);
    }
}
```

运行结果如下：

```
指定 id 获取  =com.javadevmap.bean.Product@3d921e20
指定 name 属性获取 = com.javadevmap.bean.Product@36b4cef0
指定 id 和 name 获取= com.javadevmap.bean.Product@fad74ee
```

注意上面例子，如果使用注释中的 ctx.getBean(Product.class)方法获取，会报告以下错误：

```
org.springframework.beans.factory.NoUniqueBeanDefinitionException: No qualifying bean of type [com.javadevmap.bean.Product] is defined: expected single matching bean but found 4: com.javadevmap.bean.Product#0,beanId,beanName,beanId01
```

通过 class 方法获取 Bean，Spring 发现有四个实例，但并不清楚用哪个，所以会报错，在 5.3.9 节会介绍用@Qualifier 来区分各实例，这里先注释掉。

运行本例，可在控制台看到输出结果，可以发现配置的 Bean 对象已由 ClassPathXmlApplicationContext 容器管理。之后的业务操作也不需要主动创建一个实例，只需要配置 bean 和执行获取 Bean 方法即可。Bean 的整个生命周期由 Spring 进行管理。

5.3.3　Bean 定义

上一节通过 bean 标签就能实现从容器中获取 Bean 对象，bean 标签对 Bean 的容器化管理非常重要，其属性和作用见表 5-1。

表 5-1　bean 标签

属　　性	描　　述
class	这个属性是必填的，创建对象所在类的全路径
id	这个属性指定唯一的 bean 标识符。在基于 XML 的配置元数据中，可以使用 id 或 name 属性来指定 bean 标识符。
scope	这个属性用来指定 Bean 对象的作用域
constructor-arg	通过构造函数注入
property	通过 Bean 的 setter 方法注入

了解了 bean 的基本配置属性后，考虑这样一个场景，在某些对象实例化的时候进行初始化操作，例如通过构造函数或者设置属性值。以 Product 类为例，希望在初始化 Bean 时将对象的 id 和 name 属性进行初始化，这就用到了 Spring 的 bean 标签 constructor-arg 和 property 两个配置项。

Spring IoC 容器注入依赖资源主要有以下两种基本实现方式：

（1）构造器注入：通过在 bean 定义中指定构造器参数进行注入，包括实例工厂方法参数注入（静态工厂方法参数不允许注入）。

使用 constructor-arg 指定构造函数初始化 name 和 age 属性值，具体如下：

```
<?xml version="1.0" encoding="UTF-8"?>
<beans xmlns="http://www.springframework.org/schema/beans"
    xmlns:xsi="http://www.w3.org/2001/XMLSchema-instance"
    xsi:schemaLocation="http://www.springframework.org/schema/beans
    http://www.springframework.org/schema/beans/spring-beans.xsd">

    <bean id="beanNoConstructorArg" class="com.javadevmap.bean.Product"></bean>
    <bean id="beanHasConstructorArg" class="com.javadevmap.bean.Product">
        <constructor-arg name="id" value="1001"></constructor-arg>
        <constructor-arg name="name" value="java dev map"></constructor-arg>
    </bean>
</beans>
```

通过 ClassPathXmlApplicationContext 容器获取配置的 Bean 实例：

```
public class TestIocCase02Constructor {
    ApplicationContext ctx;
    @Test
    public void testCase() {
        ctx=new ClassPathXmlApplicationContext("spring-bean-constructor.xml");
```

```
            //从容器中获得 bean
            Product product=(Product)ctx.getBean("beanNoConstructorArg");
            System.out.println("beanNoConstructorArg = "+product);
            Product productCon=(Product)ctx.getBean("beanHasConstructorArg");
            System.out.println("beanHasConstructorArg = "+productCon);
        }
    }
```

运行结果如下:

```
    invoke method -- Product(int id, String name)
    beanNoConstructorArg = com.javadevmap.bean.Product@13eb8acf
    beanHasConstructorArg = com.javadevmap.bean.Product@51c8530f
```

(2) setter 注入：通过 setter 方法进行注入。

使用 Property 注入 name 和 age 属性值，XML 配置如下：

```
    <?xml version="1.0" encoding="UTF-8"?>
    <beans xmlns="http://www.springframework.org/schema/beans"
        xmlns:xsi="http://www.w3.org/2001/XMLSchema-instance"
        xsi:schemaLocation="http://www.springframework.org/schema/beans
        http://www.springframework.org/schema/beans/spring-beans.xsd">

        <bean name="beanProperty" class="com.javadevmap.bean.Product">
            <property name="id" value="1002"></property>
            <property name="name" value="java dev map "></property>
        </bean>
    </beans>
```

通过 ClassPathXmlApplicationContext 容器获取配置的 Bean 实例：

```
    public class TestIocCase03property {
        ApplicationContext ctx;
        @Test
        public void testCase() {
            ctx=new ClassPathXmlApplicationContext("spring-bean-property.xml");
            //从容器中获得 id 为 Product 的 bean
            Product pro=(Product)ctx.getBean("beanProperty");
            System.out.println("product= "+pro);
        }
    }
```

运行结果如下:

```
    invoke method -- setId
    invoke method -- setName
    product= com.javadevmap.bean.Product@2db7a79b
```

到这里，可以看出 constructor-arg 和 property 两者均可实现对象属性的初始化，只是初始化的方式不同而已。

5.3.4 Bean 的作用域

在学习本节之前，需要弄明白什么是作用域。作用域（scope），简单来说是指 Spring 容器中 POJO 的生命周期，也可以理解为对象在 Spring 容器中的创建方式。

上一节学习了如何实例化 Bean 以及如何进行注入，那么 Spring 生成的 Bean 是单例模式[一]的还是原型模式[二]的呢？

这里调用两次 getBean 方法，比较两个 Bean 是否相同，修改之前的测试用例如下：

```
public class TestIocCaseStartSingleton {
    ApplicationContext ctx;
    @Test
    public void testCase() {
        ctx=new ClassPathXmlApplicationContext("spring-bean.xml");
        //从容器中获得 id 为 beanId 的 bean
        Product product=(Product)ctx.getBean("beanId");
        Product product2=(Product)ctx.getBean("beanId");
        System.out.println("product==product2 is "+(product==product2));
    }
}
```

运行结果如下：

```
product==product2 is true
```

可以发现 Spring 默认注入的 Bean 是单例模式的。可以通过设置 bean 标签里面的 scope 属性，指定 Bean 的作用域。

```
<bean name="beanName" scope ="XXX" class="xxx"></bean>
```

scope 属性常用 singleton 和 prototype 两种属性值。对于 singleton 作用域的 Bean，每次请求该 Bean 都将获得相同的实例，即平时说的单例模式的 Bean。如果 Bean 被设置成 prototype 作用域，程序每次请求该 id 的 Bean，Spring 都会新建一个 Bean 实例，然后返回。作用域的含义见表 5-2。

表 5-2 作用域

作用域	含义
singleton	单例模式，在整个 Spring IoC 容器中，使用 singleton 定义的 Bean 将只有一个实例
prototype	原型模式，每次通过容器的 getBean 方法获取 prototype 定义的 Bean 时，都将产生一个新的 Bean 实例
request	对于每次 HTTP 请求，使用 request 定义的 Bean 都将产生一个新实例，即每次 HTTP 请求将会产生不同的 Bean 实例。只有在 Web 应用中使用 Spring 时，该作用域才有效
session	对于每次 HTTP Session，使用 session 定义的 Bean 都将产生一个新实例。同样只有在 Web 应用中使用 Spring 时，该作用域才有效
globalsession	每个全局的 HTTP Session，使用 session 定义的 Bean 都将产生一个新实例

基于定义的 Product 类，实现 singleton 和 prototype 两种作用域的 Bean。XML 配置如下：

```
<?xml version="1.0" encoding="UTF-8"?>
<beans xmlns="http://www.springframework.org/schema/beans"
    xmlns:xsi="http://www.w3.org/2001/XMLSchema-instance"
    xsi:schemaLocation="http://www.springframework.org/schema/beans
    http://www.springframework.org/schema/beans/spring-beans.xsd">
    <bean name="beanNoScope" class="com.javadevmap.bean.Product"></bean>
    <bean name="beanPrototype" scope="prototype"
```

[一] 单例模式，是一种常用的软件设计模式。在它的核心结构中只包含一个被称为单例的特殊类。通过单例模式可以保证系统中应用该模式的类只有一个实例。

[二] 原型模式：每次注入或通过上下文获取时都会创建一个新实例。

```
                class="com.javadevmap.bean.Product"></bean>
    </beans>
```

通过上面设置的两种类型的 bean 来获取 Bean 实例。测试类核心代码如下：

```
    public class TestIocCase04scope {
        Product productOne;
        Product productTwo;
        ApplicationContext ctx;

        @Test
        public void testCase() {
            ctx=new ClassPathXmlApplicationContext("spring-bean-scope.xml");
            //从容器中获得 id 为 beanNoScope 的 bean
            productOne=(Product)ctx.getBean("beanNoScope");
            productTwo=(Product)ctx.getBean("beanNoScope");
            System.out.println("scope default productOne==productTwo is "
                                                +(productOne==productTwo));
            //从容器中获得 id 为 beanPrototype 的 bean
            productOne=(Product)ctx.getBean("beanPrototype");
            productTwo=(Product)ctx.getBean("beanPrototype");
            System.out.println("scope prototype productOne==productTwo is "
                                                +(productOne==productTwo));
        }
    }
```

运行结果如下：

```
    scope default productOne==productTwo is true
    scope prototype productOne==productTwo is false
```

通过 bean 的 scope 属性，可以灵活控制 Bean 的作用域来应对不同的业务场景。

5.3.5 Bean 的生命周期

通过前面几节，基本了解了 Spring 的 bean 的使用、注入以及配置作用域。那么 Spring 是怎么实例化 Bean 的呢？被容器管理的 Bean 什么时候创建，以及什么时候释放呢？本节讲解 Spring 的 Bean 的装配和关闭的过程。

Spring 装配 Bean 的过程如下：

1）实例化 Bean。

2）设置属性值。

3）如果实现了 BeanNameAware 接口，调用 setBeanName 设置 Bean 的 ID 或者 Name。

4）如果实现了 BeanFactoryAware 接口，调用 setBeanFactory 设置 BeanFactory。

5）如果实现了 ApplicationContextAware 接口，调用 setApplicationContext 设置 ApplicationContext。

6）调用 BeanPostProcessor 的预先初始化方法。

7）调用 InitializingBean 的 afterPropertiesSet()方法。

8）调用定制 init-method 方法。

9）调用 BeanPostProcessor 的后初始化方法。

Spring 销毁 Bean 的过程如下：
1）调用 DisposableBean 的 destroy()。
2）调用定制的 destroy-method 方法。
下面重点讲解初始化回调函数和销毁回调函数，它们常用于 Bean 的初始化和销毁时进行业务处理操作。
（1）初始化回调函数，有两种实现方式：
1）通过实现 org.springframework.beans.factory.InitializingBean 接口。
使用之前的 Product 类进行演示，将此类复制一份，然后实现 InitializingBean 接口，具体如下：

```java
package com.javadevmap.bean;
import org.springframework.beans.factory.InitializingBean;
public class ProductWithInitializingBean implements InitializingBean {
    private int id;
    private String name;
    public ProductWithInitializingBean() {
    }
    public ProductWithInitializingBean(int id, String name) {
        System.out.println("invoke method — Product(int id, String name)");
        this.id = id;
        this.name = name;
    }
    // 省略 get 与 set 方法
    @Override
    public void afterPropertiesSet() throws Exception {
        System.out.println("execute afterPropertiesSet()");
    }
}
```

在 Spring 配置文件中配置 bean 标签，具体如下：

```xml
<?xml version="1.0" encoding="UTF-8"?>
<beans xmlns="http://www.springframework.org/schema/beans"
    xmlns:xsi="http://www.w3.org/2001/XMLSchema-instance"
    xsi:schemaLocation="http://www.springframework.org/schema/beans
    http://www.springframework.org/schema/beans/spring-beans.xsd">
    <!-- 配置信息 -->
    <bean id="beanId" class="com.javadevmap.bean.ProductWithInitializingBean"></bean>
</beans>
```

编写测试代码如下：

```java
public class TestIocCaseLifeCircleWithInterface {
    AbstractApplicationContext ctx;
    ProductWithInitializingBean bean;
    @Test
    public void testCase() {
        ctx=new ClassPathXmlApplicationContext("spring-bean-initializingbean.xml");
        System.out.println("execute one");
        bean =(ProductWithInitializingBean) ctx.getBean("beanId");
        System.out.println("execute two");
        bean =(ProductWithInitializingBean) ctx.getBean("beanId");
```

```
        }
    }
```

运行结果如下：

```
execute afterPropertiesSet()
execute one
execute two
```

观察运行结果可以发现，实现 InitializingBean 接口的类仅仅在初始化的时候调用一次接口方法，由于此 bean 标签配置的是单例模式，所以之后再次使用 getBean 方法时接口方法没有被调用。

2）在配置文件中使用 init-method 属性指定无参数方法

使用之前的 Product 类进行演示，将此类复制一份，定义一个方法 initMethod，当然方法名可以随便定义，具体如下：

```
package com.javadevmap.bean;
public class ProductWithInitMethod {
    private int id;
    private String name;
    public ProductWithInitMethod() {
    }
    public ProductWithInitMethod(int id, String name) {
        System.out.println("invoke method — Product(int id, String name)");
        this.id = id;
        this.name = name;
    }
    // 省略 get 与 set 方法
    public void initMethod() {
        System.out.println("execute initMethod()");
    }
}
```

之后在 XML 文件的 bean 标签中定义 init-method 属性，属性值为上面定义的 initMethod 方法的名称。具体如下：

```xml
<?xml version="1.0" encoding="UTF-8"?>
<beans xmlns="http://www.springframework.org/schema/beans"
    xmlns:xsi="http://www.w3.org/2001/XMLSchema-instance"
    xsi:schemaLocation="http://www.springframework.org/schema/beans
    http://www.springframework.org/schema/beans/spring-beans.xsd">
    <!-- 配置信息 -->
    <bean id="beanId" init-method="initMethod"
                class="com.javadevmap.bean.ProductWithInitMethod"></bean>
</beans>
```

编写测试代码如下：

```
public class TestIocCaseLifeCircleWithInitMethod {
    AbstractApplicationContext ctx;
    ProductWithInitMethod bean;
    @Test
    public void testCase() {
        ctx=new ClassPathXmlApplicationContext("spring-bean-initMethod.xml");
        System.out.println("execute one");
```

```
            bean =(ProductWithInitMethod) ctx.getBean("beanId");
            System.out.println("execute two");
            bean =(ProductWithInitMethod) ctx.getBean("beanId");
    }
}
```

运行结果如下：

```
execute initMethod()
execute one
execute two
```

观察运行结果可以发现 initMethod 方法仅仅在初始化的时候调用一次，由于此 bean 标签配置的是单例模式，所以之后再次使用 getBean 方法时 initMethod 方法没有被调用。

（2）销毁回调函数，有两种实现方式：

1）实现 org.springframework.beans.factory.DisposableBean 接口。

使用之前的 Product 类进行演示，将此类复制一份，然后实现 DisposableBean 接口，具体如下：

```java
package com.javadevmap.bean;
import org.springframework.beans.factory.DisposableBean;
public class ProductWithDisposableBean implements DisposableBean{
    private int id;
    private String name;
    public ProductWithDisposableBean() {
    }
    public ProductWithDisposableBean(int id, String name) {
        System.out.println("invoke method -- Product(int id, String name)");
        this.id = id;
        this.name = name;
    }
    // 省略 get 与 set 方法
    @Override
    public void destroy() throws Exception {
        System.out.println("execute destroy()");
    }
}
```

在 Spring 配置文件中配置 bean 标签，具体如下：

```xml
<?xml version="1.0" encoding="UTF-8"?>
<beans xmlns="http://www.springframework.org/schema/beans"
    xmlns:xsi="http://www.w3.org/2001/XMLSchema-instance"
    xsi:schemaLocation="http://www.springframework.org/schema/beans
    http://www.springframework.org/schema/beans/spring-beans.xsd">
    <!-- 配置信息 -->
    <bean id="beanId" class="com.javadevmap.bean.ProductWithDisposableBean"></bean>
</beans>
```

为了能监听到销毁回调函数，需要在 AbstractApplicationContext[一] 类中调用关闭 hook 的 registerShutdownHook() 方法。它将确保正常关闭 Bean，并且调用 destroy 方法。编写测试代码如下：

[一] AbstractApplicationContext 抽象类，Spring 中所有 ApplicationContext 的父类，实现了一些比较核心的方法。

```java
public class TestIocCaseLifeCircleWithDisposableBean {
    AbstractApplicationContext    ctx;
    ProductWithDisposableBean bean;
    @Test
    public void testCase() {
        ctx=new ClassPathXmlApplicationContext("spring-bean-disposable.xml");
        System.out.println("execute one");
        bean =( ProductWithDisposableBean) ctx.getBean("beanId");
        System.out.println("execute two");
        bean =( ProductWithDisposableBean) ctx.getBean("beanId");
        ctx.registerShutdownHook();
    }
}
```

运行结果如下：

```
execute one
execute two
execute destroy()
```

观察控制台输出内容，当对象被销毁时会调用 DisposableBean 接口的 destroy()方法。

2）在配置文件中使用 destroy-method 属性来指定无参数方法。

使用之前的 Product 类进行演示，将此类复制一份，定义一个方法 destroyMethod，具体如下：

```java
public class ProductWithDestroyMethod {
    private int id;
    private String name;
    public ProductWithDestroyMethod() {
    }
    public ProductWithDestroyMethod(int id, String name) {
        System.out.println("invoke method -- Product(int id, String name)");
        this.id = id;
        this.name = name;
    }
    // 省略 get 与 set 方法
    public void destroyMethod() {
        System.out.println("execute destroyMethod()");
    }
}
```

在 XML 配置中指定 bean 标签 destroy-method 属性对应类中的方法名称：

```xml
<?xml version="1.0" encoding="UTF-8"?>
<beans xmlns="http://www.springframework.org/schema/beans"
    xmlns:xsi="http://www.w3.org/2001/XMLSchema-instance"
    xsi:schemaLocation="http://www.springframework.org/schema/beans
    http://www.springframework.org/schema/beans/spring-beans.xsd">
    <!-- 配置信息 -->
    <bean id="beanId" destroy-method="destroyMethod"
                class="com.javadevmap.bean.ProductWithDestroyMethod"></bean>
</beans>
```

编写测试代码如下：

```java
public class TestIocCaseLifeCircleWithDestroyMethod {
```

```
                AbstractApplicationContext    ctx;
                ProductWithDestroyMethod bean;
                @Test
                public void testCase() {
                    ctx=new ClassPathXmlApplicationContext("spring-bean-destroyMethod.xml");
                    System.out.println("execute one");
                    bean =(ProductWithDestroyMethod) ctx.getBean("beanId");
                    System.out.println("execute two");
                    bean =(ProductWithDestroyMethod) ctx.getBean("beanId");
                    ctx.registerShutdownHook();
                }
            }
```

运行结果如下：

```
execute one
execute two
execute destroyMethod()
```

这里注册了 registerShutdownHook()方法，在 Bean 销毁的时候，会执行对应配置的销毁方法。

通过上面的例子，将方法关联到 Bean 的生命周期，以便在恰当的生命周期中实现业务逻辑。例如在容器初始化时预加载缓存，或者销毁时进行日志统计等。

5.3.6 注解实现 IoC

使用 XML 方式实现 IoC，每次增加业务类都需要在 XML 中配置，非常烦琐。这里可使用注解来减轻工作量，需要在 Spring 的配置文件中设置开启注解。从 Spring 2.5 开始就可以使用注解来配置依赖注入，将 bean 的配置移动到类本身。

通过使用@Repository、@Component、@Service、@Controller 注解，Spring 会自动创建相应的 BeanDefinition 对象，并注册到 ApplicationContext 中。使用这些注解的类就成了 Spring 受管组件。

上面所说的四个注解的具体使用场景见表 5-3。

表 5-3 IoC 注解

注　解	含　义
@Repository	用于标注数据访问组件，即 DAO 组件
@Service	用于标注业务层组件
@Controller	用于标注控制层组件（类似 struts 中的 action）
@Component	泛指组件，当组件不好归属的时候，可以使用这个标注

当然，在业务类中添加上面的注解还不够，还需要在 Spring 文件中配置扫描组件，可以让 Spring 自动发现含有注解的类，进行相应的注入，即在<context:component-scan>标签中配置扫描的包路径，如果有多个可以用逗号隔开。

本项目注解的类都在 com.javadevmap 包下面，具体配置如下：

```
<?xml version="1.0" encoding="UTF-8"?>
<beans xmlns="http://www.springframework.org/schema/beans"
    xmlns:xsi="http://www.w3.org/2001/XMLSchema-instance"
```

```xml
        xmlns:p="http://www.springframework.org/schema/p"
        xmlns:context="http://www.springframework.org/schema/context"
        xsi:schemaLocation="http://www.springframework.org/schema/beans
            http://www.springframework.org/schema/beans/spring-beans.xsd
            http://www.springframework.org/schema/context
            http://www.springframework.org/schema/context/spring-context-4.3.xsd">

    <context:component-scan base-package="com.javadevmap"> </context:component-scan>
</beans>
```

下面使用一个添加商品的接口类及其实现来演示注解实现的 IoC：

```java
public interface IProductDao {
    /**
     * 添加商品接口
     */
    public String addProduct(String id,String name);

}
```

ProductAnnoDaoImpl 实现类如下：

```java
@Repository ("productDaoImpl")
public class ProductDaoImpl implements IProductAnnoDao {
    public String addProduct(String id, String name) {
        String result=String.format("添加商品 id=%s, 商品名称为 %s，成功！", id,name);
        return result;
    }
}
```

编写测试代码如下：

```java
public class TestAnnoCaseComponent {
    IProductDao productDao;
    ApplicationContext ctx;
    @Test
    public void testCase() {
        ctx=new ClassPathXmlApplicationContext("spring-bean-scan.xml");
        productDao=(IProductDao)ctx.getBean("productDaoImpl");⊖
        System.out.println("productDao is "+productDao);
        String result = productDao.addProduct("2", "javaDevmap anno");
        System.out.println(result);
    }
}
```

运行结果如下：

```
productDao is com.javadevmap.dao.ProductDaoImpl@78dd667e
添加商品 id=2，商品名称为 javaDevmap anno，成功！
```

本例中通过<context:component-scan>标签配置扫描路径，同时在扫描路径里面的类中添加注解，就可以实现类的构建，使用相比 XML 便利很多。

⊖ Spring 对注解形式的 bean 的名字默认处理就是将首字母小写，再拼接后面的字符，当类的名字是以两个或两个以上的大写字母开头时，bean 的名字会与类名保持一致。

5.3.7 注解的作用域 scope

注解的作用域需要通过使用@Scope 来指定。功能与之前在<bean>标签中配置 scope 属性一样。@Scope 里面的属性值，具体可以参考<bean>中 scope 配置的属性值。使用之前的商品服务 ProductDaoImpl①，在类中增加 Scope 注解。具体代码如下：

```
@Repository("productDaoImpl")
@Scope("prototype")
public class ProductDaoImpl implements IProductDao {

    public String addProduct(String id, String name) {
        String result=String.format("添加商品 id=%s，商品名称为 %s，成功！", id,name);
        return result;
    }
}
```

5.3.8 自动装配

在学习 Spring 自动装配之前，需要弄明白什么是装配。在 Spring 中，对象无需自己查找或创建与之关联的其他对象，容器负责把需要互相调用的对象引用赋予各个对象。而创建对象之间协作关系的行为通常称为装配。常用的装配注解见表 5-4。

表 5-4 装配注解

注 解	含 义
@Resource	默认是按照名称来装配注入的，只有当找不到与名称匹配的 Bean 才会按照类型来装配注入
@Autowired	默认是按照类型装配注入的，如果想按照名称来装配注入，则需要结合@Qualifier 一起使用

@Resource 和@Autowired 均可在 Bean 注入时使用，@Resource 并不是 Spring 的注解，它的包是 javax.annotation.Resource，需要导入，但是 Spring 支持该注解的注入。@Resource 和@Autowired 都可以标注在字段或者该字段的 setter 方法上。

下面以产品服务 ProductService 内部注入 IProductDao 为例，演示@Autowired 的使用方法：

```
@Service
public class ProductService {
    @Autowired
    IProductDao productDaoImpl;
    public void addProduct(String id,String name) {
        System.out.println("execute addProduct method()");
        String result = productDaoImpl.addProduct(id, name);
        System.out.println(result);
    }
}
```

5.3.9 @Autowired 与@Qualifier

当容器中存在多个 Bean 实现同一个接口，那么注入此接口的地方仅使用@Autowired，注入将不能执行，会抛出异常。解决办法是给@Autowired 增加一个候选值，在@Autowired 后面

---

① 这里仅作为 prototype 演示，实际使用中这个类一般都为单例模式。

增加一个@Qualifier 注解，提供一个 String 类型的值作为候选的 Bean 的名字。例如，产品增加接口针对国内和国外业务场景，有两个类实现此接口，具体如下：

```java
public interface IProductDao {
    public String addProduct(String id,String name);
}
```

ProductDaoForBusiOneImpl 实现如下：

```java
@Repository("productDaoOne")
public class ProductDaoForBusiOneImpl implements IProductDao {
    @Override
    public String addProduct(String id, String name) {
        String result = String.format(ProductDaoForBusiOneImpl.class.getSimpleName()
                +" 添加商品 id=%s，商品名称为 %s，成功！", id, name);
        return result;
    }
}
```

ProductDaoForBusiTwoImpl 实现如下：

```java
@Repository("productDaoTwo")
public class ProductDaoForBusiTwoImpl implements IProductDao {
    @Override
    public String addProduct(String id, String name) {
        String result = String.format(ProductDaoForBusiTwoImpl.class.getSimpleName()
                +" 添加商品 id=%s，商品名称为 %s，成功！", id, name);
        return result;
    }
}
```

新建类 ProductServiceBoth 来注入以上两个类实例，具体实现如下：

```java
@Service
public class ProductServiceBoth {
    @Autowired
    @Qualifier("productDaoOne")
    IProductDao productDao01;

    @Autowired
    @Qualifier("productDaoTwo")
    IProductDao productDao02;

    public void addProduct(String id,String name) {
        System.out.println("execute addProduct method()");
        String result= productDao01.addProduct(id, name);
        System.out.println(result);
        result = productDao02.addProduct(id, name);
        System.out.println(result);
    }
}
```

注意@Qualifier 注解里面的名称要与 Bean 的注解名称保持一致，否则自动装配会匹配失败。
如果在 ProductServiceBoth 中引入的 IProductDao 未增加@Qualifier 注解，则会报告如下

异常：

> Caused by: org.springframework.beans.factory.NoUniqueBeanDefinitionException: No qualifying bean of type [com.javadevmap.dao.IProductDao] is defined: expected single matching bean but found 3: productDaoOne,productDaoTwo,productDaoImpl
>   at org.springframework.beans.factory.config.DependencyDescriptor.resolveNotUnique(DependencyDescriptor.java:172)
>   at org.springframework.beans.factory.support.DefaultListableBeanFactory.doResolveDependency(DefaultListableBeanFactory.java:1106)
>   at org.springframework.beans.factory.support.DefaultListableBeanFactory.resolveDependency(DefaultListableBeanFactory.java:1056)
>   at org.springframework.beans.factory.annotation.AutowiredAnnotationBeanPostProcessor$AutowiredFieldElement.inject(AutowiredAnnotationBeanPostProcessor.java:566)
>   ... 38 more

Spring 根据接口 IProductDao 查找到了多个实现类，但无法确定该用哪个，于是报告上面的错误。需要使用@Qualifier 来明确告诉 Spring 要使用哪个实现类。

5.4 Aop

考虑这样一个业务场景：在核心业务中做日志记录，传统的做法是写好工具类，然后在业务方法的前后部分增加日志记录代码。还有一种做法是将核心业务接口抽取出来，在执行抽象接口的前后增加日志记录，这种方式更加高明。但是如果要做的业务比较多，而且部分代码不能随意修改，这样的日志记录功能就很难完成。

面向方面编程（AOP），也可称为面向切面编程，是一种编程范式，从另一个角度来考虑程序结构，从而完善面向对象编程（OOP）。

AOP 的诞生就是为了解决类似上面的业务问题，即定义好日志组件，日志组件横切业务逻辑，这样既不耦合现有的业务，还能完成需要的日志记录功能。

5.4.1 AOP 的核心概念

AOP 是通过预编译方式和运行期动态代理实现程序功能的统一维护的一种技术。AOP 是 OOP 的延续，是软件开发中的一个热点，也是 Spring 框架中的重要内容，是函数式编程的一种衍生范型。利用 AOP 可以对业务逻辑的各个部分进行隔离，从而使业务逻辑各部分之间的耦合度降低，提高程序的可重用性，同时提高开发效率。AOP 相关概念见表 5-5。

表 5-5 AOP 概念

概　　念	含　　义
切面（aspect）	类是对物体特征的抽象，切面就是对横切关注点的抽象
横切关注点	对哪些方法进行拦截，拦截后怎么处理，这些关注点称为横切关注点
连接点（joinpoint）	被拦截到的点
切入点（pointcut）	对连接点进行拦截的定义
通知（advice）	拦截到连接点之后要执行的代码，通知分为前置、后置、异常、最终、环绕通知五类
织入（weave）	将切面应用到目标对象并导致代理对象创建的过程
引入（introduction）	在不修改代码的前提下，引入可以在运行期为类动态地添加一些方法或字段

5.4.2 AOP 的代理机制

Spring AOP 是用动态代理的方式实现的，动态代理是在运行期为目标类添加增强生成子类的方式。动态代理又分为 JDK 动态代理（通过接口创建代理）、CGLib 动态代理（通过类创建代理）。Spring 优先使用 JDK 动态代理[⊖]。

Java 的代理[⊖]实现分为静态代理和动态代理。

静态代理通常是对原有业务逻辑的扩充，即通过代理类持有真实对象，然后在业务代码中调用代理类的方法，而代理类里面的方法才会调用真实对象的方法，在不改变原有业务代码的前提下，增加其他业务逻辑。

动态代理，代理类并不是在 Java 代码中实现，而是在运行期间生成，相比静态代理，动态代理可以在运行期间动态生成一个持有真实对象并实现代理接口的 Proxy，同时注入需要的扩展逻辑。

5.4.3 基于 Schema 的 AOP 使用

要了解基于 Schema 的 AOP 方式，需要先了解配置 AOP 的常用标签，见表 5-6。

表 5-6 AOP 常用标签

标 签	含 义
<aop:advisor>	定义一个 AOP 通知者
<aop:after>	后通知
<aop:after-returning>	返回后通知
<aop:after-throwing>	抛出异常后通知
<aop:around>	周围通知
<aop:aspect>	定义一个切面
<aop:before>	前通知
<aop:config>	顶级配置元素，类似于<beans>
<aop:pointcut>	定义一个切入点

配置中常用的通配符的含义如下：

- *：匹配任何数量字符。
- ..：匹配任何数量字符的重复，如在类型模式中匹配任何数量层级，在方法参数模式中匹配任何数量参数。
- +：匹配指定类型的子类型，仅能作为后缀放在类型模式后边。

这里通过 AOP 实现在调用 ProductServcie 类中方法的前后打印日志。AOP 编程其实是很简单的事情，纵观 AOP 编程，只需要实现三个部分：

（1）定义普通业务组件。

这里使用前面的 ProductService 业务类。

（2）定义切入点，一个切入点可能横切多个业务组件。

```
<?xml version="1.0" encoding="UTF-8"?>
```

[⊖] 默认使用 JDK 动态代理来创建 AOP 代理，可以为任何接口实例创建代理，当需要代理的类不是代理接口的时候，Spring 会切换为使用 CGLib 代理，也可强制使用 CGLib。

[⊖] 代理模式：为其他对象提供一种代理以控制对这个对象的访问。在某些情况下，一个对象不适合或者不能直接引用另一个对象，而代理对象可以在客户端和目标对象之间起到中介的作用。

```xml
<beans xmlns="http://www.springframework.org/schema/beans"
    xmlns:xsi="http://www.w3.org/2001/XMLSchema-instance"
    xmlns:p="http://www.springframework.org/schema/p"
    xmlns:aop="http://www.springframework.org/schema/aop"
    xmlns:context="http://www.springframework.org/schema/context"
    xsi:schemaLocation="http://www.springframework.org/schema/beans
    http://www.springframework.org/schema/beans/spring-beans.xsd
                http://www.springframework.org/schema/context
    http://www.springframework.org/schema/context/spring-context-4.3.xsd
                http://www.springframework.org/schema/aop   http://www.springframework.org/schema/aop/spring-aop-4.3.xsd">
    <!--指定要扫描的包,如果有多个可以用逗号隔开-->
    <context:component-scan base-package="com.javadevmap"></context:component-scan>
    <!-- 通知 -->
    <bean id="advice" class="com.javadevmap.bean.AdvicesBean"></bean>
    <!-- AOP 配置 -->
    <!-- proxy-target-class 属性表示被代理的类是否为一个没有实现接口的类,Spring 会根据当前被
    代理的类是否实现接口来选择代理方式。如果实现了接口则使用 JDK 内置动态代理,如果未实现接口则使用
    CGLIB 动态代理 -->
    <aop:config proxy-target-class="true">
        <!-- 切面配置 -->
        <!--ref 表示通知对象的引用 -->
        <aop:aspect ref="advice">
            <!-- 配置切入点(横切逻辑将注入的精确位置) -->
            <aop:pointcut
                expression="execution(* com.javadevmap.service.ProductService.*(..))"
                id="pointcut1" />
            <!--声明通知, method 指定通知类型, pointcut 指定切点, 就是通知应该注入哪些方法
中 -->
            <aop:before method="before" pointcut-ref="pointcut1" />
            <aop:after method="after" pointcut-ref="pointcut1" />
            <aop:around method="around"
                pointcut="execution(* com.javadevmap.service.ProductService.*(..))" />
            <aop:after-throwing method="afterThrowing"
                pointcut="execution(* com.javadevmap.service.ProductService.*(..))"
                throwing="exp" />
            <aop:after-returning method="afterReturning"
                pointcut="execution(* com.javadevmap.service.ProductService.*(..))"
                returning="result" />
        </aop:aspect>
    </aop:config>
</beans>
```

3)定义增强处理,增强处理就是通过 AOP 框架为普通业务组件织入的处理动作。

```java
public class AdvicesBean {
    //前置通知
    public void before(JoinPoint jp)
    {
        System.out.println("---------->前置通知<----------");
        System.out.println("方法名:"+jp.getSignature().getName()
                +",参数长度:"+jp.getArgs().length+",被代理对象:
            "+jp.getTarget().getClass().getName());
    }
```

```java
        //后置通知
        public void after(JoinPoint jp){
            System.out.println("---------->后置通知<----------");
        }
        //环绕通知
        public Object around(ProceedingJoinPoint pjd) throws Throwable{
            System.out.println("---------->环绕开始<----------");
            Object object=pjd.proceed();
            System.out.println("---------->环绕结束<----------");
            return object;
        }
        //异常后通知
        public void afterThrowing(JoinPoint jp,Exception exp)
        {
            System.out.println("---------->异常后通知,发生了异常: "+exp.getMessage()+"<---------- ");
        }
        //返回结果后通知
        public void afterReturning(JoinPoint joinPoint, Object result)
        {
            System.out.println("---------->返回结果后通知<----------");
            System.out.println("结果是: "+result);
        }
}
        编写测试代码如下:
        public class TestAopCase {
            ProductService proService;
            ApplicationContext ctx;
            @Test
            public void testCase() {
                ctx=new ClassPathXmlApplicationContext("spring-bean-aop.xml");
                //从容器中获得 bean
                proService=(ProductService)ctx.getBean(ProductService.class);
                proService.addProduct("4001", "java dev map");
            }
        }
```

运行结果如下:

```
---------->前置通知<----------
方法名: addProduct,参数长度: 2,被代理对象: com.javadevmap.service.ProductService
20180412_07:59:45.028        [org.springframework.beans.factory.support.DefaultListableBeanFactory][main]
[DEBUG] Returning cached instance of singleton bean 'advice'
---------->环绕开始<----------
execute addProduct method()
添加商品 id=4001,商品名称为 287af598-2c2d-46dc-89f6-d3f3003be2d4-java dev map,成功!
20180412_07:59:45.225        [org.springframework.beans.factory.support.DefaultListableBeanFactory][main]
[DEBUG] Returning cached instance of singleton bean 'advice'
---------->返回结果后通知<----------
结果是: null
---------->环绕结束<----------
20180412_07:59:45.225        [org.springframework.beans.factory.support.DefaultListableBeanFactory][main]
[DEBUG] Returning cached instance of singleton bean 'advice'
---------->后置通知<----------
```

5.4.4 基于@AspectJ 的 AOP 使用

Spring 除了支持 Schema 方式配置 AOP，还支持注解方式，即使用@AspectJ 风格的切面声明。为了支持@AspectJ，需要在 Spring 配置文件中使用如下配置：

```
<aop:aspectj-autoproxy/>
```

具体使用流程：

（1）在配置文件中添加<aop:aspectj-autoproxy/>配置。
（2）创建 Bean，使用@Aspect 注解修饰该类。
（3）创建方法，使用@Before、@After、@Around 等进行修饰，在注解中写上切入点的表达式。

这里演示通过 AOP 注解的方式实现在调用 ProductServcie 中 addProduct 方法的前后打印日志。定义一个 AspectjBean 类，具体如下：

```java
@Aspect
public class AspectjBean {
    @Pointcut("execution(* com.javadevmap.service.ProductService.*(..))") // expression
    private void businessService() {
    }
    @Before("businessService()")
    public void beforeAdvice() {
        System.out.println("beforeAdvice() --> Going to exec addProduct.");
    }
    @After("businessService()")
    public void afterAdvice() {
        System.out.println("afterAdvice() --> addProduct has been done.");
    }
    @AfterReturning(pointcut = "businessService()", returning = "retVal")
    public void afterReturningAdvice(Object retVal) {
        System.out.println("afterReturningAdvice() -->Returning");
    }
    @AfterThrowing(pointcut = "businessService()", throwing = "ex")
    public void AfterThrowingAdvice(IllegalArgumentException ex) {
        System.out.println("AfterThrowingAdvice--> There has been an exception: " + ex.toString());
    }
}
```

在 Spring 配置文件中增加如下核心配置：

```xml
<!--指定要扫描的包，如果有多个可以用逗号隔开 -->
<context:component-scan base-package="com.javadevmap"></context:component-scan>
<aop:aspectj-autoproxy />
<!-- aspect 通知 -->
<bean id="aspectJ" class="com.javadevmap.service.AspectjBean"></bean>
```

为了演示抛出异常的通知，在 ProductService 中增加 doThrowException 方法：

```java
@Service
public class ProductService {
    //省略之前业务代码逻辑
    ...
    public void doThrowException() {
        System.out.println("Exception raised");
        throw new IllegalArgumentException();
```

编写测试代码如下：

```
public class TestAopCaseAspectJ {
    ApplicationContext ctx;
    @Test
    public void testCase() {
        ctx = new ClassPathXmlApplicationContext("spring-bean-aspectj.xml");
        productService=(ProductService)ctx.getBean(ProductService.class);
        productService.addProduct("4001", "java dev map");
        productService.doThrowException();
    }
}
```

运行结果如下：

```
beforeAdvice() --> Going to exec addProduct.
execute addProduct method()
添加商品 id=4001，商品名称为 java dev map，成功！
afterAdvice() --> addProduct has been done.
afterReturningAdvice() -->Returning
beforeAdvice() --> Going to exec addProduct.
Exception raised
afterAdvice() --> addProduct has been done.
AfterThrowingAdvice--> There has been an exception: java.lang.IllegalArgumentException
```

通过前面的例子，可见在未修改自己的业务类的前提下，利用 Spring 的 AOP 特性即可在方法的前后增加一些其他业务逻辑。

5.5 集成 Logback

Logback 是一个开源的日志组件，本节将介绍在 Spring 工程中集成 Logback 实现日志打印的方法。

5.5.1 SLF4J 简介

SLF4J（Simple Logging Facade For Java）是一个针对各类 Java 日志框架的统一抽象。Java 日志框架众多，常用的有 java.util.logging、log4j、logback、commons-logging。Spring 框架使用的是 Jakarta Commons Logging API（JCL）。在使用 SLF4J 的时候，不需要在代码或配置文件中指定打算使用哪个具体的日志系统。SLF4J 提供了统一的记录日志的接口，只要按照其提供的方法记录即可，最终日志的格式、记录级别、输出方式等通过具体日志系统的配置来实现，因此可以在应用中灵活切换日志系统。

5.5.2 Logback 概述

Logback[⊖]是一个开源日志组件。Logback 当前分成三个模块：logback-core、logback-classic

---

⊖ Logback 官网是 http://logback.qos.ch。

和 logback-access。logback-core 是其他两个模块的基础模块。logback-classic 是 log4j 的一个改良版本。此外 logback-classic 完整实现了 SLF4J API，可以很方便地更换成其他日志系统，例如 log4j。logback-access 访问模块与 Servlet 容器集成，提供了通过 http 来访问日志的功能。

Logback 包含的配置内容具体如下：

（1）logger、appender 及 layout

logger 是日志的记录器，把它关联到应用的对应的 context 上后，主要用于存放日志对象，也可以定义日志类型、级别。

appender 主要用于指定日志输出的目的地，目的地可以是控制台、文件、MySQL、Oracle 和其他数据库等。

layout 负责把事件转换成字符串，格式化日志信息的输出。

（2）loggerContext

各个 logger 都被关联到一个 LoggerContext，LoggerContext 负责制造 logger，也负责以树结构排列各 logger。

（3）有效级别及级别的继承

logger 可以被分配级别。级别有：TRACE、DEBUG、INFO、WARN 和 ERROR。如果 logger 没有被分配级别，那么它将从被分配级别的最近的祖先那里继承级别。root logger 默认级别是 DEBUG。

（4）打印方法与基本的选择规则

打印方法决定了记录请求的级别。级别排序为：TRACE < DEBUG < INFO < WARN < ERROR。

5.5.3　Logback 的集成

下面将 Logback 集成到 Spring 项目中来，实现让 Logback 用一个固定格式将日志输出到控制台的功能。

（1）添加依赖

在 pom 文件中添加如下依赖：

```xml
<dependency>
    <groupId>ch.qos.logback</groupId>
    <artifactId>logback-classic</artifactId>
    <version>1.2.3</version>
</dependency>
<dependency>
    <groupId>ch.qos.logback</groupId>
    <artifactId>logback-core</artifactId>
    <version>1.2.3</version>
</dependency>
<dependency>
    <groupId>org.logback-extensions</groupId>
    <artifactId>logback-ext-spring</artifactId>
    <version>0.1.4</version>
</dependency>
<dependency>
    <groupId>org.slf4j</groupId>
    <artifactId>slf4j-api</artifactId>
```

```xml
            <version>1.7.25</version>
        </dependency>
        <dependency>
            <groupId>org.slf4j</groupId>
            <artifactId>jcl-over-slf4j</artifactId>
            <version>1.7.25</version>
        </dependency>
```

（2）添加配置文件 logback.xml

Logback 配置文件的基本结构：以<configuration>开头，后面有零个或多个<appender>元素，有零个或多个<logger>元素，有最多一个<root>元素。

在 resources 目录新建一个名为 logback.xml 的文件。定义 logback 的日志输出格式为："日期 线程名 日志级别 日志消息内容"。文件具体内容如下：

```xml
<?xml version="1.0" encoding="UTF-8"?>
<configuration>
    <!--格式化输出：%d 表示日期，%thread 表示线程名，%level：日志级别，%msg：日志消息，%n 是换行符-->
    <property name="consoleLayoutPattern"
        value="%-20(%d{yyyyMMdd_HH:mm:ss.SSS} [%logger][%thread]) [%level] %msg%n" />
    <!-- 控制台输出 -->
    <appender name="CONSOLE" class="ch.qos.logback.core.ConsoleAppender">
        <layout name="StandardFormat" class="ch.qos.logback.classic.PatternLayout">
            <pattern>${consoleLayoutPattern}</pattern>
        </layout>
    </appender>

    <root level="DEBUG">
        <appender-ref ref="CONSOLE" /> <!-- 控制台输出 -->
    </root>
</configuration>
```

可根据业务需求打印不同级别的日志信息，下面简单说明如何使用：

```java
public class TestLog4jCase {
    private Logger logger = LoggerFactory.getLogger(TestLog4jCase.class);
    ApplicationContext context=null;
    @Test
    public void testCase() {
        context = new ClassPathXmlApplicationContext("spring-bean.xml");
        logger.debug("debug");
        logger.info("test");
        logger.warn("warn");
        logger.error("error");
    }
}
```

运行结果如下：

```
20180410_15:25:34.450 [com.javadevmap.logback.SpringLogbackApp][main] [DEBUG] debug
20180410_15:25:34.450 [com.javadevmap.logback.SpringLogbackApp][main] [INFO] test
20180410_15:25:34.450 [com.javadevmap.logback.SpringLogbackApp][main] [WARN] warn
20180410_15:25:34.450 [com.javadevmap.logback.SpringLogbackApp][main] [ERROR] error
```

通过上面的步骤，轻松实现了 Spring 集成 Logback 框架。

5.5.4 输出日志到文件

在实际业务中，不仅要在控制台输出日志，也要保存日志信息到文件，以便日后系统出现问题可以根据日志文件查找原因。

当然记录日志信息的同时，还要考虑日志文件大小以及如何存放，需要制定日志文件存储策略，才能在日志信息记录的同时让日志文件大小保持在可控范围内。

基于这些前提，下面配置 Logback 文件，让日志信息记录到文件，并且控制记录的日志级别在 ERROR 级别，单个日志文件的大小控制在 10MB，隔日的日志文件打成 zip 包按日期命名。

在上一节项目的基础上，修改 logback.xml 配置文件，具体如下：

```xml
<?xml version="1.0" encoding="UTF-8"?>
<configuration>
    <property name="SYS_LOG_DIR" value="d:/logbackDir" />
    <property name="LOG_FILE" value="app.log" />
    <!--格式化输出：%d 表示日期，%thread 表示线程名，%level：日志级别，%msg：日志消息，%n 是换行符-->
    <property name="fileLayoutPattern" value=
"%-20(%d{yyyyMMdd_HH:mm:ss.SSS} [%logger{10}][%thread]) [%level] %msg%n" />
    <property name="consoleLayoutPattern"
value="%-20(%d{yyyyMMdd_HH:mm:ss.SSS} [%logger][%thread]) [%level] %msg%n" />

    <appender name="LOG_ROLLING"
        class="ch.qos.logback.core.rolling.RollingFileAppender">
        <file>${SYS_LOG_DIR}/${LOG_FILE}</file>
        <filter class="ch.qos.logback.classic.filter.ThresholdFilter">
            <level>ERROR</level>
        </filter>
        <rollingPolicy class="ch.qos.logback.core.rolling.TimeBasedRollingPolicy">
            <fileNamePattern>
                ${SYS_LOG_DIR}/%d{yyyy-MM-dd}/${LOG_FILE}_%d{yyyy-MM-dd}_%i.zip
            </fileNamePattern>
            <maxHistory>7</maxHistory>
            <timeBasedFileNamingAndTriggeringPolicy
                class="ch.qos.logback.core.rolling.SizeAndTimeBasedFNATP">
                <maxFileSize>10MB</maxFileSize>
            </timeBasedFileNamingAndTriggeringPolicy>
        </rollingPolicy>
        <layout>
            <pattern>${fileLayoutPattern}</pattern>
        </layout>
    </appender>

    <!-- 控制台输出 -->
    <appender name="CONSOLE" class="ch.qos.logback.core.ConsoleAppender">
        <layout name="StandardFormat" class="ch.qos.logback.classic.PatternLayout">
            <pattern>${consoleLayoutPattern}</pattern>
        </layout>
    </appender>
```

```xml
        <root level="DEBUG">
            <appender-ref ref="CONSOLE" /><!-- 控制台输出 -->
            <appender-ref ref="LOG_ROLLING" /><!-- 文件输出 -->
        </root>
</configuration>
```

这里为了演示日志存放的效果,临时将单个日志大小设置为 1MB,测试代码如下:

```java
public class SpringLogbackAppFile {
    private static final Logger logger = LoggerFactory.getLogger(SpringLogbackAppFile.class);
    static ApplicationContext context = null;
    public static void main(String[] args) {
        // IoC 获取 beans 的上下文
        context = new ClassPathXmlApplicationContext("spring-bean.xml");
        for (int i = 0; i < 20000; i++) {
            logger.debug("debug" + i);
            logger.info("test");
            logger.warn("warn " + i);
            logger.error("error " + i);
        }
    }
}
```

运行效果如图 5-1 和图 5-2 所示。

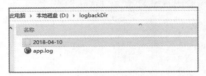

图 5-1　日志运行效果

图 5-2　日志文件打包效果

由输出可见,本例实现了指定日志位置、设置日志的级别、控制单个日志文件大小、隔天日志保存等功能。

5.6　集成 MyBatis

MyBatis 是一款优秀的持久层框架,它支持定制化 SQL、存储过程以及高级映射。MyBatis 避免了几乎所有的 JDBC 代码和手动设置参数以及获取结果集。MyBatis 可以使用简单的 XML 或注解来配置和映射原生信息,将接口和 Java 的 POJOs 映射成数据库中的记录——MyBatis-Spring[一],可将 MyBatis 代码无缝整合到 Spring 中。持久层的 mapper 需要由 Spring 进行管理。

5.6.1　数据准备

下面用 MyBatis 实现一个针对 MySQL 数据库中单表的增删改查功能。先准备好测试环境,本节数据库采用 MySQL。MySQL 的安装将在第 19 章介绍。在 MySQL 中创建一个名为 javadevmap 的数据库,创建一个 product 表。里面有 id、product_name、product_desc 和 price 四个字段。具体 SQL 语句如下:

```
CREATE DATABASE `javadevmap` DEFAULT CHARACTER SET utf8 COLLATE utf8_general_ci;
```

[一] Mybatis-Spring 的官网是 http://www.mybatis.org/spring/zh/index.html。

```
DROP TABLE IF EXISTS `product`;
CREATE TABLE `product` (
  `id` int(11) NOT NULL AUTO_INCREMENT,
  `product_name` varchar(150) COLLATE utf8_unicode_ci DEFAULT NULL,
  `price` int(11) DEFAULT NULL,
  `product_desc` varchar(500) COLLATE utf8_unicode_ci DEFAULT NULL,
  PRIMARY KEY (`id`)
) ENGINE=InnoDB AUTO_INCREMENT=3 DEFAULT CHARSET=utf8 COLLATE=utf8_unicode_ci;
```

在 MySQL 客户端执行上述 SQL 语句，就能生成数据库以及对应的表。

5.6.2　添加 Spring 与 Mybatis 集成相关依赖

在 pom 文件中添加如下依赖，其中使用 druid[①]作为数据库连接池。

```xml
<dependencies>
    <!-- 添加 spring-jdbc 包 -->
    <dependency>
        <groupId>org.springframework</groupId>
        <artifactId>spring-jdbc</artifactId>
        <version>4.3.3.RELEASE</version>
    </dependency>
    <!-- 添加 mybatis 的核心包 -->
    <dependency>
        <groupId>org.mybatis</groupId>
        <artifactId>mybatis</artifactId>
        <version>3.2.8</version>
    </dependency>
    <!-- 添加 mybatis 与 Spring 整合的核心包 -->
    <dependency>
        <groupId>org.mybatis</groupId>
        <artifactId>mybatis-spring</artifactId>
        <version>1.2.2</version>
    </dependency>
    <!-- 添加 mysql 驱动包 -->
    <dependency>
        <groupId>mysql</groupId>
        <artifactId>mysql-connector-java</artifactId>
        <version>5.1.34</version>
    </dependency>
    <!-- 添加 druid 连接池包 -->
    <dependency>
        <groupId>com.alibaba</groupId>
        <artifactId>druid</artifactId>
        <version>1.0.29</version>
    </dependency>
</dependencies>
```

5.6.3　编写相关配置文件

在 src/main/resources 目录下创建三个 Spring 与 MyBatis 整合的配置文件。这三个文件分别

[①] druid 是目前比较优秀的数据库连接池，在功能、性能、扩展性方面表现良好。

是 dbconfig.properties、spring-mybatis-base.xml 和 spring-mybatis.xml。

（1）配置数据库 dbconfig.properties

基于 Spring 开发应用的时候，一般会将数据库的配置放置在 properties 文件中，dbconfig.properties 文件用来存放数据库相关的链接以及账号信息。具体内容如下：

```
driverClassName=com.mysql.jdbc.Driver
validationQuery=SELECT 1
jdbc_url=jdbc:mysql://localhost:3306/javadevmap_db?useUnicode=true&characterEncoding=UTF-8&zeroDateTimeBehavior=convertToNull
jdbc_username=root
jdbc_password=root
```

（2）配置扫描

spring-mybatis-base.xml 配置了数据库相关配置文件以及扫描包的路径。内容如下：

```xml
<?xml version="1.0" encoding="UTF-8"?>
<beans
...
    <!-- 引入 dbconfig.properties 属性文件 -->
    <context:property-placeholder location="classpath:dbconfig.properties" />
    <context:component-scan base-package="com.javadevmap.mybatis" />
</beans>
```

（3）配置数据源

spring-mybatis.xml 配置了数据源连接属性、sqlSessionFactory、扫描器。内容如下：

```xml
<?xml version="1.0" encoding="UTF-8"?>
<beans xmlns="http://www.springframework.org/schema/beans" xmlns:xsi="http://www.w3.org/2001/XML Schema-instance" xmlns:tx="http://www.springframework.org/schema/tx" xmlns:aop="http://www.springframework.org/schema/aop" xsi:schemaLocation="
    http://www.springframework.org/schema/beans
    http://www.springframework.org/schema/beans/spring-beans-3.0.xsd
    http://www.springframework.org/schema/tx
    http://www.springframework.org/schema/tx/spring-tx-3.0.xsd
    http://www.springframework.org/schema/aop
    http://www.springframework.org/schema/aop/spring-aop-3.0.xsd
    ">

    <!-- ==================配置数据源================== -->
    <!-- 配置数据源，使用的是 alibaba 的 Druid -->
    <bean name="dataSource" class="com.alibaba.druid.pool.DruidDataSource" init-method="init" destroy-method="close">
        <property name="url" value="${jdbc_url}" />
        <property name="username" value="${jdbc_username}" />
        <property name="password" value="${jdbc_password}" />
        <!-- 初始化连接大小 -->
        <property name="initialSize" value="0" />
        <!-- 连接池最大使用连接数量 -->
        <property name="maxActive" value="20" />
        <!-- 连接池最大空闲 -->
        <property name="maxIdle" value="20" />
        <!-- 连接池最小空闲 -->
        <property name="minIdle" value="0" />
```

```xml
        <!-- 获取连接最大等待时间 -->
        <property name="maxWait" value="60000" />
        <property name="validationQuery" value="${validationQuery}" />
        <property name="testOnBorrow" value="false" />
        <property name="testOnReturn" value="false" />
        <property name="testWhileIdle" value="true" />
        <!-- 配置间隔多久才进行一次检测，检测需要关闭的空闲连接，单位是毫秒 -->
        <property name="timeBetweenEvictionRunsMillis" value="60000" />
        <!-- 配置一个连接在池中最小生存的时间，单位是毫秒 -->
        <property name="minEvictableIdleTimeMillis" value="25200000" />
        <!-- 监控数据库 -->
        <!-- <property name="filters" value="stat" /> -->
        <property name="filters" value="mergeStat" />
    </bean>

    <!-- ===================针对 myBatis 的配置项=================== -->
    <!-- 配置 sqlSessionFactory -->
    <bean id="sqlSessionFactory" class="org.mybatis.spring.SqlSessionFactoryBean">
        <!-- 实例化 sqlSessionFactory 时需要使用上述配置好的数据源以及 SQL 映射文件 -->
        <property name="dataSource" ref="dataSource" />
        <property name="mapperLocations">
            <array>
                <value>classpath:mappers/*.xml</value>
                <value>classpath:mappers/manul/*.xml</value>
            </array>
        </property>
    </bean>
    <!-- 配置扫描器 -->
    <bean class="org.mybatis.spring.mapper.MapperScannerConfigurer">
        <!-- 扫描 com.javadevmap.mybatis.dao 这个包以及它的子包下的所有映射接口类 -->
        <property name="basePackage" value="com.javadevmap.mybatis.dao" />
        <property name="sqlSessionFactoryBeanName" value="sqlSessionFactory" />
    </bean>

    <!-- 配置 Spring 的事务管理器 -->
    <bean id="transactionManager"
            class="org.springframework.jdbc.datasource.DataSourceTransactionManager">
        <property name="dataSource" ref="dataSource" />
    </bean>
</beans>
```

配置 sqlSessionFactory 的时候，mapperLocations 属性值配置在 resources 里面的 mapper 文件夹中。注意路径前缀的 classpath。classpath 本质是 JVM 的根路径，JVM 获取资源都是从该根路径下搜索的，这个根路径是个逻辑路径，并不是磁盘路径。

5.6.4 使用 generator 生成单表增删改查代码

MyBatis Generator 是一个 MyBatis 的代码生成器，可以根据数据库中表的设计生成对应的实体类、XML Mapper 文件、接口，从而实现简单数据库操作能力，但是如果需要联合查询和存储过程，仍需手写 SQL 和对象。

本节主要介绍基于 Maven plugin 方式实现代码生成。生成代码遵循以下操作步骤：配置

generatorMybatisCode.xml 文件；在 pom.xml 文件中添加 mybatis-generator-maven-plugin 插件；执行生成代码的 Maven 命令。

（1）配置 generatorMybatisCode.xml 文件

主要填写连接数据库的配置和生成的文件配置信息以及要生成的实体类所对应的表或视图。在 src/main/resources 文件夹下面创建一个 mybatis 文件夹，创建一个 generatorMybatisCode.xml 文件，具体内容如下：

```xml
<?xml version="1.0" encoding="UTF-8"?>
<!DOCTYPE generatorConfiguration PUBLIC "-//mybatis.org//DTD MyBatis Generator Configuration 1.0//EN" "http://mybatis.org/dtd/mybatis-generator-config_1_0.dtd">
<generatorConfiguration>
    <!-- 数据库驱动包位置 -->
    <classPathEntry
    location="C:\Users\Administrator\git\JavaDeveloperMap\SpringBasic\libsNeed\mysql-connector-java-5.1.17.jar" />
    <context id="DB2Tables" targetRuntime="MyBatis3">
        <commentGenerator>
            <property name="suppressAllComments" value="true" />
        </commentGenerator>
        <!-- 数据库链接 URL、用户名、密码 -->
        <jdbcConnection driverClass="com.mysql.jdbc.Driver"
            connectionURL="jdbc:mysql://localhost:3306/javadevmap"
            userId="root"
            password="root">
        </jdbcConnection>
        <javaTypeResolver>
            <property name="forceBigDecimals" value="false" />
        </javaTypeResolver>
        <!-- 生成实体类的包名和位置-->
        <javaModelGenerator targetPackage="com.javadevmap.mybatis.domain"
            targetProject="C:\Users\Administrator\git\JavaDeveloperMap\SpringBasic\target\">
            <property name="enableSubPackages" value="true" />
            <property name="trimStrings" value="true" />
        </javaModelGenerator>
        <!-- 生成的 SQL 映射文件包名和位置-->
        <sqlMapGenerator targetPackage="com.javadevmap.mybatis.mapping"
            targetProject="C:\Users\Administrator\git\JavaDeveloperMap\SpringBasic\target\">
            <property name="enableSubPackages" value="true" />
        </sqlMapGenerator>
        <!-- 生成 DAO 的包名和位置-->
        <javaClientGenerator type="XMLMAPPER"
            targetPackage="com.javadevmap.mybatis.dao"
            targetProject="C:\Users\Administrator\git\JavaDeveloperMap\SpringBasic\target\">
            <property name="enableSubPackages" value="true" />
        </javaClientGenerator>
        <!-- 要生成哪些表(更改 tableName 和 domainObjectName 就可以) -->
        <table tableName="product" domainObjectName="ProductBean"
            enableCountByExample="false" enableUpdateByExample="false"
```

```xml
        enableDeleteByExample="false" enableSelectByExample="false"
        selectByExampleQueryId="false" />
    </context>
</generatorConfiguration>
```

（2）添加 mybatis-generator-maven-plugin 插件

在 pom.xml 中添加依赖以及指定前面自定义的 generatorMybatisCode.xml 文件路径。具体如下：

```xml
<build>
    <plugins>
        <plugin>
            <groupId>org.mybatis.generator</groupId>
            <artifactId>mybatis-generator-maven-plugin</artifactId>
            <version>1.3.5</version>
            <configuration>
                <!-- 在控制台打印执行日志 -->
                <verbose>true</verbose>
                <!-- 重复生成时会覆盖之前的文件 -->
                <overwrite>true</overwrite>
<configurationFile>src\main\resources\mybatis\genratorMybatisCode.xml</configurationFile>
            </configuration>
        </plugin>
    </plugins>
</build>
```

（3）执行 Maven 生成命令

配置完上面的插件后，在工程上执行"鼠标右键单击->run as->Maven build..."，在弹出窗口的 Goals 输入框中输入"mybatis-generator:generate"，就会在 target 文件夹中生成对应的文件，复制生成的文件到对应的包路径下面即可。

现在已经完成 product 表的增删改查工具类的生成，接下来使用生成的代码对数据库进行基本操作，具体测试代码如下：

```java
@RunWith(SpringJUnit4ClassRunner.class)
@ContextConfiguration(locations = { "classpath:spring-mybatis-base.xml", "classpath:spring-mybatis.xml" })
public class SpringTest {
    // 注入
    @Autowired
    private ProductBeanMapper productBeanMapper;

    @Test
    public void insert() {
        ProductBean record=new ProductBean();
        record.setPrice(99);
        record.setProductName("java dev map");
        record.setProductDesc("产品描述，产品描述");
        int affectsNums = productBeanMapper.insert(record);
        System.out.println("insert affects row num is "+affectsNums);
    }
    @Test
    public void selectByPrimaryKey() {
        int id=100;
```

```java
        ProductBean bean = productBeanMapper.selectByPrimaryKey(id);
        System.out.println(bean.getId() +"   ");
        System.out.println(bean.getProductName() +"   ");
        System.out.println(bean.getProductDesc() +"   ");
        System.out.println(bean.getPrice() +"   ");
    }
    @Test
    public void updateByPrimaryKeySelective() {
        ProductBean record=new ProductBean();
        record.setId(100);
        record.setPrice(80);
        record.setProductName("java dev map -update");
        int affectsNums = productBeanMapper.updateByPrimaryKey(record);
        System.out.println("updateByPrimaryKeySelective affects row num is "+affectsNums);
    }
    @Test
    public void deleteByPrimaryKey() {
        int affectsNums = productBeanMapper.deleteByPrimaryKey(1);
        System.out.println("deleteByPrimaryKey affects row num is "+affectsNums);
    }
}
```

通过上面的三步，可轻松生成单表的增删改查的相关类以及 mapper 文件，节省了大量编写 SQL 语句的时间，大大提升了工作效率。

第 6 章　Spring MVC

Spring MVC 是传统框架 SSM 的组成部分，本章将介绍 Spring MVC 框架的特性以及此框架在工程中的作用和用法。

6.1　Spring MVC 概述

Spring MVC 是一种基于请求驱动类型的轻量级 Web 框架，根据 MVC 架构模式的思想，将 Web 层进行职责解耦，基于请求驱动（使用请求—响应模型），简化开发，同时 Spring MVC 分离了控制器、模型对象、过滤器以及处理程序对象的角色，这种分离让它们更容易进行定制。

6.1.1　MVC

MVC 是一种设计模式，它强制性地把应用程序的数据展示、数据处理和流程控制分开。MVC 将应用程序分成 3 个核心模块：模型、视图、控制器。它们相互联接又分别担当不同的职责，如图 6-1 所示。

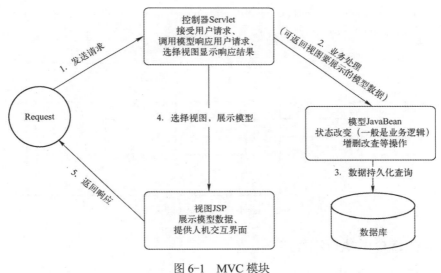

图 6-1　MVC 模块

- 模型：数据模型，提供要展示的数据，可认为是 Bean，一个模型可为多个视图提供数据。
- 视图：负责模型的展示，一般指用户界面。
- 控制器：控制器负责应用的流程控制，所谓流程控制就是接受用户请求，委托给模型进行处理，获取模型数据交由视图处理。

6.1.2　HTTP 请求处理流程

在学习 Spring MVC 之前，需要了解网络请求的原理。

通过浏览器输入网址来访问服务器，例如访问百度，在浏览器中输入 www.baidu.com，浏览器就会显示百度的首页。那么整个过程执行了怎样的操作呢？HTTP 请求和响应流程如下：

（1）域名解析，例如解析 www.baidu.com。
（2）发起 TCP 的 3 次握手。
（3）建立 TCP 连接后发起 HTTP 请求。
（4）服务器端响应 HTTP 请求，浏览器得到 html 代码。
（5）浏览器解析 html 代码，并请求 html 代码中的资源。
（6）浏览器对页面进行渲染，呈现给用户。

以上是 HTTP 请求和响应的流程。那么服务器是怎么处理的呢？处理请求和发送响应的过程是由一种叫作 Servlet 的程序来完成的，Servlet 是为了实现动态页面而衍生出来的。

6.1.3 Servlet 与 Tomcat 的关系

Java Servlet 是运行在 Web 服务器或应用服务器上的程序，它是 Web 浏览器或其他 HTTP 请求的客户端和 HTTP 服务器上的数据库或应用程序之间的中间层。

使用 Servlet，可以收集来自网页表单的用户输入，将数据库或者其他数据源的记录展示给用户，还可以动态创建网页。

Tomcat 是 Web 应用服务器，也是常用的一个 Servlet 容器。Tomcat 作为 Servlet 容器，负责处理客户端请求，把请求传送给 Servlet，并将 Servlet 的响应传送回给客户端。而 Servlet 是一种运行在支持 Java 语言的服务器上的组件。 Servlet 最常见的用途是扩展 Java Web 服务器功能，提供非常安全的、可移植的、易于使用的 CGI⊖替代品。

Servlet 与 Tomcat 的关系如图 6-2 所示。

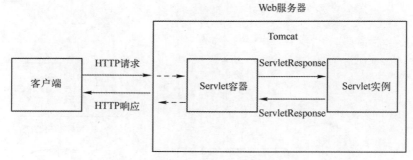

图 6-2 Servlet 与 Tomcat 的关系

6.1.4 Spring MVC 的执行流程

Spring MVC 是基于请求驱动的 Web 框架，并且也使用了前端控制器模式⊖来进行设计，再根据请求映射规则分发给相应的页面控制器（动作/处理器）进行处理。

Spring MVC 的执行步骤如下：

（1）Spring MVC 将所有的请求都提交给 DispatcherServlet，它会委托应用系统的其他模块

⊖ CGI 是 Web 服务器运行时外部程序的规范，按 CGI 编写的程序可以扩展服务器功能。
⊖ 前端控制器模式提供了一个集中的请求处理机制，所有的请求都将由一个单一的处理程序处理。该处理程序可以做认证/授权/记录日志，或者跟踪请求，然后把请求传给相应的处理程序。

第 6 章 Spring MVC

负责对请求进行真正的处理工作。

（2）DispatcherServlet 查询一个或多个 HandlerMapping，找到处理请求的 Controller。

（3）DispatcherServlet 把请求提交到目标 Controller。

（4）Controller 进行业务逻辑处理后，会返回一个 ModelAndView 对象。

（5）Dispatcher 查询一个或多个 ViewResolver 视图解析器，找到 ModelAndView 对象指定的视图对象。

（6）视图对象负责渲染返回给客户端。

在业务开发时只涉及 Handler 处理器和视图（View）层的编写，其他的组件不需要开发，但还是有必要了解一下 Spring MVC 的常用组件，见表 6-1。

表 6-1 Spring MVC 常用组件

组件名称	作　用
前端控制器（DispatcherServlet）	接收请求，响应结果，相当于转发器，使用时通过配置实现
处理器映射器（HandlerMapping）	根据请求的 url 查找 Handler
Handler 处理器	按照 HandlerAdapter 的要求编写，并且处理实际业务
处理器适配器（HandlerAdapter）	按照特定规则（HandlerAdapter 要求的规则）执行 Handler
视图解析器（ViewResolver）	进行视图解析，根据逻辑视图解析成真正的视图（View）
视图（View）	View 是一个接口，实现类支持不同的 View 类型（jsp，framemark，pdf…）

6.2 构建第一个 Spring MVC 项目

本节搭建一个 Spring MVC 项目，实现一个简单的页面展示功能。打开 Eclipse，选择 "new->Maven Project"，在弹出的对话框中选择 "Maven-archetype-webapp"，然后点击 next，在弹出的输入框中输入 Group Id 和 Artifact Id，然后点击 Finish。具体坐标信息如下：

```
<groupId>com.javadevmap</groupId>
<artifactId>SpringMVCBasic</artifactId>
<packaging>war</packaging>
<version>0.0.1-SNAPSHOT</version>
```

使用 maven build 工程，会发现项目报错，原因是由于 Maven webapp 项目默认会创建 index.jsp 页面，此页面引用了 HttpServlet 对象，当前 pom 文件没有此依赖，所以项目报错，如图 6-3 所示。

图 6-3 编译结果

打开 pom.xml 文件，添加 servlet 依赖，具体如下：

```
<dependency>
    <groupId>javax.servlet</groupId>
    <artifactId>javax.servlet-api</artifactId>
```

```xml
        <version>3.0.1</version>
        <scope>provided</scope>
    </dependency>
```

然后再次 maven build 工程即可。

创建常用 package 及存放页面的文件夹。如图 6-4 所示。
图中目录的含义如下：

- com.javadevmap.controller：存放 Spring MVC 的 controller 层代码。
- com.javadevmap.converter：存放自定义转换器代码，例如日期转换器。
- com.javadevmap.domain：存放业务对象 POJO。
- com.javadevmap.exception：存放自定义异常以及全局异常处理器。
- com.javadevmap.interceptor：存放自定义拦截器。
- com.javadevmap.mybatis.dao：存放数据库操作的 dao 层类。
- com. javadevmap.mybatis.model：存放数据库实体类。
- src/main/webapp：存放网页及配置属性等。
- web.xml：用来初始化配置信息，例如 Welcome 页面、Servlet、Servlet-mapping、filter、listener、启动加载级别等。

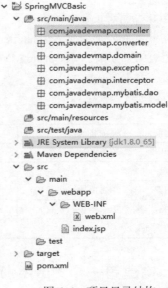

图 6-4　项目目录结构

6.2.1　添加依赖

创建完项目后，在 pom 文件中添加 Spring MVC 所需要的依赖，具体如下：

```xml
<dependency>
    <groupId>org.springframework</groupId>
    <artifactId>spring-web</artifactId>
    <version>4.3.3.RELEASE</version>
</dependency>
<dependency>
    <groupId>org.springframework</groupId>
    <artifactId>spring-webmvc</artifactId>
    <version>4.3.3.RELEASE</version>
</dependency>
```

6.2.2　配置相关文件

根据之前 Spring MVC 请求处理的流程，需要在 web.xml 中配置前端控制器 (DispatcherServlet)，打开"webapp->WEB-INFO->web.xml"文件，添加如下信息：

```xml
<servlet>
    <servlet-name>spring</servlet-name>
    <servlet-class>org.springframework.web.servlet.DispatcherServlet</servlet-class>
    <init-param>
```

```xml
            <param-name>contextConfigLocation</param-name>
            <param-value>classpath:spring/spring-*.xml</param-value>
        </init-param>
        <load-on-startup>1</load-on-startup>
    </servlet>

    <servlet-mapping>
        <servlet-name>spring</servlet-name>
        <url-pattern>/</url-pattern>
    </servlet-mapping>
```

<servlet-name> 属性随意，只要上下一致即可，url-pattern 中的 "/" 表示拦截所有请求，所有访问的地址由 DispatcherServlet 进行解析，使用此种方式可实现 RESTful 风格的 url。

上面的 contextConfigLocation 用来指定配置文件具体位置，这里配置的内容为 classpath:spring/spring-*.xml，*号为通配符，即可匹配多个以 spring 开头的配置文件。

在 resources 文件夹下新建一个 spring 文件夹，然后新创建一个 spring-servlet.xml 的配置文件，用来存放 Spring MVC 相关的配置信息。spring-servlet.xml 具体配置如下：

```xml
<?xml version="1.0" encoding="UTF-8"?>
<beans xmlns="http://www.springframework.org/schema/beans"
    xmlns:xsi="http://www.w3.org/2001/XMLSchema-instance"
    xmlns:context="http://www.springframework.org/schema/context"
    xmlns:mvc="http://www.springframework.org/schema/mvc"
    xsi:schemaLocation="http://www.springframework.org/schema/beans
        http://www.springframework.org/schema/beans/spring-beans.xsd
        http://www.springframework.org/schema/context
        http://www.springframework.org/schema/context/spring-context.xsd
        http://www.springframework.org/schema/mvc
        http://www.springframework.org/schema/mvc/spring-mvc.xsd">
    <!-- 配置扫描的包 -->
    <context:component-scan base-package="com.javadevmap.*" />
    <!-- 注册 HandlerMapper、HandlerAdapter 两个映射类 -->
    <mvc:annotation-driven />
    <!-- 访问静态资源 -->
    <mvc:default-servlet-handler />
    <!-- 视图解析器 -->
    <bean
        class="org.springframework.web.servlet.view.InternalResourceViewResolver">
        <property name="prefix" value="/WEB-INF/view/"></property>
        <property name="suffix" value=".jsp"></property>
    </bean>
</beans>
```

上面配置用到的标签含义如下：

（1）<context:component-scan...>：用于激活 Spring MVC 注解扫描功能，该功能允许使用注解，如@Controller 和@RequestMapping 等。

（2）InternalResourceViewResolver：bean 视图解析器。

（3）<mvc:default-servlet-handler />：访问静态资源。对进入 DispatcherServlet 的 URL 进行筛查，如果发现是静态资源的请求，就将该请求转给 Web 应用服务器默认的 Servlet 处理，如果不是静态资源的请求，才由 DispatcherServlet 继续处理。

（4）<mvc:annotation-driven />：注册 HandlerMapper、HandlerAdapter 两个映射类。

使用如上配置会自动扫描 com.javadevmap 下的所有包中的含有注解（如@Controller, @ Service 等）的类；<mvc:annotation-driven />会注册两个映射类，负责将请求映射到类的方法中。

6.2.3 基本页面展示

为了演示一个简单的 Spring MVC 请求的完整流程，在 WEB-INF 文件夹下面创建一个 view 文件夹，并在其中创建一个 demo.jsp 页面。demo.jsp 具体内容如下：

```jsp
<%@ page language="java" contentType="text/html; charset=UTF-8"
    pageEncoding="UTF-8"%>
<!DOCTYPE html PUBLIC "-//W3C//DTD HTML 4.01 Transitional//EN" "http://www.w3.org/TR/html4/loose.dtd">
<html>
<head>
<meta http-equiv="Content-Type" content="text/html; charset=UTF-8">
<title>首页</title>
</head>
<body>
    <h1>This is SpringMVC Demo</h1>
</body>
</html>
```

创建 com.javadevmap.controller 包（包路径需被扫描到），在其中创建 WebController 类，具体如下：

```java
import org.springframework.stereotype.Controller;
@Controller
@RequestMapping("/demo")
public class WebController   {
    @RequestMapping(value="/index", method = RequestMethod.GET)
    public String index() throws CustomException {
        return "demo";
    }
}
```

@Controller 注解定义该类作为一个 Spring MVC 控制器。@RequestMapping 表明在该控制器中处理的所有方法都是相对于/demo 路径的。Index 方法上的注解@RequestMapping(value="/index",method = RequestMethod.GET)用于匹配方法和请求路径，处理/demo/index 的 Http GET 请求。

当请求/demo/index 路径时会映射到此方法中，返回的字符串 demo 会被配置拼接为 WEB-INF/view/demo.jsp，并展示出来。把此工程生成 war 包，在 Tomcat 中部署运行，访问 http://localhost:8080/demo/index，页面展示如图 6-5 所示，说明配置成功。

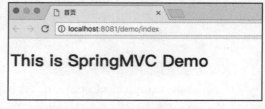

图 6-5　页面效果

6.3　Spring MVC Restful 实现

软件风格其实就是一种约定，统一的软件风格可以在研发中提高代码的可读性，从而在需要协作才能完成的工程中，保持风格的统一，提高协作的效率。本节将介绍 Spring MVC 的 Restful 实现。

6.3.1 REST 概述

REST 即表述性状态传递，是一种软件架构风格。REST 是面向资源的，将资源的状态以最适合客户端或服务端的形式从服务端转移到客户端。

在 REST 中，资源通过 URL 进行识别和定位。REST 中的行为是通过 Http 方法定义的。这些 Http 方法通常会匹配如下动作：

- Create：POST
- Read：GET
- Update：PUT 或 PATCH
- Delete：DELETE

这里用商品添加的功能举例，在使用 Restful 之前的请求为：

- http://127.0.0.1/product/query/1　　　GET　　　根据商品 id 查询商品数据
- http://127.0.0.1/product/save　　　　　POST　　　新增商品
- http://127.0.0.1/product/update　　　　POST　　　修改商品信息
- http://127.0.0.1/product/delete　　　　GET/POST　删除商品信息

使用 RESTful 后的用法：

- http://127.0.0.1/product/1　　　　　　GET　　　根据商品 id 查询商品数据
- http://127.0.0.1/product　　　　　　　POST　　　新增商品
- http://127.0.0.1/product　　　　　　　PUT　　　修改商品信息
- http://127.0.0.1/product　　　　　　　DELETE　删除商品信息

6.3.2 创建 REST 风格的 Controller

根据上一节对 REST 的理解，基于 REST 风格实现商品模块的增删改查功能。使用 MyBatis 框架将数据存放到 MySQL 中，MyBatis 集成步骤可以参看第 5 章。

创建一个返回数据的实体 ResultBean，方便封装请求返回数据、通用的状态码和描述信息。

```
package com.javadevmap.domain;
public class ResultBean {
    private int code;// 状态码
    private Product data;// 业务数据 product
    private String msg;// 业务信息

    public ResultBean() {
    }
    public ResultBean(int code, Product data, String desc) {
        this.code = code;
        this.data = data;
        this.msg = desc;
    }

    public static ResultBean ofSuccess(Product obj){
        ResultBean ret=new ResultBean();
```

```
            ret.setCode(200);
            ret.setData(obj);
            return ret;
        }
        public static ResultBean ofSuccess(String msg){
            ResultBean ret=new ResultBean();
            ret.setCode(200);
            ret.setMsg(msg);
            return ret;
        }
        public static ResultBean ofSuccess(Product obj,String msg){
            ResultBean ret=new ResultBean();
            ret.setCode(200);
            ret.setData(obj);
            ret.setMsg(msg);
            return ret;
        }
        public static ResultBean ofSuccess(){
            ResultBean ret=new ResultBean();
            ret.setCode(200);
            return ret;
        }
        public static ResultBean ofFail(int code,String desc){
            ResultBean ret=new ResultBean();
            ret.setCode(code);
            ret.setMsg(desc);
            return ret;
        }
    …..//省略 get 与 set 方法
}
```

以产品信息的增删改查为例，实现一组 REST 风格的接口。在 com.javadevmap.controller 下新建一个名为 RestController[⊖]的 Controller，具体内容如下：

```
@Controller
@RequestMapping("/rest")
public class RestController {
    @Autowired
    ProductModelMapper productModelMapper;
}
```

（1）GET 方法查询商品信息

```
@RequestMapping(value="/product/{proId}",method = RequestMethod.GET)
public @ResponseBody ResultBean getProductInfo(@PathVariable("proId")Integer proId){
    System.out.println( "getProductInfo() called with: proId = [" + proId + "]");

    ProductModel productModel = productModelMapper.selectByPrimaryKey(proId);
    if(null==productModel){
        return ResultBean.ofFail(501,"未查询到此商品");
    }
```

⊖ 这里为了演示方便，在 Controller 中直接使用了 Mapper，而在实际业务中，应该在 Controller 中注入 Service，在 Service 中注入 DAO，在 DAO 中使用 Mapper。

```
        Product product=new Product();
        BeanUtils.copyProperties(productModel,product);
        return ResultBean.ofSuccess(product,"查询成功");
    }
```

通过 Postman① 工具进行测试，执行效果如图 6-6 所示。

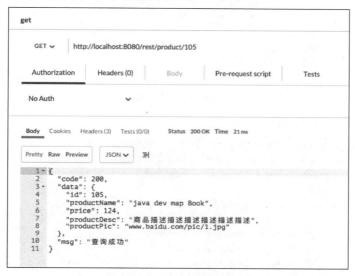

图 6-6　Get 请求效果

（2）POST 方法增加单个商品

```
    @RequestMapping(value="/product",method = RequestMethod.POST)
    public @ResponseBody ResultBean addProduct(@RequestBody Product product){
        System.out.println("addProduct() called with: product = [" + product + "]");
        //业务逻辑代码 …..
        ProductModel productModel=new ProductModel();
        BeanUtils.copyProperties(product,productModel);
        productModelMapper.insert(productModel);
        return ResultBean.ofSuccess("增加成功");
    }
```

通过 Postman 工具提交请求并返回，结果如图 6-7、图 6-8 所示。

图 6-7　Post 请求参数

---

① Postman 是一款功能强大的网页调试与发送网页 HTTP 请求的 Chrome 插件。

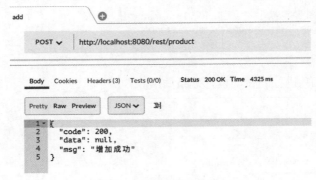

图 6-8　Post 请求结果

（3）PUT 方法更新商品信息

```
@RequestMapping(value="/product",method = RequestMethod.PUT)
public @ResponseBody ResultBean updateProduct(@RequestBody Product product){
    System.out.println("updateProduct() called with: product = [" + product + "]");
    if(product.getId()==null){
        return ResultBean.ofFail(502,"参数不正确");
    }
    ProductModel productModel=new ProductModel();
    BeanUtils.copyProperties(product,productModel);
    productModelMapper.updateByPrimaryKeySelective(productModel);
    return ResultBean.ofSuccess("更新商品成功");
};
```

通过 Postman 工具提交请求并返回，结果如图 6-9、图 6-10 所示。

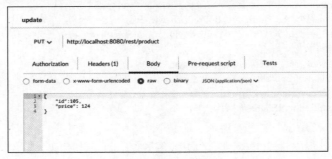

图 6-9　Put 请求参数

图 6-10　Put 请求结果

（4）DELETE 方法删除单个商品信息

```
@RequestMapping(value="/product/{proId}",method = RequestMethod.DELETE)
public @ResponseBody ResultBean delProduct(@PathVariable("proId")Integer proId){
    System.out.println("delProduct() called with: proId = [" + proId + "]");
    int i = productModelMapper.deleteByPrimaryKey(proId);
    if(i<=0){
        return ResultBean.ofFail(503,"删除失败");
    }
    return ResultBean.ofSuccess("删除成功");
};
```

通过 Postman 工具测试，结果如图 6-11 所示。

图 6-11　Delete 请求结果

通过上面的配置，基本实现了针对商品模块 REST 风格的接口 API。

6.4　Spring MVC 拦截器

Spring MVC 的处理器拦截器，类似于 Servlet 开发中的过滤器 Filter，用于对处理器进行预处理和后处理。应用场景有：

- 日志记录：记录请求日志，进行信息监控、统计、计算 PV 等。
- 权限检查：请求进入业务逻辑处理前，进行一些权限验证等操作。
- 性能监控：通过进入拦截器和执行完拦截器后记录时间戳，计算请求处理的耗时。

6.4.1　拦截器

在 Spring MVC 中实现拦截器，需要实现 HandlerInterceptor 接口，即 Spring 里面的 org.springframework.web.servlet.HandlerInterceptor。接口中提供三个方法，分别为：

（1）public boolean preHandle(HttpServletRequest request, HttpServletResponse response, Object handler) throws Exception

进入 Handler 方法之前执行。

（2）public void postHandle(HttpServletRequest request, HttpServletResponse response, Object handler, ModelAndView modelAndView) throws Exception

进入 Handler 方法之后，返回 ModelAndView 之前执行。

(3) public void afterCompletion(HttpServletRequest request, HttpServletResponse response, Object handler, Exception ex) throws Exception

执行 Handler 完成后执行此方法。

6.4.2 自定义拦截器

上一节介绍了 Spring MVC 拦截器的核心接口 HandlerInterceptor，本节通过拦截器实现请求处理耗时的性能监控功能。定义类 HandlerInterceptorTimeConsume，实现 HandlerInterceptor 接口，通过在请求处理前和请求处理后记录时间戳①的方式实现功能耗时的计算。此类内容如下：

```java
package com.javadevmap.interceptor;
import org.springframework.core.NamedThreadLocal;
import org.springframework.web.servlet.HandlerInterceptor;
import org.springframework.web.servlet.ModelAndView;
import javax.servlet.http.HttpServletRequest;
import javax.servlet.http.HttpServletResponse;
/**
 * 请求耗时统计拦截器
 */
public class HandlerInterceptorTimeConsume implements HandlerInterceptor {
    private NamedThreadLocal<Long> threadLocal =
                           new NamedThreadLocal<Long>("execConsumeTime");

    public boolean preHandle(HttpServletRequest request, HttpServletResponse response, Object object) throws Exception {
        System.out.println("Thread.currentThread().getId() = " + Thread.currentThread().getId());
        long beginTime = System.currentTimeMillis();           //开始时间
        threadLocal.set(beginTime);
        return true;
    }

    public void afterCompletion(HttpServletRequest request,
                                HttpServletResponse response,
                                Object handler,
                                Exception exception) throws Exception {
        long endTime = System.currentTimeMillis();
        long consumeMills = endTime - threadLocal.get();
        System.out.println(String.format("请求 RequestURI: %s ; method = %s -> 耗时（毫秒）%s ", request.getRequestURI(), request.getMethod(), consumeMills));
    }

    public void postHandle(HttpServletRequest request,
                           HttpServletResponse response,
                           Object handler,
                           ModelAndView modelAndView)
            throws Exception {
    }
}
```

将定义的拦截器类在配置文件中进行配置，这里针对所有的请求进行耗时统计。在配置文

---

① 此处记录时间戳，是使用 ThreadLocal 记录开始时间，然后在同一线程内，和结束时间进行相减，从而计算出来的。

件中增加了一个<mvc:interceptors>标签，通过<mvc:mapping>的 patch 属性来匹配拦截规则，将定义的拦截器通过<bean>映射进来。具体如下：

```xml
<!--拦截器 -->
<mvc:interceptors>
    <!--多个拦截器,顺序执行 -->
    <mvc:interceptor>
        <!-- /**表示所有 url 包括子 url 路径 -->
        <mvc:mapping path="/**"/>
        <bean
          class="com.javamapdev.interceptor.HandlerInterceptorTimeConsume"></bean>
    </mvc:interceptor>
</mvc:interceptors>
```

接下来访问之前定义的请求接口：

http://localhost:8080/rest/product/107

控制台输出如下信息：

```
Thread.currentThread().getId() = 31
getProductInfo() called with: proId = [107]
请求 RequestURI: /rest/product/107 ; method = GET    -> 耗时（毫秒）7
```

通过上面的输出信息，可见拦截器能够计算一个请求的响应时间，如果想更加完善，可以自定义一个阈值进行预警等。

6.4.3 拦截器执行规则

接下来通过定义两个拦截器来观察拦截器的执行规则，定义拦截器 HandlerInterceptor1 和 HandlerInterceptor2，两个类具体内容大致相同，都是在方法执行时打印自己的拦截器名称加方法名。HandlerInterceptor1 具体如下：

```java
package com.javadevmap.interceptor;
import org.springframework.web.servlet.HandlerInterceptor;
import org.springframework.web.servlet.ModelAndView;
import javax.servlet.http.HttpServletRequest;
import javax.servlet.http.HttpServletResponse;
/**
 * 拦截器
 */
public class HandlerInterceptor1 implements HandlerInterceptor {
    //进入 Handler 方法之前执行
    public boolean preHandle(HttpServletRequest request, HttpServletResponse response, Object handler) throws Exception {
        System.out.println("HandlerInterceptor1...preHandle ");
        //return false 表示拦截，不向下执行
        //return true 表示放行
        return true;
    }

    //进入 Handler 方法之后，返回 modelAndView 之前执行
    public void postHandle(HttpServletRequest request, HttpServletResponse response, Object handler, ModelAndView modelAndView) throws Exception {
```

```
            System.out.println("HandlerInterceptor1...postHandle");
        }

        //执行 Handler 完成执行此方法
        public void afterCompletion(HttpServletRequest request, HttpServletResponse response, Object handler, Exception ex) throws Exception {
            System.out.println("HandlerInterceptor1...afterCompletion");
        }
    }
```

HandlerInterceptor2 具体如下：

```
package com.javadevmap.interceptor;
import org.springframework.web.servlet.HandlerInterceptor;
import org.springframework.web.servlet.ModelAndView;
import javax.servlet.http.HttpServletRequest;
import javax.servlet.http.HttpServletResponse;
/**
 * 拦截器
 */
public class HandlerInterceptor2 implements HandlerInterceptor {
    //进入 Handler 方法之前执行
    public boolean preHandle(HttpServletRequest request, HttpServletResponse response, Object handler) throws Exception {
        System.out.println("HandlerInterceptor2 ---> preHandle");
        //return false 表示拦截，不向下执行
        //return true 表示放行
        return true;
    }

    //进入 Handler 方法之后，返回 modelAndView 之前执行
    public void postHandle(HttpServletRequest request, HttpServletResponse response, Object handler, ModelAndView modelAndView) throws Exception {
        System.out.println("HandlerInterceptor2 ---> postHandle");
    }

    //执行 Handler 完成执行此方法
    public void afterCompletion(HttpServletRequest request, HttpServletResponse response, Object handler, Exception ex) throws Exception {
        System.out.println("HandlerInterceptor2 ---> afterCompletion");
    }
}
```

然后在 spring-servlet.xml 文件中配置此拦截器。具体如下：

```
<!--拦截器 -->
<mvc:interceptors>
    <!--多个拦截器,顺序执行 -->
    <mvc:interceptor>
        <!-- /**表示所有 url 包括子 url 路径 -->
        <mvc:mapping path="/**"/>
        <bean class="com.javamapdev.interceptor.HandlerInterceptor1"></bean>
    </mvc:interceptor>
    <mvc:interceptor>
        <mvc:mapping path="/**"/>
```

```
            <bean class="com.javamapdev.interceptor.HandlerInterceptor2"></bean>
        </mvc:interceptor>
    </mvc:interceptors>
```

这里 HandlerInterceptor1 和 HandlerInterceptor2 两个拦截器的拦截路径规则配置一致。请求之前定义的页面，控制台输出如下信息：

```
HandlerInterceptor1...preHandle
HandlerInterceptor2 ---> preHandle
getProductInfo() called with: proId = [107]
HandlerInterceptor2 ---> postHandle
HandlerInterceptor1...postHandle
HandlerInterceptor2 ---> afterCompletion
HandlerInterceptor1...afterCompletion
```

可以自行设置拦截器方法返回的拦截标志 true（放行）或者 false（不放行），得出如下规律：
- 拦截器 1 放行，拦截器 2 preHandle 才会执行。
- 拦截器 2 preHandle 不放行，拦截器 2 postHandle 和 afterCompletion 不会执行。
- 只要有一个拦截器不放行，postHandle 都不会执行。

6.5 Spring MVC 异常处理器

开发 Web 应用的时候，请求处理过程中的错误和异常是经常发生的。例如 Ajax 方式发起请求，后台服务如果发生异常，前台则无法获得任何请求结果。如果希望前台能够获得请求的返回数据，可以使用 Spring MVC 提供的全局异常拦截器，保证当后台业务处理抛出异常后，会转到该拦截器中处理，根据异常信息向前台返回不同的页面或者数据。

6.5.1 Spring MVC 异常处理方式

Spring MVC 处理异常有 3 种方式：
- 使用 Spring MVC 提供的简单异常处理器 SimpleMappingExceptionResolver。
- 实现 Spring 的异常处理接口 HandlerExceptionResolver 自定义自己的异常处理器。
- 使用@ExceptionHandler 注解实现异常处理。

本节使用 HandlerExceptionResolver 方式来处理全局异常。

在处理异常之前，需要了解一下 HTTP 请求常见的响应状态码：

400：请求错误

401：校验失败

403：资源不可用

404：找不到匹配的资源

405：不支持的请求方法

406：没有和请求 accept 匹配的 MIME 类型

415：不支持的 MIME 类型

500：服务内部错误

6.5.2 实现自定义异常处理类

平时业务开发，有的请求返回页面，有的返回 Json 数据，所以针对异常也要分情况进行处理。

下面实现的异常处理功能，既能处理普通请求 Web 页面的异常，也能处理 Ajax 请求的异常。实现这个功能的核心是根据请求头进行判断，普通请求与 Ajax 请求区别在于 Ajax 请求头会有 x-requested-with，通过区分不同请求头从而反馈给客户端不同形式的异常信息。

首先定义一个自定义异常，用于业务出现问题时主动抛出此异常。具体如下：

```
package com.javadevmap.exception;
public class CustomException extends Exception {
    public CustomException() {
    }
    public CustomException(String message) {
        super(message);
    }
}
```

本节使用 HandlerExceptionResolver 方式来处理全局异常，所以需要在包 com.javadevmap.exception 下新建一个 CustomExceptionResolver 类并且实现 HandlerExceptionResolver 接口。具体内容如下：

```
public class CustomExceptionResolver implements HandlerExceptionResolver {
    protected static final String DEFAULT_ERROR_MESSAGE = "系统忙，请稍后再试";
    /**
     * @param ex 系统抛出的异常
     */
    public ModelAndView resolveException(HttpServletRequest request, HttpServletResponse response, Object handler, Exception ex) {
        String errorMsg = ex instanceof CustomException ?
                                    ex.getMessage() : DEFAULT_ERROR_MESSAGE;
        String errorStack = Throwables.getStackTraceAsString(ex);
        System.out.printf("Request: %s raised %s", request.getRequestURI(), errorStack);
        ModelAndView modelAndView =null;
        ////如果是 Ajax 请求响应头会有 x-requested-with
        if (request.getHeader("x-requested-with") != null &&
                request.getHeader("x-requested-with").equalsIgnoreCase("XMLHttpRequest")){
            modelAndView = handleAjaxError(response,errorMsg);
        }else{
            //非 Ajax 请求时，session 失效的处理
            modelAndView = handleViewError(request.getRequestURI(),ex.getMessage(),"error");
        }
        return modelAndView;
    }
}
```

针对 Web 页面的异常处理由 handleViewError 方法具体实现，此方法会返回一个错误页面。

```
protected ModelAndView handleViewError(String url,
                        String errorMessage,
                        String viewName) {
    ModelAndView mav = new ModelAndView();
    mav.addObject("url", url);
    //将错误信息传到页面
    mav.addObject("message", errorMessage);
    //指向错误页面
    mav.setViewName(viewName);
```

```
        return mav;
    }
```

定义一个错误页面,在 WEB-INF/view 文件夹下面定义一个 error.jsp 页面,用来显示错误信息。具体内容如下:

```
<%@ page contentType="text/html;charset=UTF-8" language="java" %>
<html>
<head>
    <title>错误页面提示</title>
</head>
<body>
错误信息:  ${message}
</body>
</html>
```

而针对 Json 接口异常由 handleAjaxError 方法具体实现,此方法会在 response 中返回错误信息:

```
protected ModelAndView handleAjaxError(HttpServletResponse rsp, String errorMessage) {
    try {
        rsp.setCharacterEncoding("UTF-8");
        rsp.setContentType("application/json;charset=utf-8");
        rsp.setStatus(HttpStatus.OK.value());
        PrintWriter writer = rsp.getWriter();
        ResultBean bean=ResultBean.ofFail(500,errorMessage);
        ObjectMapper mapper = new ObjectMapper();
        //ResultBean 类转 JSON
        String json = mapper.writeValueAsString(bean);
        System.out.println(json);
        writer.write(json);// 输出
        writer.flush();
        writer.close();
    }catch (Exception e){
        e.printStackTrace();
    }
    return null;
}
```

为了演示异常处理效果,定义了一个 Controller 类 ExceptionController,分别为普通请求和 Ajax 请求主动抛出异常。

```
@Controller
@RequestMapping("/exception")
public class ExceptionController {
    @RequestMapping("/web")
    public String index() throws CustomException {
        //业务处理
        if(1==1){// 此条件判断是为了抛出自定义异常
            throw new CustomException("页面请求出现了异常");
        }
        return "demo";
    }
```

```
@RequestMapping(value="/json",method = RequestMethod.POST)
public @ResponseBody ResultBean addProduct(@RequestBody Product product)throws
CustomException {
    System.out.println("addProduct() called with: product = [" + product + "]");
    //业务逻辑代码 …..
    if(1==1){// 抛出自定义异常
        throw new CustomException("Ajax 请求出现了异常");
    }
    return ResultBean.ofSuccess();
}
```

普通页面抛出异常，测试结果如图 6-12 所示。

图 6-12 异常页面

Ajax 页面抛出异常，测试结果如图 6-13 所示。

图 6-13 接口异常

本节实现了既能处理普通请求也能处理 Ajax 请求的异常处理器。实际业务代码开发的时候，建议与业务功能相关的异常，在 Service 中抛出，而与业务无关的异常在 Controller 层抛出，然后在异常处理器中进行统一捕获并返回数据。

6.6 Spring MVC 上传和下载文件

在平时业务开发过程中，文件的上传和下载是很常见的场景。例如前面实现的商品模块，商品中有对应图片资源，此功能就涉及图片上传和下载的功能。Spring MVC 为文件上传提供了直接的支持，这种支持是由 MultipartResolver 实现的。Spring MVC 使用 Apache Commons FileUpload 技术实现了一个 MultipartResolver 实现类——CommonsMultipartResolver。因此，Spring MVC 的文件上传需依赖 Apache Commons FileUpload 的组件。本节实现基于 Spring MVC 的文件上传和下载功能。

6.6.1 MultipartFile 对象

Spring MVC 会将上传的文件绑定到 MultipartFile 对象中。MultipartFile 提供了获取上传文件内容、文件名等方法。通过 transferTo()方法还可以将文件存储到本地。MultipartFile 对象中的常用方法见表 6-2。

表 6-2 MultipartFile 常用方法

方 法 名	作 用
byte[] getBytes()	获取文件数据
String getContentType[]	获取文件 MIME 类型，如 image/jpeg 等
InputStream getInputStream()	获取文件流
String getName()	获取表单中文件组件的名字
String getOriginalFilename()	获取上传文件的原名
Long getSize()	获取文件的字节大小，单位为 byte
boolean isEmpty()	检测是否有上传文件
void transferTo(File dest)	将上传文件保存到一个目录文件中

了解并掌握 MultipartFile 的常用方法，可轻松实现文件上传。

6.6.2 上传文件

在之前商品模块基础上，增加商品图片上传功能，将前端传递的图片信息以及图片资源存放到服务器，并把文件的路径信息保存到数据库中。

上传文件，必须将页面中表单的 method 方法设置为 POST，并将 enctype 设置为 multipart/form-data。这样，浏览器才会把用户选择的文件以二进制数据的形式发送给服务端。一旦设置了 enctype 为 multipart/form-data，浏览器就会采用二进制流的方式来处理表单数据。

Spring MVC 上下文默认没有装配 MultipartResolver，因此不能直接处理文件上传，需要先在 spring-servlet.xml 中配置 multipart 类型解析器，具体内容如下：

```xml
<!-- 文件上传 -->
<bean id="multipartResolver"
      class="org.springframework.web.multipart.commons.CommonsMultipartResolver">
    <!-- 设置上传文件的最大尺寸为 5MB -->
    <property name="maxUploadSize">
        <value>5242880</value>
    </property>
</bean>
```

在 WEB-INFO/view 文件夹新建一个图片上传页面 uploadFile.jsp：

```jsp
<%@ page language="java" contentType="text/html; charset=UTF-8"
    pageEncoding="UTF-8"%>
<!DOCTYPE html PUBLIC "-//W3C//DTD HTML 4.01 Transitional//EN" "http://www.w3.org/TR/ html4/ loose.dtd">
<html>
<head>
    <meta http-equiv="Content-Type" content="text/html; charset=UTF-8">
    <title>文件上传</title>
</head>
<body>
```

```html
        <h2>文件上传</h2>
        <form action="upload" enctype="multipart/form-data" method="post">
            <table>
                <tr>
                    <td>商品ID:</td>
                    <td><input type="text" name="productId"></td>
                </tr>
                <tr>
                    <td>文件描述:</td>
                    <td><input type="text" name="description"></td>
                </tr>
                <tr>
                    <td>请选择文件:</td>
                    <td><input type="file" name="file"></td>
                </tr>
                <tr>
                    <td><input type="submit" value="上传"></td>
                </tr>
            </table>
        </form>
    </body>
</html>
```

在工程包 com.javadevmap.controller 下面创建一个 FileUploadController 类,用来接收请求处理业务,具体内容如下:

```java
@Controller
@RequestMapping("/file")
public class FileUploadController {
    @RequestMapping(value = "/uploadFile", method = {RequestMethod.POST,
                                                     RequestMethod.GET})
    public String uploadFile() throws Exception {
        return "uploadFile";
    }
}
```

部署服务到 Tomcat 上,访问路径为:

http://localhost:8080/file/uploadFile

页面效果如图 6-14 所示。

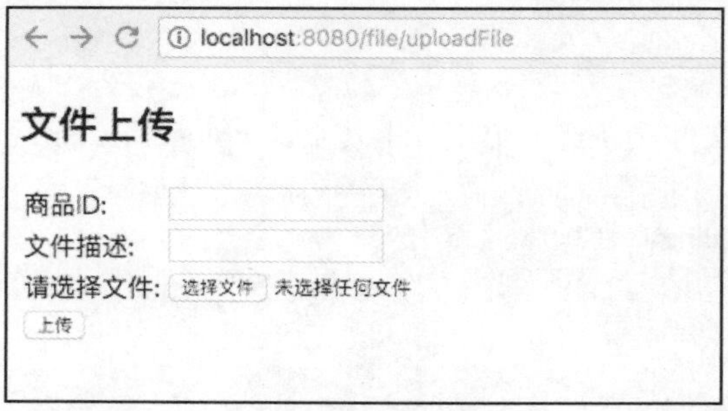

图 6-14 文件上传页面

在 FileUploadController 中添加 uploadProductPic 方法用来接收文件上传请求。

```java
@Autowired
ProductModelMapper productModelMapper;//dao

@RequestMapping(value = "/upload", method = RequestMethod.POST)
public String uploadProductPic(
        HttpServletRequest request,
        Model model,
        @RequestParam("productId") int productId,
        @RequestParam("description") String description,
        MultipartFile file) throws Exception {
    System.out.println(description);
    //原始名称
    String originalFilename = file.getOriginalFilename();
    String newFileName =null;
    //上传图片
    if (file != null && originalFilename != null && originalFilename.length() > 0) {
        //存储图片的物理路径
        String path = request.getServletContext().getRealPath(".");
        System.out.println(path);
        //新的图片名称
        newFileName = "javadevmap" +
                        originalFilename.substring(originalFilename.lastIndexOf("."));
        System.out.println("newFileName = " + newFileName);
        //新图片
        File newFile = new File(path ,newFileName);

        //将内存中的数据写入磁盘
        file.transferTo(newFile);
        model.addAttribute("title", "文件上传");
        model.addAttribute("message", "文件上传成功");
        model.addAttribute("originalFilename", originalFilename);
        model.addAttribute("newFileName", newFileName);
    }else{
        model.addAttribute("title", "文件上传");
        model.addAttribute("message", "文件上传失败");
    }
    // 更新商品信息
    ProductModel productModel = productModelMapper.selectByPrimaryKey(productId);
    if(null != productModel){
        model.addAttribute("productName", productModel.getProductName());
        productModel.setProductPic("download?filename="+newFileName);
        productModel.setProductDesc(description);
        productModelMapper.updateByPrimaryKeySelective(productModel);
    }
    return "status";
}
```

根据前端传递过来的图片，将文件重命名并保存到服务器指定位置，然后跳转到指定的结果页。在WEB-INFO/view文件夹新建一个status.jsp文件用来展示上传结果：

```jsp
<%@ page contentType="text/html;charset=UTF-8" language="java" %>
<html>
<head>
    <title>${title}</title>
```

```
            </head>
            <body>
            <div>
                商品：
                ${productName}
            </div>
            <div>
                文件上传结果：
                ${message}
            </div>
            </body>
            </html>
```

上传文件后，页面效果如图 6-15 所示。

6.6.3 下载文件

除了文件上传功能，平时业务下载功能也必不可少。接下来，实现商品图片下载功能。文件下载相对简单，直接在页面给出一个超链接，该链接的 href 属性设置成下载文件路径，点击超链接浏览器将执行文件下载。

图 6-15 文件上传成功页面

Spring MVC 提供了一个 ResponseEntity 类型，使用它可以很方便地定义返回的 HttpHeaders 和 HttpStatus。在 FileUploadController 中增加下载方法，代码如下：

```
@RequestMapping(value = "/download")
public ResponseEntity<byte[]> download(HttpServletRequest request,
                    @RequestParam("filename") String filename,
                    Model model) throws Exception {
    //下载文件路径
    String path = request.getServletContext().getRealPath(".");
    File file = new File(path + File.separator + filename);
    HttpHeaders headers = new HttpHeaders();
    //下载显示的文件名，解决中文名称乱码问题
    String downloadFileName = new String(filename.getBytes("UTF-8"), "iso-8859-1");
    //通知浏览器以 attachment（下载方式）打开图片
    headers.setContentDispositionFormData("attachment", downloadFileName);
    //application/octet-stream ： 二进制流数据（最常见的文件下载）。
    headers.setContentType(MediaType.APPLICATION_OCTET_STREAM);
    return new ResponseEntity<byte[]>(FileUtils.readFileToByteArray(file),
                    headers, HttpStatus.CREATED);
}
```

代码中定义了 download 方法，此方法接收页面传递的文件名后，使用 Apache Commons FileUpload 组件的 FileUtils 读取服务器本地文件，并将其构建成 ResponseEntity 对象返回客户端下载。

使用 ResponseEntity 对象，可以很方便地定义返回的 HttpHeaders 和 HttpStatus。上面代码中的 MediaType⊖，代表的是 Internet Media Type，即互联网媒体类型，也叫作 MIME 类型。在 HTTP 协议消息头中，使用 Content-Type 来表示具体请求中的媒体类型信息。HttpStatus 类型代

⊖ Internet Media Type，有时在一些协议的消息头中叫作"Content-Type"。它使用两部分标识符来确定一个类型。

表的是 HTTP 协议中的状态。

为了查看效果，修改 status.jsp 页面，具体如下：

```jsp
<%@ page contentType="text/html;charset=UTF-8" language="java" %>
<html>
<head>
    <title>${title}</title>
</head>
<body>
<div>
    商品：
    ${productName}
</div>
<div>
    文件上传结果：
    ${message}
</div>
<hr>
<img src="download?filename=${newFileName}"/>
<h3>文件下载</h3>
<a href="download?filename=${newFileName}">
    ${originalFilename}
</a>
</body>
</html>
```

部署服务到 Tomcat 上，访问路径为：

```
http://localhost:8080/file/uploadFile
```

上传图片后运行效果如图 6-16 所示，当点击图片名称时，会进行图片下载操作。

图 6-16　文件下载页面

本节实现了 Spring MVC 文件的上传和下载功能，实际业务开发时，考虑到性能、访问速度等因素，一般会把文件存放到 FastDFS⊖（第 14 章会有介绍）中，或者阿里的 OSS⊖服务器上。

⊖ FastDFS 是一个开源的高性能分布式文件系统，它的主要功能包括：文件存储、文件同步和文件访问，具备了大容量和负载均衡的能力。

⊖ OSS：阿里云对象存储服务（Object Storage Service，简称 OSS），是阿里云提供的海量、安全、低成本、高可靠的云存储服务。

第 7 章 Spring Boot

前面已经详细地介绍了 Spring 的能力，相信大家对 Spring 已经有了自己的理解。那么 Spring Boot 和 Spring 之间存在什么联系呢？

Spring 的核心理念是让研发者专注于业务的逻辑，而不过分考虑框架的治理，基于此思想，Spring 确实做出了很多改进，例如使用 XML 进行配置和后期的使用注解进行配置。但是即使如此，Spring 还是没有逃脱大量的配置工作，例如引入外部工程依赖时的配置等；在工程中管理大量的工程依赖时，各个依赖版本间的兼容性和配置属性烦琐等问题变得更为明显。这些问题与 Spring 的初衷相悖，所以 Spring Boot 的出现就是 Spring 初衷继续贯彻的升级版。因此，可以把 Spring Boot 理解为简化并且丰富了的 Spring。

Spring Boot 的使用会让编程更加简单，更加专注于业务，如果对比上面 Spring MVC 的配置，你会发现 Spring Boot 的改进到底有多大。这些改进基于 Spring Boot 的自动配置和起步依赖。本书对 Spring Boot 做最精要的提炼，目的是让大家快速了解 Spring Boot 的神奇之处。

7.1 构建第一个 Spring Boot 工程

正如前面所说，Spring Boot 让研发者更专注于业务，那么工程的构建必须方便快速，并且具备非常简单的框架能力，下面创建一个 Spring Boot 工程，看看 Spring Boot 初始工程已经具备了什么能力。

7.1.1 IDE 搭建及特性

（1）用 Spring Initializr 创建工程

通过浏览器打开http://start.spring.io 网址，会显示如图 7-1所示页面，在页面中可以选择工程依赖管理工具、选择语言和 Spring Boot 版本这些基础选项⊖，然后只要键入自己的工程属性例如 Group 等，就可以轻松创建一个 Spring Boot 工程，最后点击 Generate Project 进行下载，整个流程连 IDE 都没有使用就完成了一个 Spring Boot 工程的创建。

（2）用 STS 创建工程

STS（spring tool suite）是一个定制版的 Eclipse，专为 Spring 开发而定制，在 STS 中可以轻松创建 Spring Boot 工程。可以访问https://spring.io/tools/sts/all下载 STS，具体安装过程很简单，这里不再介绍。在 STS 中创建 Spring Boot 工程时其实是依赖 Spring Initializr 的，所以必须保证网络连接的畅通。

下面使用 STS 创建第一个 Spring Boot 工程。执行 "File->New->Spring Start Project"，可见如图 7-2 所示页面。

⊖ 在后面的演示中选择 Maven 进行工程管理，使用 Java 作为开发语言，Spring Boot 版本使用 1.5.10+。

第 7 章　Spring Boot

图 7-1　Initializr 页面

图 7-2　创建工程

在页面中，逐项填写工程信息，点击 Next，即进入起步依赖选择页面，如图 7-3 所示，这里选择 Web 依赖作为该项目的起步依赖。

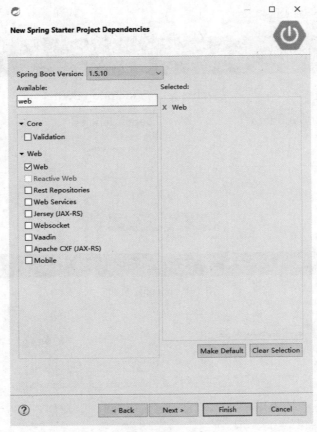

图 7-3 添加依赖

到此，一个简单的 Spring Boot 工程创建完毕，虽然这仅仅是一个新建的工程，但是它已经具备了 Web 能力，只要简单地添加一个方法即可完成一个 Web 接口的编写工作，而不需要像之前那样进行烦琐的配置。

7.1.2 工程目录

本节将介绍 Spring Boot 工程的目录结构，并且创建第一个 Spring Boot 工程的 Web 接口。工程的目录结构如图 7-4 所示。

各个模块的分工较为明确，每个根目录都有自己的职责划分，具体如下：

- src/main/java：用于业务逻辑代码的编写，Java 代码就放在这里。
- SpringBootBasicApplication.java 文件：工程的启动文件，承担了非常重要的职责，它会通过 @SpringBootApplication 注解进行自动配置并扫描它及以下层级的目录里的文件进行自动配置。
- src/main/resources：用于存放资源，其中 static 用

图 7-4 工程目录

于存放静态资源；templates 用于存放页面模板；application.properties 是工程配置文件，可以把其改名为 application.yml 文件，从而采用其他配置书写格式。
- src/test/java：用于存放测试代码，SpringBootBasicApplicationTests.java 可以获得工程上下文并且进行测试。
- pom：Mavan 依赖管理文件，特殊的地方就是包含了 spring-boot-starter-parent 这个父依赖并且自动添加了 spring-boot-maven-plugin 打包工具和 test 依赖。

了解目录结构和用处后，下面实现第一个 Spring Boot 接口。只要编写一个类，并且完成一个方法即可。

首先创建 package 为 com.javadevmap.basic.controllers。然后创建类 BasicController，类中具体内容如下：

```
@RestController
@RequestMapping("/SpringBoot")
public class BasicController {
    @RequestMapping(value="/hello",method=RequestMethod.GET)
    public String hello() {
        return "******hello Spring Boot******";
    }
}
```

这样，一个 Web 接口即完成了研发，虽然它不具备什么业务能力，但是确实是免除了之前大量的配置工作。通过 STS 的 "Run As->Spring Boot APP" 即可启动这个工程。通过浏览器访问 http://localhost:8080/SpringBoot/hello，能看到输出为 "******hello Spring Boot******"。

7.2 起步依赖

目前，仅仅创建了一个基础工程，编写了一段不到十行的代码，就完成了一个接口的编写，这是怎么做到的呢？相对于前面 Spring MVC 的大量配置工作，上一节的工程中并没有做这些工作。那么到底是谁默默地做了这些事情？这就是 Spring Boot 的功劳。

上一节工程中使用了 Spring Boot 的起步依赖 spring-boot-starter-web，可以说它是 Web 相关依赖的合集，仅使用这一个起步依赖就完成了所有 Web 依赖包的引入。查看 Web 起步依赖的 pom 文件，可以发现它具体引入的依赖⊖。

```
<dependencies>
    <dependency>
        <groupId>org.springframework.boot</groupId>
        <artifactId>spring-boot-starter</artifactId>
    </dependency>
    <dependency>
        <groupId>org.springframework.boot</groupId>
        <artifactId>spring-boot-starter-tomcat</artifactId>
    </dependency>
    <dependency>
        <groupId>org.hibernate</groupId>
        <artifactId>hibernate-validator</artifactId>
```

⊖ 手动分别引入依赖时可能会发生版本冲突的问题，使用起步依赖会保证所引入的依赖版本之间的可用性。

```xml
        </dependency>
        <dependency>
            <groupId>com.fasterxml.jackson.core</groupId>
            <artifactId>jackson-databind</artifactId>
        </dependency>
        <dependency>
            <groupId>org.springframework</groupId>
            <artifactId>spring-web</artifactId>
        </dependency>
        <dependency>
            <groupId>org.springframework</groupId>
            <artifactId>spring-webmvc</artifactId>
        </dependency>
    </dependencies>
```

起步依赖一般以 spring-boot-starter 开头，Spring Boot 的起步依赖有很多种，后面会接触到很多常用的依赖，所以这里就不一一介绍各种起步依赖的能力，仅仅介绍起步依赖所做的事情。

起步依赖帮助工程完成依赖包的集合引入，但是好像距离像之前那样完成所有配置工作还差一些，那么具体的属性配置工作就是由 Spring Boot 的自动配置能力完成的。

7.3 配置

Spring Boot 的配置能力非常灵活，在什么都不写的情况下，它会自动检测工程内的依赖引用情况然后完成默认值的配置工作。也可以编写 application.yml 文件完成自定义属性值的配置。由于 Spring Boot 的配置优先级原则，可以用高优先级的配置覆盖低优先级的配置，例如用启动命令覆盖自定义配置，还可以编写不同环境的配置文件，通过 Spring Boot 的多环境配置选择不同的环境启动服务。

7.3.1 自动配置

在前面的例子中，启动了一个 Spring Boot 服务，从控制台可以看到服务启动过程中的页面输出，下面截取部分输出数据观察工程的启动流程和相应组件的自动配置。

```
Starting SpringBootBasicApplication…
Tomcat initialized with port(s): 8080 (http)…
Starting Servlet Engine: Apache Tomcat/8.5.27…
Mapping servlet: 'dispatcherServlet' to [/]
Mapped "{[/SpringBoot/hello],methods=[GET]}" onto public…
Tomcat started on port(s): 8080 (http)…
Started SpringBootBasicApplication in 2.334 seconds (JVM running for 3.625)
```

这个项目启动的流程为：
（1）工程分配进程，并且开始启动。
（2）起步依赖中的 Tomcat 配置端口号。
（3）启动 Servlet 引擎。
（4）配置 Servlet 的前端控制器 dispatcherServlet。
（5）进行接口和方法的映射。
（6）通报 Tomcat 容器启动完成。

（7）整个工程启动完成，统计总体耗时。

以上仅用自动配置就完成了一个非常基础的 Web 服务的启动流程。Spring Boot 的自动配置项非常多，只要把起步依赖加入工程，就执行自动配置。如果对自动配置的原理感到好奇，可以先注释掉 SpringBootBasicApplication 类中的 @SpringBootApplication 注解[注]，再启动工程观察具体的变化。

7.3.2 设置配置值

在 application.properties 文件中，可以设置服务启动的端口号，只要在文件中添加 server.port=18080，然后启动服务就可以在控制台看到 Tomcat 的端口号改为了 18080（Tomcat started on port(s): 18080 (http)）。这种配置的格式很常见，但是本节介绍另一种配置格式 YAML，此种格式在层次划分上更加清晰。

YAML 语言是专门用来书写配置文件的语言，它的格式简洁，并且配置块能够明显区分开来，同一类配置放到一起便于阅读并且非常美观。下面简单演示 YAML 的配置方式，以后的大部分实例配置都采用此种格式。

在 resources 目录下，删除 application.properties 文件，创建 application.yml 文件，在文件中添加如下内容。

```
server:
  port: 18088
  tomcat:
    max-threads: 64
    min-spare-threads: 16
```

这时启动服务，端口变为了 18088，可见配置生效。

YAML 语言的格式，简单概括起来就是用键值对进行赋值，用缩进表示所属层级。这样理解起来就会非常容易。

Spring Boot 服务的配置项非常多，所以只能先熟悉一些常用的配置项，而其他配置项要在实际项目中进行摸索和理解。如果非常想了解所有的配置项，那么推荐《Spring Boot 实战》这本书，里面有非常详细的配置项说明。

7.3.3 配置优先级

配置项只能写在同一个地方吗？答案当然是否定的。Spring Boot 允许在多处进行配置，并且配置之间可以根据优先级进行覆盖。下面演示一种命令行启动服务的配置方式。

执行"鼠标右键单击工程->Run As->Run Configurations…"进入配置页面，设置 Arguments 中的 Program arguments 的内容为 --server.port=18089，如图 7-5 所示。

点击下方的 Run，启动的工程端口为：Tomcat started on port(s): 18089 (http)。可见在 yml 文件中的设置被新设置所覆盖。Spring Boot 的属性源有很多，优先级由高到低分别是：命令行参数（如本例）、java:comp/env 里的 JNDI 属性、JVM 系统属性、操作系统环境变量、随机生成的带 random.* 前缀的属性、应用程序外的配置文件（application.yml）、应用程序内的配置文件、默认属性。

[注] 它由以下几个注解组成：@SpringBootConfiguration、@EnableAutoConfiguration、@ComponentScan，这三个合起来形成了 Spring Boot 的扫描、自动配置等能力。

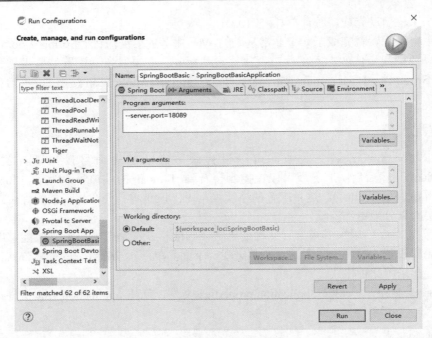

图 7-5 启动配置

一般来讲需要研发人员编写的是配置文件里的属性，而系统和环境里的属性会对程序造成影响，需要注意。命令行属性的作用一般是临时配置或者做配置选择。

7.3.4 多环境配置

在研发过程中，一般会面临最少两个环境，一个是研发和测试的环境，称为功能环境；另一个是实际业务运行的环境，称为生产环境。在这两个环境中，服务所使用的端口，所连接的数据库地址可能是不同的，这就需要有两套配置应用于两种环境。使用 Spring Boot 可以方便地进行多环境的部署工作。这里介绍两种多环境配置方法。

（1）同一文件不同 profile

把不同的环境配置写入同一个文件，然后通过启动命令选择不同的环境进行启动。Application.yml 文件修改如下：

```yaml
server:
  port: 18088
  tomcat:
    max-threads: 64
    min-spare-threads: 16
---
spring:
  profiles: dev

server:
  port: 18001
---
spring:
  profiles: prod
```

```
server:
    port: 18002
```

在这里，不同的环境用"---"隔开，针对每一个环境用 profiles 设置环境名称，一些通用的配置可以放在默认环境中。在启动工程时，选择不同的环境即可。如果选择的环境中缺少某些默认环境已经配置的配置项，则使用默认环境的配置，如果选择的环境中包含默认环境的配置项，则覆盖默认环境。进入启动配置页，在 Profile 中填入环境名称[①]。如图 7-6 所示。

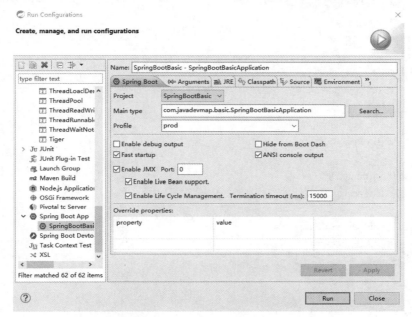

图 7-6　多环境选择

这样，通过选择不同的 profile，即可根据不同的环境配置启动工程。

（2）不同文件环境配置

通过创建不同文件名的文件，达到多环境配置的目的。

在 resources 文件夹中创建两个文件，分别为 application-dev.yml 和 application-prod.yml，在文件中设置自己的属性值，启动时也可以通过 profile 选择不同的环境进行启动。Profile 名称为 application-后面至扩展名之前的名字。如图 7-7 所示，文件的 profile 就是 dev 和 prod。

图 7-7　多配置文件

作者推荐使用多文件的配置方式，这种方式可以使配置间的分隔更加明确，并且避免了提交冲突等问题。

7.3.5　自定义类的注入

前面已经讲述了 Spring Boot 自带组件的配置方法，那么如果想自定义一个类，并且希望通过 Spring Boot 管理，就像通过 Spring 管理一样，应该怎么编写呢？

（1）首先引入 pom 文件依赖：

[①] 记得去掉之前在 Arguments 中设置的端口号，否则配置文件中的值会被覆盖。

```xml
<dependency>
    <groupId>org.springframework.boot</groupId>
    <artifactId>spring-boot-configuration-processor</artifactId>
    <optional>true</optional>
</dependency>
```

（2）编写自定义的类，这里使用一个抽象基类和两个派生类（后面会演示两种加载方法）：

```java
public abstract class IocAnimal {
    private int weight;
    private String desc;

    public int getWeight() {
        return weight;
    }
    public void setWeight(int weight) {
        this.weight = weight;
    }
    public String getDesc() {
        return desc;
    }
    public void setDesc(String desc) {
        this.desc = desc;
    }
}

@Component
@ConfigurationProperties(prefix="iocfish")
public class IocFish extends IocAnimal{

}

@Component
@ConfigurationProperties(prefix="ioctiger")
public class IocTiger extends IocAnimal{

}
```

（3）编写 yml 配置文件：

```yaml
ioctiger:
  weight: 500
  desc: i am tiger

iocfish:
  weight: 500
  desc: i am fish
```

（4）在 Controller 类中注入：

```java
@RestController
```

```
@RequestMapping("/SpringBoot")
public class BasicController {
    @Autowired
    private IocAnimal iocFish;

    @Autowired
    @Qualifier("iocTiger")
    private IocAnimal iocAnimal;

    @RequestMapping(value="/tiger",method=RequestMethod.GET)
    public IocAnimal getTiger() {
        return iocAnimal;
    }

    @RequestMapping(value="/fish",method=RequestMethod.GET)
    public IocAnimal getFish() {
        return iocFish;
    }
}
```

代码中使用了两种注入方式，两种方式的原理在第 5 章已经有了详细的介绍，这里仅演示自定义类在 Spring Boot 工程中的创建及注入方法。

7.4　使用 Thymeleaf 构建页面

通过前面的学习，读者对 Spring Boot 的主要特性有了较多的认识。下面可以基于 Spring Boot 工程实现自己的业务站点。最基础的业务站点需要有页面和数据存储，虽然本书不会把页面部分作为重点，但是绘制简单页面可以为后台开发带来一些额外的成就感，毕竟只返回 JSON 格式的数据看起来没有页面那么美观。

7.4.1　Thymeleaf 基本使用

Thymeleaf 是一款用于渲染 XML/XHTML/HTML5 内容的模板引擎，也是 Spring Boot 官方推荐使用的模板引擎，可以用它替代 JSP 进行页面绘制。

（1）添加依赖

可以在 STS 中选择 pom 文件，然后进入 pom 文件的 Dependencies 选项栏，执行"鼠标右键单击->Spring->Edit Starters"即可进入如图 7-8 所示页面，在页面中可以方便地查找想要使用的依赖，而不必使用 pom 的文本编辑输入依赖项。

在搜索栏中输入 thyme 就可以搜出 Thymeleaf 依赖，选择即可。在 pom 文件中会增加如下依赖：

```
<dependency>
    <groupId>org.springframework.boot</groupId>
    <artifactId>spring-boot-starter-thymeleaf</artifactId>
</dependency>
```

图 7-8 依赖选择页面

（2）添加模板页面

在 templates 目录下，添加 welcome.html 文件，完成一个最简单的页面模板。

```
<html xmlns:th="http://www.thymeleaf.org">
<head>
<title>welcome spring boot</title>
</head>
<body>
Welcome Spring Boot!
</body>
</html>
```

（3）添加 Controller

在 controllers 包下新建一个类 PageController，目前这个类仅实现一个方法，这个方法的目的是对访问路径和模版页面进行映射。

```
@Controller
@RequestMapping(value="/page")
public class PageController {
    @RequestMapping(value="/welcome")
    public ModelAndView welcome() {
        ModelAndView modelAndView = new ModelAndView("welcome");
        return modelAndView;
    }
}
```

注意，PageController 类中使用@Controller 作为注解，方法返回的是 ModelAndView 类型。

通过浏览器访问 http://localhost:18088/page/welcome 地址，可以查看这个页面，页面上有"Welcome Spring Boot!"的输出。

（4）添加静态资源

前面已经通过 Spring Boot 完成了一个简单的页面，但是这个页面确实有些太简单了，可以稍微对页面进行优化，添加一些静态图片，让它稍微美观一些。在 static 中添加静态资源 spring_boot.jpg 文件，然后修改 welcome.html 文件，更改样式和插入图片。

```
<html xmlns:th="http://www.thymeleaf.org">
    <head>
        <title>welcome spring boot</title>
    </head>
    <body>
        <div style="width:600px;border:1px solid black">
            <img style="height:200px;margin:0 auto;display:block" alt="" src="/spring_boot.jpg"></img>
            <div>
                <label style="font-size:48px;display:block;width:100%;text-align:center" th:text="'Welcome Spring Boot!'"></label>
            </div>
        </div>
    </body>
</html>
```

通过浏览器访问此网址，可以看到如图 7-9 所示页面[⊖]。

图 7-9　页面运行效果

7.4.2　添加页面逻辑

现在虽然已经可以绘制简单的页面了，但是这种页面明显与实现基本的业务逻辑相差甚远。下面给程序添加页面和数据交互的逻辑，让页面变得更有意义。

设计一个页面，可以通过输入框输入用户信息，然后提交给后台程序。

（1）添加输入页面模板

添加页面模板 createuser.html，把页面输入数据的部分放入 form 中，并且通过 action 标记数据处理的方法，这个方法会在 submit 时被调用。

⊖ 在调试程序时，可能针对某些修改要重启程序，这会对研发速度造成一定的影响，可以通过引入 spring-boot-devtools 依赖从而实现动态更新程序的目的，而不必不断手动启动服务。

```html
<html xmlns:th="http://www.thymeleaf.org">
<head>
<title>create user</title>
<meta http-equiv="Content-Type" content="text/html; charset=UTF-8" />
</head>
<body>
    <div>
        <label style="font-size: 18px" th:text="'创建用户'"></label>
    </div>
    <form th:action="@{/page/save}" method="post" th:object="${user}" style="width:600px">
        <fieldset>
            <p>
                <label style="font-size: 18px" th:text="'名字：'"></label> <input
                    type="text" id="name" name="name" tabindex="1"></input>
            </p>
            <p>
                <label style="font-size: 18px" th:text="'年龄：'"></label> <input
                    type="text" id="age" name="age" tabindex="2"></input>
            </p>
            <p>
                <label style="font-size: 18px" th:text="'电话号码：'"></label> <input
                    type="text" id="phoneNum" name="phoneNum" tabindex="3"></input>
            </p>
            <p>
                <label style="font-size: 18px" th:text="'住址：'"></label> <input
                    type="text" id="address" name="address" tabindex="4"></input>
            </p>
            <p id="buttons">
                <input id="submit" type="submit" tabindex="5" value="创建"></input> <input
                    id="reset" type="reset" tabindex="6" value="取消"></input>
            </p>
        </fieldset>
    </form>
</body>
</html>
```

（2）添加页面映射 Controller 类方法

```java
@RequestMapping(value="/createuser")
public ModelAndView createUser() {
    ModelAndView modelAndView = new ModelAndView("createuser");
    return modelAndView;
}
```

（3）添加页面数据模型

此数据结构用来承载页面输入的数据。

```java
public class User {
    private int id;
    private String name;
    private int age;
    private String phoneNum;
    private String address;
    …
```

```
    @Override
    public String toString() {
        return "name = " + name + ";age = " + age + ";phoneNum = " + phoneNum + ";address = " + address;
    }
}
```

（4）页面展示

通过以上的配置，可以根据模板映射和模板页面的实现，在浏览器中显示实际的前端页面。如图 7-10 所示。

（5）添加数据交互

根据上面的页面进行输入，但是输入后的数据由谁来承接呢？这就需要提供一个接口，用来承接页面数据。

```
@RequestMapping(value="/save")
public ModelAndView save(@ModelAttribute User user) {
    ModelAndView modelAndView = new ModelAndView("save");
    modelAndView.addObject("info", user.toString());
    return modelAndView;
}
```

以上为承接页面数据的接口。然后编写一个 save.html 模板页面把 toString 生成的数据显示出来，具体如下。

```
<html xmlns:th="http://www.thymeleaf.org">
<head>
<title>welcome spring boot</title>
<meta http-equiv="Content-Type" content="text/html; charset=UTF-8" />
</head>
<body>
<label style="font-size:18px;display:block" th:text="${info}"></label>
</body>
</html>
```

（6）页面显示

在创建用户的页面输入一些 User 信息，如图 7-11 所示，然后观察 save 页面的显示。

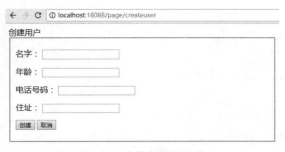
图 7-10　用户信息输入页面

图 7-11　输入用户信息

点击"创建"后，显示的页面内容如图 7-12 所示。

现在虽然可以在后台程序接收前台页面的输入了，但是这么直白地显示出来好像也没有什么意义，而且一旦程序退出，之前输入的数据也会丢失，所以需要把数据持久化。

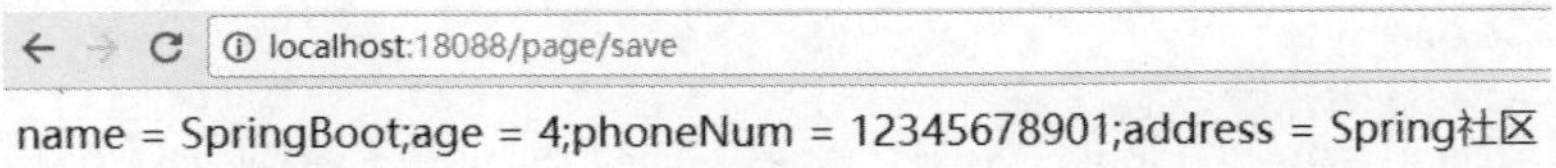

name = SpringBoot;age = 4;phoneNum = 12345678901;address = Spring社区

图 7-12　保存结果页面

7.5　使用 JPA 构建持久化存储

Spring Data JPA 是 Spring 在 ORM[⊖]框架、JPA[⊖]规范的基础上封装的一套 JPA 应用框架，可使开发者用极简的代码实现对数据的访问和操作。

本节将要用 JPA 连接一个 MySQL 数据库，如果读者之前对数据库一点都不了解，那么使用 JPA 倒是非常简单的，就像给方法起一些有意义的名字，就可以操作数据库了；如果读者习惯了用 SQL 的方式使用数据库，那么反倒可能造成一些影响，JPA 的写法还是要适应一段时间的。

7.5.1　JPA 基本使用

这里用最原始的 JPA 来实现数据操作。可以把业务数据存入数据库中，并且在需要的时候取出。

（1）添加依赖

在 pom 文件中引入如下依赖，因为要连接 MySQL 数据库，所以除了引入 JPA 依赖还要引入 MySQL 连接的依赖。

```xml
<dependency>
    <groupId>org.springframework.boot</groupId>
    <artifactId>spring-boot-starter-data-jpa</artifactId>
</dependency>
<dependency>
    <groupId>mysql</groupId>
    <artifactId>mysql-connector-java</artifactId>
    <scope>runtime</scope>
</dependency>
```

（2）添加数据库配置

```yaml
spring:
  datasource:
    driver-class-name: com.mysql.jdbc.Driver
    url: jdbc:mysql://localhost:3306/javadevmap?characterEncoding=utf-8
    username: root
    password: mypass
  jpa:
    database-platform: org.hibernate.dialect.MySQL5Dialect
    hibernate:
      ddl-auto: update
```

[⊖] ORM 即对象关系映射（英语：Object Relational Mapping，简称 ORM，或 O/RM，或 O/R mapping），是一种程序技术，用于实现面向对象编程语言里不同类型系统的数据之间的转换。

[⊖] JPA 是 Java Persistence API 的简称，中文名为 Java 持久层 API，通过 JDK 5.0 注解或 XML 描述"对象－关系表"的映射关系，并将运行期的实体对象持久化到数据库中。

在上面的配置中，对 datasource 设置了驱动 com.mysql.jdbc.Driver，并且通过 url 设置了数据库的地址，通过 username 和 password 设置数据库的账号和密码。

对 jpa 设置了目标数据库，并且设置了自动更新库表。如果读者不熟悉 JPA，那么自动更新这个设置非常重要，当数据库中不存在表时它会帮你创建，当你添加映射类的字段时它会帮你在表中添加列。需要注意的是，如果在 MySQL 数据库中手动创建表，那么 MySQL 中的表和 JPA 中的表的字段匹配万一出现问题会导致 JPA 无法使用表，排查这个问题会比较困难。

(3) 添加实体映射类

使用之前的 User 类，只是把这个类通过注解标记，使之成为实体映射类，这样 JPA 就可以通过这个类去创建数据库表。

```java
@Entity
@Table(name="User")
public class User {
    @Id
    @GeneratedValue(strategy=GenerationType.IDENTITY)
    private int id;
    private String name;
    private int age;
    private String phoneNum;
    private String address;
    …
}
```

通过它，JPA 会在数据库中生成一个 user 表，并且以 id 为自增的主键。

(4) 定义数据访问接口

使用原生的数据库访问接口的方法，所以只要用一个接口类继承 JpaRepository 就可以了。

```java
public interface UserRepository extends JpaRepository<User, Integer>{

}
```

这种写法还真的适合懒人，因为就是什么都没写。可以向类中添加自定义的方法，从而实现不同的能力。

(5) 实现存储和展示逻辑

下面，要做的事情就是把页面传过来的数据放入数据库中存储，并且从数据库中取部分数据用于展示。

新建一个 service 接口类和实现类，用于业务逻辑的实现。这个类实现了两个方法，一个调用 repository 把数据存入数据库；另一个倒序取出数据库中的最后 10 条数据，方法中用到了 PageRequest 的分页能力，设置的是每页显示 10 条数据，显示第一页，通过 Sort 实现倒序。

```java
public interface UserService {
    public void add(User user);
    public List<User> getList();
}

@Service
public class UserServiceImpl implements UserService{
    @Autowired
    private UserRepository repository;
```

```java
    public void add(User user) {
        repository.save(user);
    }

    public List<User> getList(){
        Sort sort = new Sort(Sort.Direction.DESC, "id");
        return repository.findAll(new PageRequest(0, 10, sort)).getContent();
    }
}
```

在 Controller 类中注入 service 类实例，进行业务逻辑操作，并且把要显示的数据传入页面。

```java
@Autowired
private UserService userService;
@RequestMapping(value="/save")
public ModelAndView save(@ModelAttribute User user) {
    userService.add(user);
    List<User> list = userService.getList();
    ModelAndView modelAndView = new ModelAndView("save");
    modelAndView.addObject("list", list);
    return modelAndView;
}
```

修改 save.html 页面，用于显示用户 list。

```html
<html xmlns:th="http://www.thymeleaf.org">
<head>
<style>
table {
    border-collapse: collapse;
    border-spacing: 0;
    border-left: 1px solid #888;
    border-top: 1px solid #888;
    background: #efefef;
}

th, td {
    border-right: 1px solid #888;
    border-bottom: 1px solid #888;
    padding: 5px 15px;
}

th {
    font-weight: bold;
    background: #ccc;
}
</style>
<title>welcome spring boot</title>
<meta http-equiv="Content-Type" content="text/html; charset=UTF-8" />
</head>
<body>
    <div style="width: 600px">
        <label style="font-size: 18px" th:text="'用户列表："></label>
    </div>
    <div style="width: 600px">
```

```
            <table style="width: 100%">
                <tr>
                    <td>编号</td>
                    <td>名字</td>
                    <td>年龄</td>
                    <td>手机号</td>
                    <td>住址</td>
                </tr>
                <tr th:each="user : ${list}">
                    <td th:text="${user.id}"></td>
                    <td th:text="${user.name}"></td>
                    <td th:text="${user.age}"></td>
                    <td th:text="${user.phoneNum}"></td>
                    <td th:text="${user.address}"></td>
                </tr>
            </table>
        </div>
        <label style="font-size: 18px; display: block" th:text="${info}"></label>
    </body>
</html>
```

（6）显示效果如图 7-13 所示。

localhost:18088/page/save

用户列表：

编号	名字	年龄	手机号	住址
12	mybatis	18	12345678901	地球
11	spring boot	18	12345678901	地球
10	spring MVC	18	12345678901	地球
9	spring	18	12345678901	地球
8	maven	18	12345678901	地球
7	svn	18	12345678901	地球
6	git	18	12345678901	地球
5	linux	18	12345678901	地球
4	c++	18	12345678901	地球
3	java	18	12345678901	地球

图 7-13　数据保存列表

目前的 Spring Boot 工程已经实现了前后台的打通，前台输入数据，后台存储数据。虽然这个功能看起来有些简单，但确实是一个项目的最基础和核心的内容，一个大的平台也是从这些小功能开始慢慢丰富和发展起来的。如果想了解更多，最好的学习办法就是在实际项目中接触更多的内容，解决更多的问题，想更多的解决方案。

7.5.2　定义 JPA 扩展接口

当使用 JPA 时，只要继承 JpaRepository，就继承了一些默认的方法，主要包含如下几类：delete、find、save、count、exists。上例中就使用了 save 和 find 方法来实现业务逻辑，但是这些默认方法还不能覆盖所有的 SQL 操作，需要手动扩展方法。

扩展使用方法时，就好像用英文直译要做的事情一样来定义方法名，例如想通过 id 来查找某条数据，只要定义一个方法 findById，就具备了通过 id 查找的能力。下面在 UserRepository 中定义几个方法来扩充对数据库操作的能力。

```
public interface UserRepository extends JpaRepository<User, Integer>{
    public User findById(int id);
    public User findByName(String name);
    public List<User> findByAgeBetween(int start,int end);
    public List<User> findByAgeLessThan(int age);
}
```

如上面所写,只要把想做的事按照 JPA 的规则命名一个方法,即完成了数据库操作语句的建立。具体的 JPA 规则见表 7-1。

表 7-1 JPA 规则

关键字	示例	JPQL 代码
And	findByLastnameAndFirstname	… where x.lastname = ?1 and x.firstname = ?2
Or	findByLastnameOrFirstname	… where x.lastname = ?1 or x.firstname = ?2
Is,Equals	findByFirstname,findByFirstnameIs,findByFirstnameEquals	… where x.firstname = ?1
Between	findByStartDateBetween	… where x.startDate between ?1 and ?2
LessThan	findByAgeLessThan	… where x.age < ?1
LessThanEqual	findByAgeLessThanEqual	… where x.age <= ?1
GreaterThan	findByAgeGreaterThan	… where x.age > ?1
GreaterThanEqual	findByAgeGreaterThanEqual	… where x.age >= ?1
After	findByStartDateAfter	… where x.startDate > ?1
Before	findByStartDateBefore	… where x.startDate < ?1
IsNull	findByAgeIsNull	… where x.age is null
IsNotNull,NotNull	findByAge(Is)NotNull	… where x.age not null
Like	findByFirstnameLike	… where x.firstname like ?1
NotLike	findByFirstnameNotLike	… where x.firstname not like ?1
StartingWith	findByFirstnameStartingWith	… where x.firstname like ?1(parameter bound with appended %)
EndingWith	findByFirstnameEndingWith	… where x.firstname like ?1(parameter bound with prepended %)
Containing	findByFirstnameContaining	… where x.firstname like ?1(parameter bound wrapped in %)
OrderBy	findByAgeOrderByLastnameDesc	… where x.age = ?1 order by x.lastname desc
Not	findByLastnameNot	… where x.lastname <> ?1
In	findByAgeIn(Collection ages)	… where x.age in ?1
NotIn	findByAgeNotIn(Collection age)	… where x.age not in ?1
TRUE	findByActiveTrue()	… where x.active = true
FALSE	findByActiveFalse()	… where x.active = false
IgnoreCase	findByFirstnameIgnoreCase	… where UPPER(x.firstame) = UPPER(?1)

使用以上规则,就可以自由组装接口方法来扩展数据库操作能力。

7.6 Actuator

Spring Boot Actuator 提供了众多 Web 端点,可以通过端点了解服务内部的运行情况。Actuator 提供的端点见表 7-2。

表 7-2 Actuator 端点

HTTP 方法	路径	描述	鉴权
GET	/autoconfig	查看自动配置的使用情况	TRUE
GET	/configprops	查看配置属性，包括默认配置	TRUE
GET	/beans	查看 bean 及其关系列表	TRUE
GET	/dump	打印线程栈	TRUE
GET	/env	查看所有环境变量	TRUE
GET	/env/{name}	查看具体变量值	TRUE
GET	/health	查看应用健康指标	FALSE
GET	/info	查看应用信息	FALSE
GET	/mappings	查看所有 url 映射	TRUE
GET	/metrics	查看应用基本指标	TRUE
GET	/metrics/{name}	查看具体指标	TRUE
POST	/shutdown	关闭应用	TRUE
GET	/trace	查看基本追踪信息	TRUE

按照 Spring Boot 的一贯风格，只要引入 pom 依赖，再进行一些简单的配置，就可以使用某组件了，Actuator 也是这样。

7.6.1 Actuator 的基本使用

（1）添加依赖

```xml
<dependency>
    <groupId>org.springframework.boot</groupId>
    <artifactId>spring-boot-starter-actuator</artifactId>
</dependency>
```

现在访问程序的 http://localhost:18088/health 端点，可以得到如下信息：

```
{"status":"UP","diskSpace":{"status":"UP","total":195696783360,"free":145456672768,"threshold":10485760},"db":{"status":"UP","database":"MySQL","hello":1}}
```

如果访问http://localhost:18088/metrics会发生什么呢？其实是不可访问，这就是上表中鉴权的意思，这个端点是受保护的，如果想直接访问可以先关闭端点安全能力。

（2）关闭安全认证

在 yml 文件中添加如下设置，即可关闭端点的安全认证。

```yaml
management:
  security:
    enabled: false
```

这样访问 metrics 端点就可以得到数据。

（3）使用端点关闭程序

通过 yml 文件的配置，开启 shutdown 端点。

```yaml
endpoints:
  shutdown:
    enabled: true
```

使用 shutdown 端点可以关闭程序，它与其他端点不同的地方是它是 POST 类型的请求，使用 Postman 工具提交一个 POST 请求到端点，如图 7-14 所示。

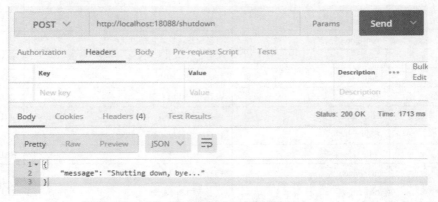

图 7-14 shutdown 请求展示

执行后可以得到如下输出结果：

{"message":"Shutting down, bye..."}

7.6.2 端点的保护

如果程序运行在网络中，那么把端点暴露出去是很危险的，所以应该想一些办法尽量不要让其他人调用到程序的端点。

（1）隐藏端点

可以用一个简单的办法，既不开启安全认证，又能对 shutdown 端点适当隐藏。通过修改 shutdown 端点的 id，把端点的调用指向另一个路径，这样可以避免熟悉 Actuator 的人调用 shutdown 端点。

```
endpoints:
  shutdown:
    enabled: true
    id: kill
```

（2）端点安全认证

可以为端点设置一个路径，然后对这个路径进行加密保护，这样端点就安全多了。

添加依赖：

```
<dependency>
    <groupId>org.springframework.boot</groupId>
    <artifactId>spring-boot-starter-security</artifactId>
</dependency>
```

修改端点配置：

```
management:
  context-path: /admin
  security:
    enabled: true
security:
  user:
```

```
      name: admin
      password: mypass
    basic:
      enabled: true
      path: /admin
endpoints:
  shutdown:
    enabled: true
    id: kill
    sensitive: true
  metrics:
    enabled: true
    sensitive: true
```

端点的访问路径通过 context-path 设置为/admin 根路径，这样访问端点时前面要加上此路径。打开安全认证组件，并且设置账号密码后，访问/admin 路径时都需要输入账号信息，所以端点安全多了。

7.7 部署

目前，调试 Spring Boot 程序时，都是在 IDE 工具中，如果程序编写完毕需要上线运行，肯定不是在 IDE 中运行，而是运行在服务器中。在服务器中运行 Spring Boot 程序的方法非常简单，按照以下流程操作即可。

（1）程序打包

执行"鼠标右键单击工程->Run As->Maven install"，这样程序会通过 Maven 打包成一个 Jar 包。Jar 包名字为 SpringBootBasic-0.0.1-SNAPSHOT.jar。如果成功打包会有如下输出。

```
[INFO] BUILD SUCCESS
[INFO] ------------------------------------------------------------------------
[INFO] Total time: 01:41 min
[INFO] Finished at: 2018-03-31T21:47:43+08:00
[INFO] Final Memory: 31M/306M
[INFO] ------------------------------------------------------------------------
```

（2）把程序传至服务器

生产环境的管理一般是运维人员进行的，他们可能会使用非常严格的安全认证才能连接上生产环境，例如使用动态密码等工具。这些对于研发来讲可能就太专业了，本节仅需要把 jar 传到服务器的某个位置即可。使用 VanDyke 公司的 SecureFX（一个可视化的工具），拖动文件即可实现文件上传，所以不再过多介绍。

（3）启动服务

前面已经介绍了 Linux 的基本命令，所以这里就不过多介绍。登录 Linux 服务器进入 jar 包的文件夹，通过如下命令启动服务。

```
$ nohup java –jar SpringBootBasic-0.0.1-SNAPSHOT.jar &
```

（4）服务运行情况如图 7-15 所示。

可见，Spring Boot 服务可以在 Linux 服务器上顺利运行。

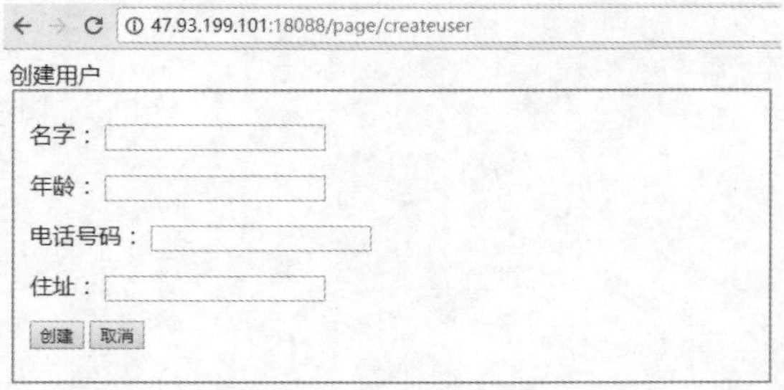

图 7-15　Linux 服务运行

7.8　参数校验

目前的 Spring Boot 工程已经实现了程序的接口与方法的映射、页面的显示、业务逻辑的处理、数据的保存，并且把程序已经运行到服务器了，好像这个程序的所有事情都已经完成了。那么请看下面的情况，如图 7-16 所示。

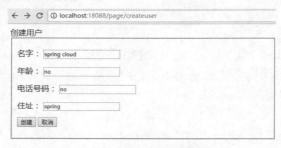

图 7-16　输入错误用户信息

如果使用之前的代码，填写上面的信息，程序会报告错误。如图 7-17 所示。

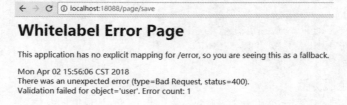

图 7-17　错误信息

错误的原因是通过表单构建 User 对象的时候，由于年龄和电话号码使用了错误的类型，所以无法正确赋值给 User 对象。因此需要修改代码。

7.8.1　前台完成基本参数校验

参数的校验一般是需要前后台配合的，前台期望得到正确的返回，所以前台会尽量矫正用户的输入错误，并且尽量保证数据的正确。后台会验证数据的合法性来保证业务的正确。对于上述错误情况，前台可以进行如下修改，在模板页面限定用户的输入类型。

```
<p>
    <label style="font-size: 18px" th:text="'年龄：'"></label> <input
        type="number" id="age" name="age" tabindex="2"></input>
```

```
</p>
<p>
    <label style="font-size: 18px" th:text="''电话号码: ''"></label> <input
    type="number" id="phoneNum" name="phoneNum" tabindex="3"></input>
</p>
```

模板中限定输入年龄和手机号的输入框必须输入数字,这样就可以保证前台输入类型的正确性。

7.8.2 前后台配合完成数据校验

设想一种情况,如果用户输入的数据不合法,例如年龄字段输入负数或者大于 200 的数字,这样的数据对于业务来讲是没有意义的[⊖]。那么可以通过后台进行用户输入的校验,并且通过前台显示提示信息,这里使用了 hibernate validator[⊖]组件。

要想实现此能力,需要修改几个地方,首先,后台的提交接口需要具备可以验证的能力;其次,针对每个字段的验证,需要设置字段规则;最后,前台可以接收验证的结果并且展示出来。优化部分就是保留用户已填内容,避免由于部分内容填写错误而导致所有信息全部重填。

(1) 后台接口修改

修改 save 接口,对前端传入的数据进行验证,并且返回页面和错误信息。修改 createuser 接口,使前台页面可以正确显示。

```
@RequestMapping(value="/createuser")
public ModelAndView createUser() {
    ModelAndView modelAndView = new ModelAndView("createuser");
    modelAndView.addObject("user",new User());
    return modelAndView;
}
@RequestMapping(value="/save")
public ModelAndView save(@Valid User user,BindingResult result,Model model) {
    if(result!=null && result.hasErrors()) {
        ModelAndView modelAndView = new ModelAndView("createuser");
        modelAndView.addObject(model);
        return modelAndView;
    }else {
        userService.add(user);
        List<User> list = userService.getList();
        ModelAndView modelAndView = new ModelAndView("save");
        modelAndView.addObject("list", list);
        return modelAndView;
    }
}
```

在 save 方法中,通过@Valid 注解验证对象数据,通过 BindingResult 实例返回页面错误信息,并且根据数据是否发生错误返回不同的页面。

(2) 数据字段验证规则

设置每个字段的校验规则,输入的字段必须符合校验的要求,否则不能通过。

⊖ 这里不考虑其他情况,例如从某些产品来讲,希望获取用户的门槛越低越好,尽管用户输入了一个不合理数据也期望用户能够使用业务。

⊖ hibernate validator 不需要单独引用,Web 起步依赖已经集成。

```
@Entity
@Table(name="User")
public class User {
    @Id
    @GeneratedValue(strategy=GenerationType.IDENTITY)
    private int id;

    @NotBlank(message="姓名不能为空")
    private String name;

    @NotNull(message="年龄不能为空")
    @Min(value=0,message="输入年龄小于最小值")
    @Max(value=150,message="输入年龄大于最大值")
    private Integer age;

    @Length(min=7,max=11,message = "输入号码错误")
    private String phoneNum;

    @NotBlank(message="地址不能为空")
    private String address;
}
```

代码中修改了 age 字段的类型，由 int 改为 Integer，这样可以避免前台页面默认显示 int 的初始化数字 0，改为 Integer 后前台没有默认输出内容。

User 类中针对每个字段都做了限定，例如设置了年龄的最大/最小值，输入手机号码的长度限制等，其实 validator 还有很多其他注解可以限定字段校验，见表 7-3。

表 7-3 validator 注解

注 解	作 用
@Null	限制只能为 null
@NotNull	限制不能为 null
@AssertFalse	限制必须为 false
@AssertTrue	限制必须为 true
@DecimalMax(value)	限制必须为一个不大于指定值的数字
@DecimalMin(value)	限制必须为一个不小于指定值的数字
@Digits(integer,fraction)	限制必须为一个小数，且整数部分的位数不能超过 integer，小数部分的位数不能超过 fraction
@Future	限制必须是一个将来的日期
@Max(value)	限制必须为一个不大于指定值的数字
@Min(value)	限制必须为一个不小于指定值的数字
@Past	限制必须是一个过去的日期
@Pattern(value)	限制必须符合指定的正则表达式
@Size(max,min)	限制字符长度必须在 min 到 max 之间
@Past	验证注解的元素值（日期类型）比当前时间早
@NotEmpty	验证注解的元素值不为 null 且不为空（字符串长度不为 0、集合大小不为 0）
@NotBlank	验证注解的元素值不为空（不为 null、去除首位空格后长度为 0），不同于@NotEmpty，@NotBlank 只应用于字符串且在比较时会去除字符串的空格
@Email	验证注解的元素值是 Email，也可以通过正则表达式和 flag 指定自定义的 Email 格式

（3）修改前台显示模板

前台显示模板的修改较为简单，主要是两点，分别是错误信息显示和已填信息的回填。createuser.html 文件具体如下。

```html
<html xmlns:th="http://www.thymeleaf.org">
<head>
<title>create user</title>
<meta http-equiv="Content-Type" content="text/html; charset=UTF-8" />
</head>
<body>
    <div>
        <label style="font-size: 18px" th:text="'创建用户'"></label>
    </div>
    <form th:action="@{/page/save}" method="post" th:object="${user}"
        style="width: 600px">
        <fieldset>
            <p>
                <label style="font-size: 18px" th:text="'名字："></label> <input
                    type="text" id="$name" name="name" tabindex="1" th:field= "*{name}"> </input>
                <td th:if="${#fields.hasErrors('name')}" th:errors="*{name}" />
            </p>
            <p>
                <label style="font-size: 18px" th:text="'年龄："></label> <input
                    type="number" id="age" name="age" tabindex="2" th:field= "*{age}"> </input>
                <td th:if="${#fields.hasErrors('age')}" th:errors="*{age}" />
            </p>
            <p>
                <label style="font-size: 18px" th:text="'电话号码："></label> <input
                    type="number" id="phoneNum" name="phoneNum" tabindex="3"
                    th:field="*{phoneNum}"></input>
                <td th:if="${#fields.hasErrors('phoneNum')}" th:errors="*{phoneNum}" />
            </p>
            <p>
                <label style="font-size: 18px" th:text="'住址："></label> <input
                    type="text" id="address" name="address" tabindex="4"
                    th:field="*{address}"></input>
                <td th:if="${#fields.hasErrors('address')}" th:errors="*{address}" />
            </p>
            <p id="buttons">
                <input id="submit" type="submit" tabindex="5" value="创建"></input> <input
                    id="reset" type="reset" tabindex="6" value="取消"></input>
            </p>
        </fieldset>
    </form>
</body>
</html>
```

页面模板在 input 标签中添加 th:field="*{name}"，来做已填数据的回填；通过<td th:if="${#fields.hasErrors('name')}" th:errors="*{name}" />来检测错误信息，如果有错误则显示。

（4）效果展示

现在打开 createuser 页面，然后任何信息都不填写，点击"创建"，观察页面的变化。如图 7-18

所示。

现在，这个 Spring Boot 工程既可以执行基本的业务逻辑，又可以通过前台的限制和后台的校验保证业务的准确，并且人性化地提示了错误的原因。对于后台研发人员来讲，前台的页面部分了解这些已经可以了。

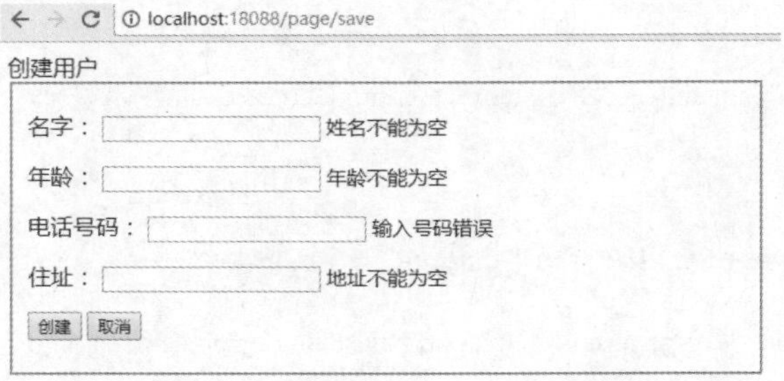

图 7-18　输入错误提示

7.9　MyBatis 的框架整合及数据校验

对于很多后台研发人员来讲，可能不太习惯 JPA 的使用方式，大家可能更习惯使用 MyBatis 操作数据库。下面就介绍 MyBatis 的 Spring Boot 工程集成方法，并且通过 validator 对纯后台接口进行数据校验和错误返回。

7.9.1　整合 MyBatis

在讲解 Spring 的时候，已经添加了 MyBatis 的整合，所以本节简单介绍 Spring Boot 的 MyBatis 整合。新建一个工程 SpringBootMybatis。

（1）添加依赖

由于 Spring Boot 的起步依赖较为完善，所以此工程添加的依赖主要是以下 3 个。

```
<dependency>
    <groupId>org.springframework.boot</groupId>
    <artifactId>spring-boot-starter-web</artifactId>
</dependency>
<dependency>
    <groupId>org.mybatis.spring.boot</groupId>
    <artifactId>mybatis-spring-boot-starter</artifactId>
    <version>1.3.2</version>
</dependency>
<dependency>
    <groupId>mysql</groupId>
    <artifactId>mysql-connector-java</artifactId>
    <scope>runtime</scope>
</dependency>
```

（2）配置 yml 文件

在配置文件中，配置数据库连接 datasource 和 Mybatis 的相关路径。

```
server:
  port: 18089
```

```yaml
spring:
  application:
    name: SpringBootMybatis
  datasource:
    driver-class-name: com.mysql.jdbc.Driver
    url: jdbc:mysql://39.106.10.196:3306/javadevmap?characterEncoding=utf-8
    username: root
    password: mypass
mybatis:
  type-aliases-package: com.javadevmap.mybatis.*.model
  mapper-locations: classpath:/mybatis/sqlmap/*.xml
  config-location: classpath:/mybatis/mybatis-config.xml
```

(3) 添加 Mybatis 配置文件

在 mybatis-config.xml 文件中添加如下配置。

```xml
<?xml version="1.0" encoding="UTF-8"?>
<!DOCTYPE configuration
        PUBLIC "-//www.mybatis.org//DTD Config 3.0//EN"
        "http://mybatis.org/dtd/mybatis-3-config.dtd">
<configuration>
    <settings>
        <setting name="cacheEnabled" value="true" />
        <setting name="lazyLoadingEnabled" value="true" />
        <setting name="multipleResultSetsEnabled" value="true" />
        <setting name="useColumnLabel" value="true" />
        <setting name="useGeneratedKeys" value="false" />
        <setting name="defaultExecutorType" value="SIMPLE" />
        <setting name="defaultStatementTimeout" value="25000" />
    </settings>
</configuration>
```

(4) 配置生成文件

在 generatorConfig.xml 文件中添加如下配置，这里设置直接将生成的代码添加至工程中。

```xml
<?xml version="1.0" encoding="UTF-8"?>
<!DOCTYPE generatorConfiguration
        PUBLIC "-//mybatis.org//DTD MyBatis Generator Configuration 1.0//EN"
        "http://mybatis.org/dtd/mybatis-generator-config_1_0.dtd">

<generatorConfiguration>
<classPathEntry location="C:\Users\T460\.m2\repository\mysql\
                    mysql-connector-java\5.1.44\mysql-connector-java-5.1.44.jar" />
    <context id="mysqlStepyee" targetRuntime="MyBatis3" >
        <commentGenerator>
            <property name="suppressAllComments" value="true" />
            <property name="suppressDate" value="true" />
        </commentGenerator>
        <jdbcConnection driverClass="com.mysql.jdbc.Driver"
                        connectionURL="jdbc:mysql://39.106.10.196:3306/javadevmap?
                        autoReconnect=true&amp;useUnicode=true&amp;characterEncoding=utf8"
                userId="root" password="mypass">
        </jdbcConnection>
        <javaTypeResolver>
```

```xml
                <property name="forceBigDecimals" value="false"/>
            </javaTypeResolver>
            <javaModelGenerator targetPackage="com.javadevmap.mybatis.model"
                                targetProject="src\main\java">
                <property name="enableSubPackages" value="true"/>
                <property name="trimStrings" value="true"/>
            </javaModelGenerator>
            <sqlMapGenerator targetPackage="mybatis.sqlmap" targetProject="src\main\resources">
                <property name="enableSubPackages" value="true"/>
            </sqlMapGenerator>
            <javaClientGenerator type="MIXEDMAPPER"
                                 targetPackage="com.javadevmap.mybatis.model.mapper"
                                 targetProject="src\main\java">
                <property name="enableSubPackages" value="true"/>
            </javaClientGenerator>
            <table tableName="user" schema="javadevmap"/>
        </context>
</generatorConfiguration>
```

（5）添加插件并生成

在 pom 文件中，添加如下生成插件，然后运行生成命令"Run As->Maven build…"，输入框中填写 mybatis-generator:generate，点击 Run 即可生成代码。

```xml
<build>
    <plugins>
        …
        <plugin>
            <groupId>org.mybatis.generator</groupId>
            <artifactId>mybatis-generator-maven-plugin</artifactId>
            <version>1.3.2</version>
            <configuration>
                <configurationFile>src/main/resources/mybatis/generatorConfig.xml
                </configurationFile>
                <overwrite>true</overwrite>
            </configuration>
        </plugin>
    </plugins>
</build>
```

（6）启动类配置

在启动类中添加@MapperScan("com.javadevmap.mybatis.model.mapper")注解，用于扫描 mapper 文件。

（7）添加 DAO

添加 DAO 类的接口和实现，用于操作数据库，类中添加两个方法，一个是查询，一个是保存。

```java
public interface UserDao {
    public User getUser(int id);
    public int saveUser(User user);
}

@Repository
```

```java
public class UserDaoImpl implements UserDao{
    @Autowired
    private UserMapper mapper;

    @Override
    public User getUser(int id) {
        return mapper.selectByPrimaryKey(id);
    }

    @Override
    public int saveUser(User user) {
        return mapper.insert(user);
    }
}
```

（8）添加一个承接请求的 User 类

虽然 MyBatis 自动生成时，会生成一个 User 类，但是为了不破坏自动生成的代码，或者避免修改 User 文件后又被自动生成所覆盖，所以这里定义一个用于承接 Web 请求的 DomainUser 类。

```java
public class DomainUser {
    private Integer id;
    private String address;
    private Integer age;
    private String name;
    private String phoneNum;
    //…getset
}
```

（9）添加 Service

添加 Service 类的接口和实现，用于业务逻辑的处理，在 Service 中注入 Dao 类用以操作数据库。

```java
public interface UserService {
    public DomainUser getUser(int id);
    public int saveUser(DomainUser user);
}

@Service
public class UserServiceImpl implements UserService{
    @Autowired
    private UserDao dao;

    @Override
    public DomainUser getUser(int id) {
        User user = dao.getUser(id);
        if(user!=null) {
            DomainUser domainUser = new DomainUser();
            BeanUtils.copyProperties(user, domainUser);
            return domainUser;
        }
        return null;
    }
```

```
    @Override
    public int saveUser(DomainUser domainUser) {
        User user = new User();
        BeanUtils.copyProperties(domainUser, user);
        return dao.saveUser(user);
    }
}
```

（10）添加 Controller 接口

在 Controller 中，注入 Service 实现，并且通过 Service 进行相应的业务操作。接口使用 REST 的方式获取用户信息，通过 post 请求添加用户。

```
@RestController
@RequestMapping("/user")
public class UserController {
    @Autowired
    private UserService service;

    @RequestMapping(value="/{Id}",method=RequestMethod.GET)
    public DomainUser getUser(@PathVariable("Id") int id) {
        return service.getUser(id);
    }

    @RequestMapping(value="/add",method=RequestMethod.POST)
    public void addUser(@RequestBody DomainUser user) {
        service.saveUser(user);
    }
}
```

（11）获取用户信息

使用 Postman，可以轻松地通过 REST 接口得到用户的信息。如图 7-19 所示。

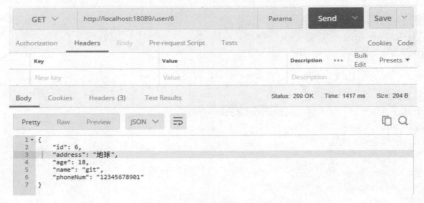

图 7-19　请求用户信息

（12）添加用户

可以使用 Postman 工具提交一个 POST 请求到服务器，模拟真实的前台请求操作。

Postman 的参数设置如图 7-20 所示。

第 7 章　Spring Boot

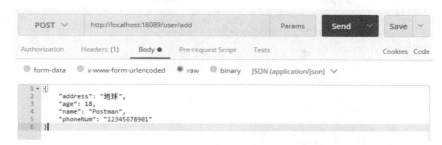

图 7-20　Postman 参数设置

如果正确添加用户数据，逻辑肯定会正常执行；如果非法添加用户数据，并且希望后台服务能够正确检查出非法的原因，那么就又涉及了参数校验。

7.9.2　后台接口请求校验

对于前台传过来的数据，可以在 Controller 的方法中用 if 语句，逐字段判断其是否符合业务规则，然后向前台反馈错误信息，例如

```
if(user.getAddress()==null) {
    return "address can not be null!";
}
```

这种方法完全可行，而且已经成为很多编程人员的习惯用法，此处不再赘述。至于这个返回数据仅仅是个 String 类型，不够规范的问题，会在下一小节中介绍返回数据规范化。

此工程继续使用 validator 请求数据校验。校验的是接口数据而非前台页面传递过来的数据，所以写法上存在一些差异，但是原理基本相同。

（1）DomainUser 添加注解规范

和验证前端数据一样，从 REST 接口获得的数据如果要进行合法性校验，也要添加 validator 注解。

```
public class DomainUser {
    private Integer id;

    @NotBlank(message="地址不能为空")
    private String address;

    @NotNull(message="年龄不能为空")
    @Min(value=0,message="输入年龄小于最小值")
    @Max(value=150,message="输入年龄大于最大值")
    private Integer age;

    @NotBlank(message="姓名不能为空")
    private String name;

    @Length(min=7,max=11,message = "输入号码错误")
    private String phoneNum;
}
```

（2）Controller 类修改方法参数

在 Controller 类的 addUser 方法中添加@Valid 注解，用以标明需要验证的参数。

181

```
@RequestMapping(value="/add",method=RequestMethod.POST)
Public void addUser(@RequestBody @Valid DomainUser user) {
    service.saveUser(user);
}
```

（3）验证校验情况

通过 Postman 提交请求，故意把 address 字段的数据删掉，观察返回结果。

```
{
    "timestamp": 1522825947907,
    "status": 400,
    "error": "Bad Request",
    "exception": "org.springframework.web.bind.MethodArgumentNotValidException",
    "errors": [{
        "codes":     ["NotBlank.domainUser.address",    "NotBlank.address",    "NotBlank.java.lang.String",
"NotBlank"],
        "arguments": [{
            "codes": ["domainUser.address", "address"],
            "arguments": null,
            "defaultMessage": "address",
            "code": "address"
        }],
        "defaultMessage": "地址不能为空",
        "objectName": "domainUser",
        "field": "address",
        "rejectedValue": "",
        "bindingFailure": false,
        "code": "NotBlank"
    }],
    "message": "Validation failed for object='domainUser'. Error count: 1",
    "path": "/user/add"
}
```

validator 校验了请求参数，并且返回了大量的用于排查问题的信息，其中包含了代码中的自定义错误提示语"地址不能为空"。虽然 validator 返回的数据非常完善，但是对于前端研发人员来讲，他们期望后台返回的数据具备统一模板，而不是突然出现的 validator 风格的错误提示，所以工程面临统一返回模板的问题。

7.9.3 规范数据返回

下面规范一个最简单的返回模板，包含返回信息码、返回信息提示语和前台需要的数据对象。用 Java 泛型模板来实现返回格式的统一。

（1）定义数据统一返回模板

先定义一个枚举值，通过枚举值生成统一返回模板的错误码和返回信息。

```
public enum ResultCode{
    OK,Bad_Request,Unauthorized,Not_Found,ERROR,Unavailable
}

public class Result<T> {
    private int resultCode;
    private String msg;
```

```java
    private T data;

    public Result() {

    }

    public Result(int code,String msg) {
        this.resultCode = code;
        this.msg = msg;
    }

    public Result(int code,String msg,T data) {
        this.resultCode = code;
        this.msg = msg;
        this.data = data;
    }

    public Result(ResultCode rCode) {
        switch (rCode) {
        case OK:
            this.resultCode = 200;
            this.msg = "ok";
            break;
        case Bad_Request:
            this.resultCode = 400;
            this.msg = "Bad Request";
            break;
        case Unauthorized:
            this.resultCode = 401;
            this.msg = "Unauthorized";
            break;
        case Not_Found:
            this.resultCode = 404;
            this.msg = "Not Found";
            break;
        case ERROR:
            this.resultCode = 500;
            this.msg = "Server Error";
            break;
        case Unavailable:
            this.resultCode = 503;
            this.msg = "Service Unavailable";
            break;
        default:
            this.resultCode = 400;
            this.msg = "Bad Request";
            break;
        }
    }

    public Result(ResultCode rCode,T data) {
        this(rCode);
        this.data = data;
```

```java
        }
        public int getResultCode() {
            return resultCode;
        }
        public void setResultCode(int resultCode) {
            this.resultCode = resultCode;
        }
        public String getMsg() {
            return msg;
        }
        public void setMsg(String msg) {
            this.msg = msg;
        }
        public T getData() {
            return data;
        }
        public void setData(T data) {
            this.data = data;
        }
}
```

（2）修改 Controller 接口

对 Controller 中对外提供接口的方法进行修改，使其返回的数据格式统一为 Result。

```java
@RestController
@RequestMapping("/user")
public class UserController {
    @Autowired
    private UserService service;

    @RequestMapping(value="/{Id}",method=RequestMethod.GET)
    public Result<DomainUser> getUser(@PathVariable("Id") int id) {
        DomainUser user = service.getUser(id);
        Result<DomainUser> result = null;
        if(user!=null) {
            result = new Result<>(ResultCode.OK, user);
        }else {
            result = new Result<>(ResultCode.Not_Found);
        }
        return result;
    }

    @RequestMapping(value="/add",method=RequestMethod.POST)
    public Result<String> addUser(@RequestBody @Valid DomainUser user) {
        int ret = service.saveUser(user);
        Result<String> result = null;
        if(ret == 1) {
            result = new Result<>(ResultCode.OK);
        }else {
            result = new Result<>(ResultCode.ERROR);
        }
        return result;
```

        }
    }

（3）返回信息展示

请求 getUser 对外提供的接口，观察其数据返回情况。

成功情况：

```
{
    "resultCode": 200,
    "msg": "ok",
    "data": {
        "id": 8,
        "address": "地球",
        "age": 18,
        "name": "maven",
        "phoneNum": "12345678901"
    }
}
```

失败情况：

```
{
    "resultCode": 404,
    "msg": "not found",
    "data": null
}
```

可见，数据返回已经具备了标准统一的格式，但是对于通过 validator 校验失败的请求，其返回格式还没有被统一进来，所以需要进行如下修改。

（4）模板化 validator 输出

对 validator 的标准化，需要对 validator 抛出的异常进行统一处理，然后返回模板化错误信息，具体如下。

```java
@ControllerAdvice
public class ParamValidateControllerAdvice {
    @ExceptionHandler({MethodArgumentNotValidException.class})
    @ResponseBody
    public Result<String> handleMethodArgumentNotValidException(MethodArgumentNotValidException e) {
        return handleFieldErrors(e.getBindingResult());
    }

    public Result<String> handleFieldErrors(BindingResult bindingResult) {
        List<FieldError> fieldErrors = bindingResult.getFieldErrors();
        List<String> errors = new ArrayList();
        for(FieldError error : fieldErrors) {
            errors.add(""+error.getField()+":"+error.getDefaultMessage());
        }
        Result<String> result = new Result<>();
        result.setResultCode(400);
        result.setMsg(errors.toString());
        return result;
    }

}
```

这样，当请求参数发生错误的时候也可以具有统一的返回信息格式了。例如当提交的所有字段都不合法时，会出现如下提示信息。

```
{
    "resultCode": 400,
    "msg": "[phoneNum:输入号码错误, name:姓名不能为空, address:地址不能为空, age:输入年龄大于最大值]",
    "data": null
}
```

7.10 添加日志及记录请求信息

对于这个工程，可能还需要记录一些日志信息，这些日志信息在服务运行时会辅助研发人员调试程序。可以使用 AOP 技术进行切面逻辑的添加，在 Controller 中记录服务器收到请求和返回的数据，这样如果服务出现问题可以方便排查。

7.10.1 添加日志模块

日志模块在第 5 章介绍过，这里只是简单介绍 Spring Boot 工程的日志引入方法。

（1）使用 yml 文件配置日志

由于日志组件的依赖已经通过 Spring Boot 的起步依赖引入进来，所以这里不再单独引入日志的 pom 文件依赖。之前工程一直没有特意去配置日志，但是控制台却都有日志的打印，这是 Spring Boot 工程的自动配置做的事情。所以对于日志来讲，在还没有关注它的时候它就已经可以正确输出了。

如果只是简单地使用日志，对配置文件进行简单修改即可。

```
logging:
  level:
    root: DEBUG
  pattern:
    console: "%d - %msg%n"
  path: /logs
```

上面的配置，设置了日志的总体打印级别是 DEBUG，设置了日志的打印格式，还设置了一个日志文件的输出路径。如果启动工程，就可以在控制台和对应路径下看到日志。

如果要在程序内手动打印日志，只要在类中创建日志引用，在方法中使用 log.info 日志打印，即可打印日志。

```
import org.slf4j.Logger;
import org.slf4j.LoggerFactory;

@RestController
@RequestMapping("/user")
public class UserController {
    private static final Logger log = LoggerFactory.getLogger(UserController.class);

    @RequestMapping(value="/{Id}",method=RequestMethod.GET)
    public Result<DomainUser> getUser(@PathVariable("Id") int id) {
        DomainUser user = service.getUser(id);
```

```java
                Result<DomainUser> result = null;
                if(user!=null) {
                    result = new Result<DomainUser>(ResultCode.OK, user);
                }else {
                    result = new Result<DomainUser>(ResultCode.Not_Found);
                }
                log.info(result.toString());
                return result;
            }
            …
        }
```

（2）在 resources 文件夹下添加日志配置文件

使用上面的方法虽然可以正确打印日志，但是配置项还是相对较少，所以可以使用常规的日志配置方法打印日志。注释掉 yml 文件的日志配置，在 resources 文件夹下创建 logback-spring.xml 文件。

```xml
<?xml version="1.0" encoding="UTF-8"?>
<configuration debug="true">
    <include resource="org/springframework/boot/logging/logback/defaults.xml" />
    <property name="APP_Name" value="SpringBootMybatis" />
    <property name="LOG_HOME" value="/logs" />
    <contextName>${APP_Name}</contextName>

    <jmxConfigurator />
    <appender name="STDOUT" class="ch.qos.logback.core.ConsoleAppender">
        <encoder>
            <pattern>%d{yyyy-MM-dd} %d{HH:mm:ss.SSSZ} %-5level %logger{36}[${APP_Name}], [%15.15t] : %m%n%wEx</pattern>
        </encoder>
    </appender>

    <appender name="FILE"
        class="ch.qos.logback.core.rolling.RollingFileAppender">
        <rollingPolicy
            class="ch.qos.logback.core.rolling.SizeAndTimeBasedRollingPolicy">
            <FileNamePattern>${LOG_HOME}/${APP_Name}/%d{yyyy-MM-dd}.%i.log
            </FileNamePattern>
            <MaxHistory>10</MaxHistory>
            <maxFileSize>100MB</maxFileSize>
        </rollingPolicy>
        <layout class="ch.qos.logback.classic.PatternLayout">         <pattern>%d{yyyy-MM-dd} %d{HH:mm:ss.SSSZ} %-5level %logger{36}[${APP_Name}], [%15.15t] : %m%n%wEx</pattern>
        </layout>
    </appender>

    <root level="info">
        <appender-ref ref="STDOUT" />
        <appender-ref ref="FILE" />
    </root>
</configuration>
```

添加以上配置后，日志可以通过控制台和文件分别输出。日志文件的输出采用日期和文件大小两个维度限制，即日志记录超过一天或者日志文件大于 100MB 时，则拆分日志文件。

7.10.2　AOP 实现接口信息打印

上一节在方法内打印了请求的返回值，如果有多个请求的话则需要对每个方法逐一添加日志打印语句，这明显是重复工作，而且在批量添加时还有可能出错。最简单的办法就是使用 AOP 技术，使用 AOP 技术切入方法的开始和结束，打印方法的请求和返回值。

（1）添加 AOP 依赖

```xml
<dependency>
    <groupId>org.springframework.boot</groupId>
    <artifactId>spring-boot-starter-aop</artifactId>
</dependency>
```

（2）添加切面类

```java
@Component
@Aspect
public class AopAspect {
    private static final Logger log = LoggerFactory.getLogger(AopAspect.class);
    ObjectMapper mapper = new ObjectMapper();

    @Pointcut("execution(* com.javadevmap.mybatis.controllers..*.*(..))")
    public void AopPointCut() {

    }

    @Before(value="AopPointCut()")
    public void AopBefore(JoinPoint point) {
        try {
            StringBuilder builder = new StringBuilder();
            builder.append(point.getSignature().getDeclaringTypeName());
            builder.append(" method = ");
            builder.append(point.getSignature().getName());
            builder.append(" args = ");
            for(Object object : point.getArgs()) {
                builder.append(mapper.writeValueAsString(object));
            }
            log.info(builder.toString());
        } catch (Exception e) {
            e.printStackTrace();
        }
    }

    @AfterReturning(value="AopPointCut()",returning="ret")
    public void AopAfterReturning(JoinPoint point,Object ret) {
        try {
            StringBuilder builder = new StringBuilder();
            builder.append(point.getSignature().getDeclaringTypeName());
            builder.append(" method = ");
            builder.append(point.getSignature().getName());
            builder.append(" args = ");
            for(Object object : point.getArgs()) {
```

```
                builder.append(mapper.writeValueAsString(object));
            }
            builder.append(" ret = ");
            builder.append(mapper.writeValueAsString(ret));
            log.info(builder.toString());
        } catch (Exception e) {
            e.printStackTrace();
        }
    }
}
```

如上代码使用切面切入了 Controllers 包中的所有类，对类中方法的开始和结束插入日志打印逻辑，如果是方法的开始则打印方法名和请求参数，如果是方法的结束则再添加打印返回值。通过这么一个简单的切面类，就可以在不修改接口的情况下，完成所有请求和返回的记录。

（3）日志结果

使用浏览器访问http://localhost:18089/user/10这个 get 接口，可以看到如下日志输出。

```
2018-04-08  22:39:19.758+0800  INFO  c.j.mybatis.config.AopAspect[SpringBootMybatis], [io-18089-exec-3] : com.javadevmap.mybatis.controllers.UserController method = getUser args = 10
2018-04-08  22:39:19.777+0800  INFO  c.j.mybatis.config.AopAspect[SpringBootMybatis], [io-18089-exec-3] : com.javadevmap.mybatis.controllers.UserController method = getUser args = 10 ret = {"resultCode": 200, "msg": "ok","data":{"id":10,"address":"地球","age":18,"name":"spring MVC","phoneNum": "12345678901"}}
```

到此，本章对 Spring Boot 的使用和特性已经介绍了很多。现在看看 SpringBootMybatis 的工程目录，如图 7-21 所示，已经初具规模了。

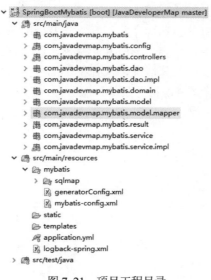

图 7-21　项目工程目录

本章讲解 Spring Boot 的自动配置和起步依赖，以这两点为基础，通过 Thymeleaf 构建了前台页面；通过 JPA 保存数据；通过端点查看程序运行情况；集成 Mybatis 用于提供第二种数据操作方法；使用 validator 对参数进行校验，统一返回数据的格式并且通过日志记录请求的具体情况。了解了这些，就可以使用 Spring Boot 开始业务的研发了。

第 8 章 服务架构

　　第一篇的几章内容，从 Java 语言开始，讲解如何使用语言以及了解语言的特性；然后讲解了 Maven 工程的管理；接下来使用 Git 和 Svn 版本控制软件来管理代码；介绍了 Linux 系统命令，并且在 Linux 服务器中运行了一个 Java 程序。在第二篇的头几章内容中，学习了使用 Spring 框架治理来管理程序，并且了解了 Spring MVC 的页面编写等内容；上一章使用 Spring Boot 更简便地管理程序。其实到目前为止，已经可以通过讲解的内容编写自己的业务了。但是本书的范围明显不限于此。通过本章的了解，可以看到一个小系统是如何一步步变大的，以及系统变大后这种复杂系统的管理办法。

　　以一个电商系统为例，电商系统必须包含的模块有：用户、商品、订单订购关系、卖家及后台。这是最核心的几个模块，是初始搭建系统的时候，必须首先实现的。如果用 Spring Boot 实现这个基础系统，那么这些模块可能分属同一个工程的不同 package，但是总体还是一个程序，程序外部连接数据库和一些静态的图片资源。这个系统可能会部署在一台服务器之内，当然了，如果这台服务器宕机了，那么电商平台也完蛋了。这个初始版的电商虽然不能保证稳定地提供业务，但是它确实具备了基础的功能业务能力，如图 8-1 所示。

　　这个系统的所有组件都部署在同一台服务器中，数据库和程序会共用 CPU 和内存，数据库和文件服务会共用磁盘，文件服务会和程序共用网络带宽。随着这个小电商系统慢慢开始有人访问，如果服务器性能不高的话，很容易形成性能瓶颈，此瓶颈是由硬件资源不足造成的。在此情况下，就需要根据硬件资源的具体情况，选择把一部分能力拆分出去，部署到另外一台服务器中，例如把数据库系统和静态资源拆分到其他服务器，那么现在就具备了 3 台服务器。如图 8-2 所示。

图 8-1　单机服务器
应用服务、数据库、文件系统

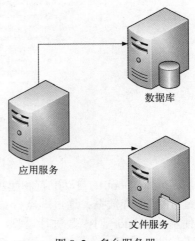

图 8-2　多台服务器

　　有一天，某台服务器宕机了，导致整个系统都瘫痪了，研发人员突然认识到单节点对业务造成了多大的风险。现在业务已经顺利开展了，不希望这种事情再次发生，于是决定把所有单节点的组件都变为双节点，这样即使其中一台瘫痪，系统还是能正常运行。这就涉及了反向代理和负载均衡，需要把系统压力合理分散到业务承载服务器上，如图 8-3 所示。

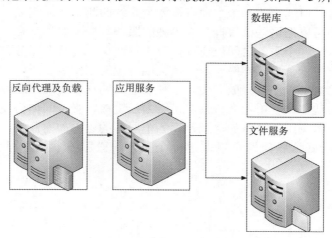

图 8-3　带冗余的服务器集群

业务的发展越来越好，用户越来越多，服务的压力也越来越大，请求响应时间开始变长。经过诊断，发现几种情况，例如程序所在的应用服务器 CPU 使用率非常高，那么应该扩容应用服务器，提供更多的程序节点；例如发现数据库查询越来越慢，可能会对数据库进行分表或者读写分离，进行多写多读，或者使用分布式缓存⊖和本地缓存，以降低数据库的压力；例如静态资源获取非常慢，可能会对静态资源配置 CDN 以提高速度，如图 8-4 所示。

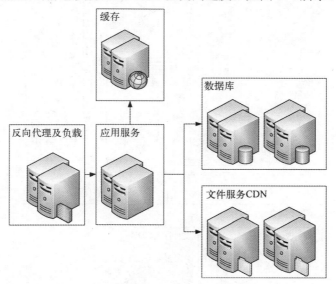

图 8-4　加入数据缓存等服务器集群

⊖　缓存的读写能力强于数据库，可以把一些经常访问的数据放入缓存中，以降低数据库的压力。本地缓存也是提高性能的好办法，只是同步问题需要解决。

业务已经运行了很长时间了，工程中代码实现的业务逻辑也越来越多，越来越复杂。例如订单模块里会有打折，打折还要根据用户的等级设置折扣比例，用户等级的提高需要达到某个条件才行，这个条件的统计可能还在订单模块里，这些相关联的模块对研发和测试工作造成了很大的复杂度，可能修改一个小地方需要全系统的测试才能完成。为了降低程序内模块间的复杂度，可不可以把这些模块拆分为单独的服务？每个服务负责一部分能力，通过接口对外提供能力支持，各个服务之间也通过接口进行通信。当某个服务更新了，只要测试这个服务的接口能力就可以了，这样降低了全量集成测试的工作量，所以微服务的理念就诞生了。微服务是把一个大的程序拆分成一堆具备独立能力的程序集合，这种拆分会面临很多问题，例如服务间如何通信、使用什么协议、是不是长链接、服务间如何发现其他依赖服务、负载策略是什么、服务调用的链路如何跟踪、各个服务的压力如何监控、如此多的服务如何进行配置更新等。还好这些问题可以通过一个完善的微服务框架进行解决，例如 Spring Cloud，如图 8-5 所示。

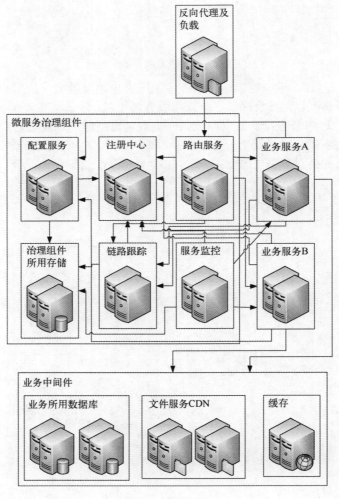

图 8-5 微服务集群

现在系统中的服务已经拆分成了微服务体系,服务间的通信可能存在一些特殊情况,例如一些比较耗时或者实时性不高的业务,可能需要使用消息队列进行处理,有一些需要定时执行的任务可能需要分布式定时任务组件进行处理,还有日志体系也需要聚合起来进行跟踪,这样系统又会引入一系列功能组件,如图8-6所示。

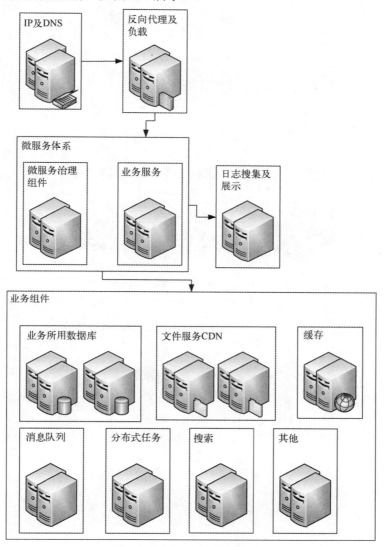

图 8-6　功能组件及微服务集群

现在系统已经非常庞大了,所有的程序是分模块的,并且每个模块都有很多个服务实例,甚至每个模块都有自己独立的数据库。每个服务使用了很多种业务组件,服务间的通信也需要 RPC 或消息队列,这对运维管理提出了很高的挑战,所以这个系统还需要添加持续集成和镜像等能力以辅助运维的工作,让系统可以实现自动化部署并且更加灵活。本书会涉及部分持续集成和镜像的内容,目的是让研发人员从运维角度理解服务集群的运行,如图 8-7 所示。

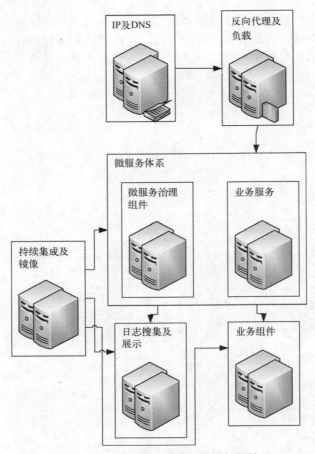

图 8-7 运维体系、组件及微服务集群

通过以上这些操作,这个系统已经做得非常强大了,并且各个模块职责清晰,单独扩展能力强,持续集成可降低人工的工作量,镜像技术使系统更加灵活。此时,研发工程师可以非常轻松地在自己所负责的模块中进行开发工作而不用担心框架的问题;而架构师则可以通过框架治理等各个可视化界面来管控这么庞大的系统集群。

第 9 章 Spring Cloud

Spring Cloud 是一套微服务治理框架，正如前面提到的，如果把一个传统架构的程序拆分为一个一个小的微服务，那么出现的治理和使用上的问题就可以通过 Spring Cloud 解决。

Spring Cloud 提供了服务发现、配置管理、消息总线、负载均衡、断路器、链路跟踪、数据监控等微服务治理能力，使微服务集群可以全面地被管理和组合起来。同时 Spring Cloud 各个组件是基于 Spring Boot 的，这些能力可以通过 Spring Boot 的简单配置实现。

本章使用 Spring Cloud 的 Edgware.SR2 版本○。Spring Cloud 的可选组件很多，书中选择其中的一部分进行演示，例如 Eureka 和 Zookeeper 都可以作为服务发现组件，书中只演示 Eureka。

9.1 Eureka

Eureka 是 Spring Cloud 的服务注册发现组件，微服务集群内的业务服务都通过 Eureka 进行注册，这样 Eureka 上就保留了业务服务的名字和地址。如果集群内的服务间需要互相调用，通过 Eureka 上已经注册的信息就能查到目标服务的地址列表○，从而实现集群内的服务间调用。

9.1.1 Eureka 基础使用

首先搭建一个单节点的 Eureka，然后创建一个业务服务，并且把业务服务注册到 Eureka 上。之后就可以通过 Eureka 的可视化页面查看服务的列表，并且通过 RestAPI 可以查看服务的详细注册信息。

（1）创建 Eureka 服务端

由于 Eureka 服务端也是一个 Spring Boot 工程，所以按照 Spring Boot 工程的方式去创建，如图 9-1 所示。

然后在接下来的组件选择中选择 Eureka Server，如图 9-2 所示。

图 9-1 创建 Eureka 工程

○ 本章介绍的组件，大部分是通过 Dalston.SR4 版本进行编写，之后又升级到 Edgware.SR2 版本的，所以在这两个版本中，程序实现基本是通用的。但是由于 Spring Cloud 版本更新很快，如果读者拿到本书时使用的 Spring Cloud 版本不是这两个，可能会发生不适配的情况。

○ 业务服务从 Eureka 获取到的服务列表信息会自己进行缓存，而不是每次调用都要请求 Eureka 获取列表。

Java 服务端研发知识图谱

图 9-2　Eureka 依赖选择

创建后工程的 pom 文件如下：

```xml
<?xml version="1.0" encoding="UTF-8"?>
<project xmlns="http://maven.apache.org/POM/4.0.0"xmlns:xsi="http://www.w3.org/2001/XMLSchema-instance"
    xsi:schemaLocation="http://maven.apache.org/POM/4.0.0
    http://maven.apache.org/xsd/maven-4.0.0.xsd">
    <modelVersion>4.0.0</modelVersion>

    <groupId>com.javadevmap</groupId>
    <artifactId>SpringCloudEureka</artifactId>
    <version>0.0.1-SNAPSHOT</version>
    <packaging>jar</packaging>

    <name>SpringCloudEureka</name>
    <description>Eureka project for Spring Cloud</description>

    <parent>
        <groupId>org.springframework.boot</groupId>
        <artifactId>spring-boot-starter-parent</artifactId>
        <version>1.5.11.RELEASE</version>
        <relativePath/> <!-- lookup parent from repository -->
    </parent>

    <properties>
        <project.build.sourceEncoding>UTF-8</project.build.sourceEncoding>
        <project.reporting.outputEncoding>UTF-8</project.reporting.outputEncoding>
        <java.version>1.8</java.version>
        <spring-cloud.version>Edgware.SR2</spring-cloud.version>
    </properties>

    <dependencies>
        <dependency>
            <groupId>org.springframework.cloud</groupId>
```

```xml
            <artifactId>spring-cloud-starter-eureka-server</artifactId>
        </dependency>

        <dependency>
            <groupId>org.springframework.boot</groupId>
            <artifactId>spring-boot-starter-test</artifactId>
            <scope>test</scope>
        </dependency>
    </dependencies>

    <dependencyManagement>
        <dependencies>
            <dependency>
                <groupId>org.springframework.cloud</groupId>
                <artifactId>spring-cloud-dependencies</artifactId>
                <version>${spring-cloud.version}</version>
                <type>pom</type>
                <scope>import</scope>
            </dependency>
        </dependencies>
    </dependencyManagement>

    <build>
        <plugins>
            <plugin>
                <groupId>org.springframework.boot</groupId>
                <artifactId>spring-boot-maven-plugin</artifactId>
            </plugin>
        </plugins>
    </build>
</project>
```

修改 Eureka 工程的配置文件，通过 server.port 和 spring.application.name 给 Eureka 服务设置端口号和服务名。eureka.instance.hostname 是服务实例主机的名字，本例是在本机演示单节点 Eureka 服务，所以使用 localhost 来指代主机名。由于是单节点，所以通过 eureka.client 下面的设置 register-with-eureka 和 fetch-registry 暂时关闭了 Eureka 服务向自己注册的能力，service-url 为 Eureka 服务的地址，这里就是 Eureka 服务本身。

```yaml
server:
  port: 18001
spring:
  application:
    name: eureka-server
eureka:
  instance:
    hostname: localhost
  client:
    register-with-eureka: false
    fetch-registry: false
    service-url:
      defaultZone: http://localhost:18001/eureka/
```

最后一步就是添加启动类的注解：

```
import org.springframework.boot.SpringApplication;
import org.springframework.boot.autoconfigure.SpringBootApplication;
import org.springframework.cloud.netflix.eureka.server.EnableEurekaServer;
@EnableEurekaServer
@SpringBootApplication
public class SpringCloudEurekaApplication {
    public static void main(String[] args) {
        SpringApplication.run(SpringCloudEurekaApplication.class, args);
    }
}
```

在程序启动类中添加了注解@EnableEurekaServer，用于开启 Eureka 的服务端能力。这样，一个 Eureka 服务端就配置好了。可以使用启动 Spring Boot 工程的方式启动此 Eureka 服务。

（2）创建 Eureka 客户端

Eureka 的客户端也是一个 Spring Boot 工程，只是工程内引用 Eureka 客户端依赖，然后配置好服务注册的相关配置，即可完成服务注册。

创建一个 Spring Boot 工程 SpringCloudServiceProvider。在工程中引入 Eureka 的客户端依赖。

```
<dependency>
    <groupId>org.springframework.cloud</groupId>
    <artifactId>spring-cloud-starter-eureka</artifactId>
</dependency>
```

修改此工程的 yml 配置文件，配置本服务的端口和名称，并且配置连接 Eureka 服务端的地址等。

```
server:
  port: 18010
spring:
  application:
    name: service-provider
eureka:
  instance:
    hostname: localhost
  client:
    service-url:
      defaultZone: http://localhost:18001/eureka/
```

在启动类中通过注解@EnableEurekaClient㊀开启服务注册能力。这样，就完成了一个业务服务的 Eureka Client 配置。

（3）观察 Eureka 服务注册信息

Eureka 服务注册中心和一个业务服务均已经配置完毕，下面启动这两个服务，然后访问 Eureka 的可视化页面，即可看见服务列表，如图 9-3 所示。

㊀ 本章用到的新依赖包和注解较多，使用注解时一般都需要 import 相应的资源。由于本书使用顺序的方式讲述每个知识点，所以 import 导入的内容和当时讲解的组件一般都有关系，当添加注解时如果书中没有明确说明引入的资源包，一般 IDE 默认提示的资源包都可以使用，但大家还要多注意分辨。

第 9 章 Spring Cloud

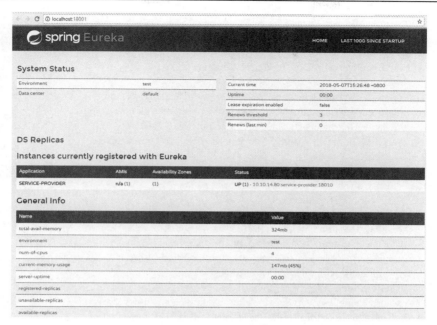

图 9-3　Eureka 页面

可见，客户端服务 SERVICE-PROVIDER 已经注册到了 Eureka Server 上。

9.1.2　配置服务注册信息

现在已经把业务服务注册到 Eureka 上了，但是如果想修改业务服务注册到 Eureka 上的信息，例如在 Eureka 页面上直接点击业务服务的链接，就可以看到业务服务的说明。

要想实现这个能力，必须弄清楚几个问题：业务服务向 Eureka 服务注册了什么信息、点击的链接实际打开的是哪个注册信息中的内容以及配置一个应用服务的说明信息。下面来实现这个能力。

（1）通过 Eureka RestAPI 获取服务信息

可以通过 Eureka 对外提供的接口，查看当前 Eureka 上的服务注册信息。访问http://localhost:18001/eureka/apps可以看到如下输出。

```
<applications>
  <versions__delta>1</versions__delta>
  <apps__hashcode>UP_1_</apps__hashcode>
  <application>
    <name>SERVICE-PROVIDER</name>
    <instance>
      <instanceId>10.10.14.80:service-provider:18010</instanceId>
      <hostName>localhost</hostName>
      <app>SERVICE-PROVIDER</app>
      <ipAddr>10.10.14.80</ipAddr>
      <status>UP</status>
      <overriddenstatus>UNKNOWN</overriddenstatus>
      <port enabled="true">18010</port>
      <securePort enabled="false">443</securePort>
      <countryId>1</countryId>
      <dataCenterInfo class="com.netflix.appinfo.InstanceInfo$DefaultDataCenterInfo">
```

```xml
            <name>MyOwn</name>
        </dataCenterInfo>
        <leaseInfo>
            <renewalIntervalInSecs>30</renewalIntervalInSecs>
            <durationInSecs>90</durationInSecs>
            <registrationTimestamp>1525678611057</registrationTimestamp>
            <lastRenewalTimestamp>1525678611057</lastRenewalTimestamp>
            <evictionTimestamp>0</evictionTimestamp>
            <serviceUpTimestamp>1525678611058</serviceUpTimestamp>
        </leaseInfo>
        <metadata>
            <management.port>18010</management.port>
            <jmx.port>55186</jmx.port>
        </metadata>
        <homePageUrl>http://localhost:18010/</homePageUrl>
        <statusPageUrl>http://localhost:18010/info</statusPageUrl>
        <healthCheckUrl>http://localhost:18010/health</healthCheckUrl>
        <vipAddress>service-provider</vipAddress>
        <secureVipAddress>service-provider</secureVipAddress>
        <isCoordinatingDiscoveryServer>false</isCoordinatingDiscoveryServer>
        <lastUpdatedTimestamp>1525678611058</lastUpdatedTimestamp>
        <lastDirtyTimestamp>1525678610862</lastDirtyTimestamp>
        <actionType>ADDED</actionType>
    </instance>
</application>
</applications>
```

在此返回信息中，最外层标签是 applications，记录了 Eureka 上所有的服务；第 2 层的 application 标签记录了某个服务的所有程序实例，其中 Instance 标签是具体的服务实例的信息，这个信息可以通过应用服务的 eureka.instance 属性进行配置。

（2）配置服务显示信息

通过对业务服务添加 actuator 依赖，可以启动业务服务的端点显示。然后在 yml 配置文件中添加服务的显示信息。

```
info:
    author: hw
    book: javadevmap
    project: service provider demo
```

这样当访问此业务服务端点 http://10.10.14.80:18010/info 时，可以得到如下输出：{"author":"hw","book":"javadevmap","project":"service provider demo"}。

（3）Eureka 页面点击服务链接显示信息

点击 Eureka 页面中的服务链接实际是打开服务配置中的 status-page-url 属性，这个地址默认是 "http://主机名:端口/info" 的形式。实现点击链接即可看到服务信息的方法有 4 种，这 4 种方法的效果相同，具体选择哪种方法根据实际情况决定。

1）如果已经正确地配置了 hostname 的解析，那么点击 Eureka 页面链接时，会自动跳转到服务的 info 端点，并且显示服务信息，本例中 hostname 使用的是 localhost，所以可以正确地打开服务信息页。

2）不配置 host 解析，而是在 eureka client 中，设置 eureka.instance.hostname 为 IP 地址，例如如下设置。

```yaml
eureka:
  instance:
    hostname: ${spring.cloud.client.ipAddress}
```

3）不使用 host 解析，也不配置 hostname 为 IP 地址，而是使用另一个配置。这个配置可以把使用 hostname 的地方全部替换为服务 IP。

```yaml
eureka:
  instance:
    hostname: service-provider
    prefer-ip-address: true
```

4）直接修改 eureka.instance.status-page-url 属性值完成配置，这里通过只修改此字段，把服务说明指向了其他的地址。

```yaml
eureka:
  instance:
    hostname: service-provider
    prefer-ip-address: true
    status-page-url: http://www.javadevmap.com
```

当然，instance 还有其他配置项用来修改服务在 Eureka 上的注册信息，例如修改 eureka.instance.instance-id 可以修改服务在 Eureka 页面上的显示等。

9.1.3 基于 Host 的高可用 Eureka

基于 Host 进行 Eureka 的高可用配置，需要提供多个可以解析的 Host 地址，在 Eureka 集群中的不同 Eureka 实例中使用不同的 Host 地址作为 hostname，并且在各自服务的 defaultZone 中配置其他 Eureka 服务的地址。

例如配置两台 Eureka 服务的集群，其中服务 A 的 hostname 为 eureka-serverA，defaultZone 地址为 http://eureka-serverB:18002/eureka/；服务 B 的 hostname 为 eureka-serverB，defaultZone 地址为 http://eureka-serverA:18001/eureka/。然后在业务服务的 defaultZone 中配置两台 Eureka 服务的地址即可。

（1）配置 Host 解析

在服务器的 hosts 文件中，使用正确的 IP 地址配置 host 解析。

```
172.17.238.237 eureka-serverA
172.17.238.237 eureka-serverB
```

（2）Eureka A 中的配置

```yaml
server:
  port: 18001
eureka:
  instance:
    hostname: eureka-serverA
    instance-id: ${spring.cloud.client.ipAddress}:${server.port}
  client:
    register-with-eureka: true
    fetch-registry: true
    service-url:
      defaultZone: http://eureka-serverB:18002/eureka/
```

（3）Eureka B 中的配置

```
server:
  port: 18002
eureka:
  instance:
    hostname: eureka-serverB
    instance-id: ${spring.cloud.client.ipAddress}:${server.port}
  client:
    register-with-eureka: true
    fetch-registry: true
    service-url:
      defaultZone: http://eureka-serverA:18001/eureka/
```

（4）业务服务的配置

```
server:
  port: 18010
spring:
  application:
    name: service-provider
eureka:
  instance:
    hostname: service-provider
    instance-id: ${spring.cloud.client.ipAddress}:${server.port}
    prefer-ip-address: true
  client:
    service-url:
      defaultZone: http://eureka-serverA:18001/eureka/,http://eureka-serverB:18002/eureka/
info:
  author: hw
  book: javadevmap
  project: service provider demo
```

这样访问其中任意一台 Eureka，即可看到业务服务 SERVICE-PROVIDER。由于开启了 Eureka 服务的自注册功能，所以还能看到两台 Eureka 服务。

9.1.4 基于 IP 的高可用 Eureka

使用上节的方法可以搭建 Eureka 集群，但是配置 Host 解析和 Eureka 集群中多实例相互配置的工作比较麻烦。下面提供一种基于 IP 的 Eureka 集群方案，该方案可以实现同样的 Eureka 集群效果，并且避免了多台 Eureka 配置属性不同的问题，前提是 Eureka 部署在不同的服务器上。

（1）Eureka 服务端配置

```
server:
  port: 18001
spring:
  application:
    name: eureka-server
eureka:
  instance:
    hostname: eureka-server
    instance-id: ${spring.cloud.client.ipAddress}:${server.port}
    prefer-ip-address: true
  client:
```

```
    fetch-registry: true
    register-with-eureka: true
    region: javadevmap
    availability-zones:
        javadevmap: map_eureka
    service-url:
        map_eureka: http://172.17.238.237:18001/eureka/,http://172.17.238.239:18001/eureka/
```

上面的配置中，开启 Eureka 的自注册能力；在配置 Eureka 集群地址时，使用了 eureka.client.region⊖的方式来替代 defaultZone 的方式配置 Eureka 服务器的地址。Eureka 的服务集群没有使用相互配置的方法，而是把全量服务地址都写到一起，这样当集群中有多台 Eureka 服务器时，避免了多个实例配置值不同的麻烦，只要在对应 IP 地址的服务器中启动 Eureka 服务即可。

（2）应用服务配置

应用服务的配置还是连接 Eureka 服务集群的地址。

```
eureka:
    instance:
        hostname: service-provider
        instance-id: ${spring.cloud.client.ipAddress}:${server.port}
        prefer-ip-address: true
    client:
        region: javadevmap
        availability-zones:
            javadevmap: map_eureka
        service-url:
            map_eureka: http://172.17.238.237:18001/eureka/,http://172.17.238.239:18001/eureka/
```

（3）页面展示如图 9-4 所示。

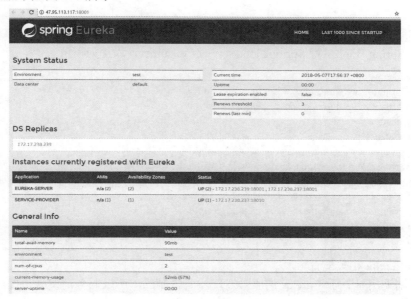

图 9-4　Eureka 页面服务列表

⊖　Region 表示区域，这里设定的区域为 javadevmap；map_eureka 表示集群分组。在这里仅仅作为演示的写法，效果和 defaultZone 是同样的。当然 region 还有其他含义，只是这里没有使用。

本节用到了 Eureka 最主要的能力，当然 Eureka 还有一些更详细的配置项，例如设置更新时间的配置项①等，这些就需要在日常工作中更详细地去了解，可以在编写代码时根据 IDE 的提示看到 Eureka 配置项的作用。

9.2 Ribbon 与 Feign

当把一个传统架构的服务进行微服务拆分后，每个微服务都负责一部分功能，但是整个系统的正常运转是需要各个微服务模块贡献其能力，组成一个业务的最终形态的，所以这就用到了服务间的调用。

服务调用还有一层意义就是服务数据的组装，例如一个服务负责用户模块，另一个服务负责订单模块，如果要查询某一用户生成了什么订单，会涉及两个模块的数据组装问题。如果直接从用户模块调用订单模块的接口，会使用户模块的业务和订单服务造成耦合，从而破坏用户模块的独立性，所以这时候提炼出来一个服务，专门负责调用业务核心服务并且完成数据的组装，从而保证最底层核心服务的纯净。

Eureka 的服务注册发现能力是服务调用的基础，服务消费者通过 Eureka 找到服务提供者的地址清单，继而实现服务间调用。可以使用 Ribbon 进行带负载均衡②的服务间调用，还可以使用 Feign 来简化服务间调用的写法，Feign 默认集成了 Ribbon。

9.2.1 Ribbon

先创建一个服务的消费者（类似数据组装层），用服务消费者单独调用服务的提供者（类似核心功能模块），完成基础的 Ribbon 调用；然后为服务提供者再添加一个实例。看一下 Ribbon 的负载情况。

（1）添加服务消费者依赖

新建一个 Spring Boot 工程 SpringCloudServiceConsumer，添加如下依赖。

```
<dependency>
    <groupId>org.springframework.boot</groupId>
    <artifactId>spring-boot-starter-web</artifactId>
</dependency>
<dependency>
    <groupId>org.springframework.cloud</groupId>
    <artifactId>spring-cloud-starter-eureka</artifactId>
</dependency>
<dependency>
    <groupId>org.springframework.cloud</groupId>
    <artifactId>spring-cloud-starter-ribbon</artifactId>
</dependency>
```

（2）添加 Eureka 相关配置

由于服务间调用是基于 Eureka 的服务发现注册机制的，所以服务消费者也要注册到 Eureka

---

① Eureka 中已注册的服务信息的移除在极端情况下的反应时间会非常长，例如使用 kill 命令杀掉业务服务节点，Eureka 可能没有那么快地更新业务服务状态变化，eureka.server.enable-self-preservation=false 配置项可以关闭 Eureka 的数据缓存保护，但是仍存在延迟时间，所以应尽量使用 shutdown 端口等方式正常关闭业务服务，因为正常关闭会有移除服务节点逻辑执行。

② 当有多个服务提供者时，负载均衡会通过其负载策略，调用多台服务提供者的实例，而不是仅仅调用其中的一台。

上，才能获取服务提供者的信息。这里可以参照上一节服务提供者的 Eureka 配置进行修改。不同的地方就是服务名字和端口号，消费者服务名使用 service-consumer，端口号使用 18020。打开 Eureka 可以看到相关注册信息，如图 9-5 所示。

Application	AMIs	Availability Zones	Status
EUREKA-SERVER	n/a (2)	(2)	UP (2) - 172.17.238.239:18001 , 172.17.238.237:18001
SERVICE-CONSUMER	n/a (1)	(1)	UP (1) - 172.17.238.237:18020
SERVICE-PROVIDER	n/a (1)	(1)	UP (1) - 172.17.238.237:18010

图 9-5　Eureka 服务列表

（3）服务提供者提供可调用接口

对服务提供者进行改造，把前面使用过的 SpringBootMybatis 工程的代码移植[一]过来，从而使 SpringCloudServiceProvider[二] 工程具备了用户查询接口、日志记录等能力，这样 Provider 工程就成为了注册到 Eureka 上的负责用户模块的微服务。

（4）服务消费者调用配置

在 SpringCloudServiceConsumer 的启动类中，添加如下代码，使调用的 RestTemplete 实例被 Spring 管理。

```
@Bean
@LoadBalanced
RestTemplate restTemplate() {
    return new RestTemplate();
}
```

（5）服务消费者逻辑添加

在 Consumer 工程中，添加 Service 接口类和它的实现类，在实现类中调用 Provider 服务的接口，从接口返回的数据中，选取需要的数据，返回给 Controller 类，然后在 Controller 类中完成数据的模板化返回[三]。同时需要在 Consumer 工程中建立一个 DomainUser 类，用于承接 Provider 服务的返回数据，这个类的数据结构和 Provider 中的相同，这里就不再赘述。

```
public interface ConsumerService {
    public DomainUser getUserFromProvider(int id);
}

import org.springframework.web.client.RestTemplate;
@Service
public class ConsumerServiceImpl implements ConsumerService{
    @Autowired
    private RestTemplate restTemplate;

    @Override
    public DomainUser getUserFromProvider(int id) {
        Result<DomainUser> result = restTemplate.
                            getForObject("http://SERVICE-PROVIDER/user/"+id, Result.class);
        if(result.getResultCode()==200) {
            return (DomainUser)result.getData();
```

[一] 移植过程中注意修改对应的包名，否则无法编译或程序运行出现错误。

[二] 在实际项目中，把此工程名改为带有 User 标记的工程名更为合适，因为那样更能直观地表达这个服务提供者负责的模块是什么。这里没有修改工程名是因为本章不会出现其他能力的服务提供者，并且用消费者和提供者作为工程名，对于新手更加直观。

[三] 模板化返回和请求的切面日志记录需要添加进工程，以后的 Spring Boot 业务服务都默认这些已经添加进工程。

```
            }else {
                return null;
            }
        }
    }
```

添加 Controller 类逻辑如下：

```
@RestController
@RequestMapping(value="/consumer")
public class ConsumerController {
    @Autowired
    private ConsumerService service;

    @RequestMapping(value="/{Id}",method=RequestMethod.GET)
    public Result<DomainUser> getUser(@PathVariable("Id") int id) {
        DomainUser user = service.getUserFromProvider(id);
        Result<DomainUser> result = null;
        if(user!=null) {
            result = new Result<>(ResultCode.OK, user);
        }else {
            result = new Result<>(ResultCode.Not_Found);
        }
        return result;
    }
}
```

上面的代码在 Service 中使用 RestTemplete 的 getForObject 方法调用服务提供者的接口，由于服务都已经注册到 Eureka 上，所以 Http 链接上可以直接使用服务的名字[一]。在方法中期望返回一个 Result 类型的数据，所以在请求参数中设置了 Result 类型，然后根据返回的信息码判断是否成功，如果成功则可以得到 DomainUser 类型的实例，然后 Consumer 服务再对数据进行封装返回。

通过用浏览器访问 Consumer 服务的接口完成调用，但并没有返回期望得到的输出结果，并且 Consumer 服务报错了。但是排查 Provider 服务的输出日志[二]发现此服务已经正确返回了数据。那么问题出在哪里？下面从问题排查的角度来逐步分析上面出现的问题，然后尝试去解决这个问题。

（6）Ribbon 获取数据的几种方法

由于查看 Provider 日志，发现 Provider 已经正确返回了数据，那么可能先怀疑，Consumer 是否正确收到了数据。所以对 Service 中的 getUserFromProvider 方法进行改造。

```
@Override
public DomainUser getUserFromProvider(int id) {
    String retString = restTemplate.getForObject("http://SERVICE-PROVIDER/user/"+id, String.class);
    log.info("consumer receive is " + retString);
    return null;
}
```

通过日志查看，会发现 Consumer 服务已经成功收到了数据。那么问题可能就出在 Json 数

[一] 可以使用 IP 端口的方式进行调用，即把例子中的服务名改为 IP 和端口，这里就不再赘述。
[二] 在你无法对程序进行断点跟踪的时候，日志会帮你的大忙。

据到类实例的转化过程。

如果读者仔细阅读了第 1 章，应该记得 Java 的泛型模板和 C++的不同，Java 的泛型会擦除类型。既然 RestTemplete 提供了根据类型获取返回数据的接口，那么如果不使用模板，而使用确定类型来获取数据呢？

所以，下面创建一个确定类型的类，来承接 getForObject 方法返回的数据，这个确定类型不使用模板，而是直接包含一个 DomainUser 对象。

```
public class ResultDomainUser {
    private int resultCode;
    private String msg;
    private DomainUser data;
    …
}

@Override
public DomainUser getUserFromProvider(int id) {
    ResultDomainUser result = restTemplate.getForObject("http://SERVICE-PROVIDER/ user/"+ id, ResultDomainUser.class);
    if(result.getResultCode()==200) {
        return result.getData();
    }else {
        return null;
    }
}
```

通过 Postman 访问获取用户接口，可以看到数据的输出。如图 9-6 所示。

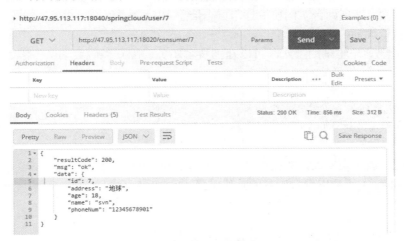

图 9-6　获取用户信息

现在数据确实可以返回了，但是这样却破坏了模板化输出的结构，调用方要针对每个提供方返回的数据提供两个类，来承接数据返回，所以还是期望能够使用模板来实现数据传递的简单化和标准化。在 Service 实现类中添加如下引用，并修改 getUserFromProvider 方法。

```
import org.springframework.core.ParameterizedTypeReference;
import org.springframework.http.HttpMethod;
import org.springframework.http.ResponseEntity;
```

```
    @Override
    public DomainUser getUserFromProvider(int id) {
        ParameterizedTypeReference responseType = new
                                ParameterizedTypeReference<Result<DomainUser>>() {};
        ResponseEntity<Result<DomainUser>> resp = restTemplate.exchange
            ("http://SERVICE-PROVIDER/user/"+id, HttpMethod.GET, null, responseType);
        Result<DomainUser> result = resp.getBody();
        if(result.getResultCode()==200) {
            return (DomainUser)result.getData();
        }else {
            return null;
        }
    }
```

模板类型应该使用 exchange 方法获取，并且要在参数中设置一个特殊的类型，不能使用普通的类型获取方法。通过这样修改，模板又可以使用了。

（7）多台负载

通过 Spring Boot 的多环境配置，用另一个端口再启动一个 Provider 实例[㊀]，如图 9-7 所示。

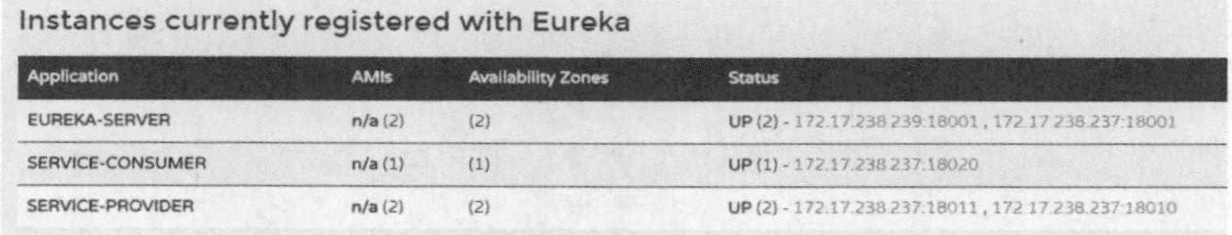
Instances currently registered with Eureka

Application	AMIs	Availability Zones	Status
EUREKA-SERVER	n/a (2)	(2)	UP (2) - 172.17.238.239:18001 , 172.17.238.237:18001
SERVICE-CONSUMER	n/a (1)	(1)	UP (1) - 172.17.238.237:18020
SERVICE-PROVIDER	n/a (2)	(2)	UP (2) - 172.17.238.237:18011 , 172.17.238.237:18010

图 9-7　Provider 服务多实例

调用 Consumer 服务接口，观察 Provider 服务的两台实例是否都得到了调用。最简单的办法就是观察日志，你会发现两台实例轮流打印日志信息。

（8）修改负载规则

默认的负载规则是轮询的方式，可以在消费者服务的配置类或启动类中添加如下代码修改负载规则：

```
@Bean
public IRule ribbonRule() {
    return new RandomRule();
}
```

这段代码把负载规则修改为随机策略。当然 Ribbon 的负载还包含其他策略，可以进入工程的 ribbon-loadbalancer 依赖中查看更多的负载策略。

9.2.2 Feign

在上一小节中，使用 Ribbon 完成了服务间的调用。但是读者可能会发现 Ribbon 的写法稍显麻烦，而且对于普遍使用的模板还要另外进行配置。下面介绍一种简单的实现服务间调用的组件。

Feign 默认集成了 Ribbon，并和 Eureka 结合默认实现负载均衡的效果。只要使用接口注解声明的方式就可以实现接口调用的能力。

（1）添加 Feign 依赖

㊀ 如果你的服务器够多，就不用这么麻烦地配置多环境，只要在另一台服务器上启动一个实例即可。

按照 Spring Boot 工程的惯例，必须在 Consumer 工程中添加依赖。

```xml
<dependency>
    <groupId>org.springframework.cloud</groupId>
    <artifactId>spring-cloud-starter-feign</artifactId>
</dependency>
```

（2）在启动类中添加注解

```java
import org.springframework.cloud.netflix.feign.EnableFeignClients;
@EnableFeignClients
```

（3）添加 Feign 接口定义

在接口类上方，使用@FeignClient 定义接口类，说明此接口类使用 Feign 方法调用 SERVICE-PROVIDER 服务；然后在接口类内部对需要调用的 Provider 服务接口定义方法，方法的注解中标明路径和请求类型，方法的参数标明传递的参数，方法的返回类型就是 Provider 服务返回的数据类型。

```java
@FeignClient(value="SERVICE-PROVIDER")
public interface ConsumerFeign {
    @RequestMapping(value="/user/{id}",method=RequestMethod.GET)
    public Result<DomainUser> getUser(@RequestParam("id") int id);

    @RequestMapping(value="/user/add",method=RequestMethod.POST)
    public Result<String> addUser(@RequestBody DomainUser user);
}
```

（4）修改 Service 类

在 Service 类中注入 Feign，并且修改 Service 类的方法实现。

```java
@Service
public class ConsumerServiceImpl implements ConsumerService{
    @Autowired
    private ConsumerFeign feign;

    @Override
    public DomainUser getUserFromProvider(int id) {
        Result<DomainUser> result = feign.getUser(id);
        if(result.getResultCode()==200) {
            return (DomainUser)result.getData();
        }else {
            return null;
        }
    }

    @Override
    public int saveUserToProvider(DomainUser user) {
        Result<String> result = feign.addUser(user);
        if(result.getResultCode()==200) {
            return 1;
        }else {
            return 0;
        }
    }
}
```

}

（5）修改 Controller 类

在 Controller 类中，添加 add 方法，然后检验其查询和添加的调用情况。

```
@RestController
@RequestMapping(value="/consumer")
public class ConsumerController {
    @Autowired
    private ConsumerService service;

    @RequestMapping(value="/{Id}",method=RequestMethod.GET)
    public Result<DomainUser> getUser(@PathVariable("Id") int id) {
        DomainUser user = service.getUserFromProvider(id);
        Result<DomainUser> result = null;
        if(user!=null) {
            result = new Result<>(ResultCode.OK, user);
        }else {
            result = new Result<>(ResultCode.Not_Found);
        }
        return result;
    }

    @RequestMapping(value="/add",method=RequestMethod.POST)
    public Result<String> addUser(@RequestBody @Valid DomainUser user) {
        Result<String> result = null;
        int ret = service.saveUserToProvider(user);
        if(ret == 1) {
            result = new Result<>(ResultCode.OK);
        }else {
            result = new Result<>(ResultCode.ERROR);
        }
        return result;
    }
}
```

（6）调用结果

通过 Postman 提交一个 Post 请求到 Consumer 服务，可以看到服务的日志为：

```
    2018-04-11 16:55:08.442+0800 INFO    c.j.serviceconsumer.config.AopAspect[SpringCloudServiceConsumer], [io-18020-exec-2] : com.javadevmap.serviceconsumer.controllers.ConsumerController method = addUser args = {"id":null,"address":"地球","age":18,"name":"svn","phoneNum":"12345678901"}
    2018-04-11 16:55:08.458+0800 INFO    c.j.serviceconsumer.config.AopAspect[SpringCloudServiceConsumer], [io-18020-exec-2] : com.javadevmap.serviceconsumer.controllers.ConsumerController method = addUser args = {"id":null,"address":"地球","age":18,"name":"svn","phoneNum":"12345678901"} ret = {"resultCode":200,"msg":"ok","data":null}
```

使用 Feign 形式的服务调用，可以让程序的代码简洁很多，本节之后的服务调用全部使用此方式。

9.3 Hystrix 与 Turbine

设想一种情况，通过 Spring Cloud 部署了一个服务集群，前面创建的 Provider 服务成为了专门负责用户信息处理的服务，它可以对外提供每秒 1000 次的访问能力。集群内同时存在其他很多业务服务，都会调用 Provider 服务获取基本的用户信息。突然有一天，由于一次运营宣传活动非常成功，整个集群访问压力突然加大，集群内其他服务对 Provider 的压力超过了每秒 1000 次，这时会发生什么？

Provider 服务会发生大量的请求积压，而单个请求的响应时间会变慢，其他依赖 Provider 的服务会等待返回数据，所以请求积压响应也开始变慢。这种情况持续一段时间整个系统可能就会瘫痪，这种效应称为"雪崩"。

Hystrix 为了防止以上问题做了很多努力，它采用线程或者信号量隔离、近乎实时的监控业务状况和指标、熔断条件配置、快速失败及默认失败返回等方法使整个集群不至于由于一点故障而全部被拖垮。例如 Provider 服务出现了业务积压，那么其他调用服务就会出现请求超时的情况，如果消费者服务在请求时配置了 Hystrix，Hystrix 会监控此请求在一定时间内失败次数是否达到 Hystrix 设定的阈值，如果达到就会触发 Hystrix 的熔断机制，从而使服务消费者停止请求 Provider 服务，这样 Provider 就有机会在非正常情况下恢复过来，保证一定的业务承载能力，而不至于由它一点而导致整个集群的瘫痪。

当然，整个系统不会由于配置了 Hystrix 而保证所有请求都正常，所有请求都正常需要服务各节点具备相应的承载能力。Hystrix 在上一情况中，仅会保证集群不会全部瘫痪，并且能够保证部分的业务承载。

本小节会介绍 Hystrix 的使用方法，包含部分 Hystrix 的配置及作用，还会介绍 HystrixDashboard 监控页面以及 Turbine 聚合监控。

9.3.1 Hystrix 基本使用

（1）添加依赖

在 Consumer 工程中，添加 Hystrix 的组件依赖。

```xml
<dependency>
    <groupId>org.springframework.cloud</groupId>
    <artifactId>spring-cloud-starter-hystrix</artifactId>
</dependency>
```

（2）添加启动类注解

在启动类中，添加注解@EnableHystrix。

（3）添加 Hystrix 快速失败逻辑

Hystrix 的快速失败只要在方法上添加一个@HystrixCommand，并在里面配置一个失败默认执行的方法即可，下面在 Service 实现类的 getUserFromProvider 方法中添加失败逻辑。

```java
@Override
@HystrixCommand(fallbackMethod="getUserFallback")
public DomainUser getUserFromProvider(int id) {
    Result<DomainUser> result = feign.getUser(id);
    if(result.getResultCode()==200) {
        return (DomainUser)result.getData();
    }else {
        return null;
    }
}

public DomainUser getUserFallback(int id) {
```

```
            DomainUser user = new DomainUser();
            user.setName("hystrix");
            user.setAge(-1);
            user.setAddress("hystrix fall back");
            return user;
        }
```

（4）效果演示

在 Provider 服务启动的情况下，调用 Consumer 服务的 get 接口http://47.95.113.117:18020/consumer/7可以得到如下输出：

{"resultCode":200,"msg":"ok","data":{"id":7,"address":"地球","age":18,"name":"svn","phoneNum": "12345678901"}}。

关闭 Provider 服务后，再次调用此接口，输出为[一]

{"resultCode":200,"msg":"ok","data":{"id":null,"address":"hystrix fall back","age":-1,"name":"hystrix","phoneNum":null}}

可见，当服务提供者不可用时，服务消费者中配置了 Hystrix 的方法可以快速失败并且返回 fallback 方法默认的数据。

（5）添加可视化页面依赖

添加 Hystrix 的可视化监控页面，需要添加如下依赖：

```xml
<dependency>
    <groupId>org.springframework.cloud</groupId>
    <artifactId>
        spring-cloud-starter-hystrix-dashboard
    </artifactId>
</dependency>
```

（6）添加启动类注解

添加可视化页面的启动类注解@EnableHystrixDashboard。

（7）查看页面

要查看 Hystrix 的监控页面，只需要在服务的 IP 和端口后添加/hystrix 路径即可。例如此服务的 Hystrix 页面访问路径是 47.95.113.117:18020/hystrix，如图 9-8 所示。

图 9-8 Hystrix 页面

[一] 这里仅为了演示，所以没有修改返回的错误码。实际业务中，应根据希望达到的效果正确使用 fallback。

第 9 章 Spring Cloud

在地址栏中输入http://127.0.0.1:18020/hystrix.stream[⊖]，Delay默认使用 2000，Title 输入框输入 Consumer，然后点击 Monitor Stream，可以看到如图 9-9 所示页面。

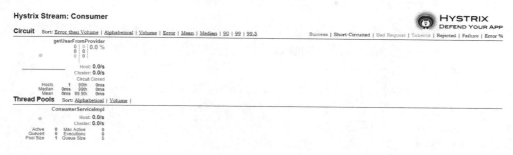

图 9-9　Hystrix 监控页面

（8）页面中数据的含义

这个页面就是 Hystrix 用于监控展示的页面，页面分为两部分，上面为被监控方法的展示，下面为线程监控展示。

重新启动 Provider 服务，然后对 Consumer 服务的这个接口进行一定的访问，观察页面变化。如图 9-10 所示。

图中的上半部分最明显的两个图形元素是实心圆和变化曲线。

- 实心圆会随着业务访问的压力情况变化，流量越大实心圆就越大。同时实心圆的颜色会表示这个方法的健康状态，绿色为最好，红色为最差。
- 变化曲线表示流量的相对变化。

图中上半部分的数字，当移动鼠标到上面的时候会有相应的提示，各个监控数据含义如图 9-11 所示。

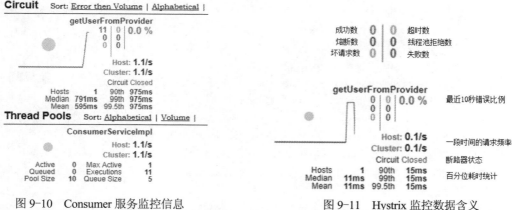

图 9-10　Consumer 服务监控信息　　　　图 9-11　Hystrix 监控数据含义

图中下半部分的数据 Thread Pools 是线程监控的情况。后面会介绍线程数量的设置，也就

⊖ 如果感兴趣，可以直接访问http:// ip:18020/hystrix.stream地址，观察页面的原始数据。另外注意服务的地址，由于演示工程的服务已经部署到外网上，所以前面页面请求的地址是一个外网地址，但是此处输入的是 127.0.0.1，是由于此服务是自我数据的访问，所以输入自己的本机地址即可。但是如果想用浏览器访问外网的 hystrix.stream，还是应该使用外网地址。

是图中的 Pool Size 以及线程的定义等。

9.3.2 Feign 与 Hystrix 结合

Feign 是自带断路器的，但是需要在配置文件中打开 Feign 的断路器开关。打开开关后，只要在 Feign 接口类的注解@FeignClient 中添加 fallback 设置，就可以实现快速失败返回。

（1）打开断路器开关

在 yml 文件中添加如下配置，即可打开 Feign 的断路器开关。

```
feign:
  hystrix:
    enabled: true
```

（2）配置 Feign 的 fallback

在 Feign 的接口类上方@FeignClient 中，配置 fallback 属性，属性指向一个负责处理快速失败的类。这个处理快速失败的类继承 Feign 接口类，所以类中的方法和 Feign 保持相同，用于实现对应的 Feign 接口类中方法的快速返回。

```
@FeignClient(value="SERVICE-PROVIDER",fallback=ConsumerFeignFallBack.class)
public interface ConsumerFeign {
    @RequestMapping(value="/user/{id}",method=RequestMethod.GET)
    public Result<DomainUser> getUser(@RequestParam("id") int id);

    @RequestMapping(value="/user/add",method=RequestMethod.POST)
    public Result<String> addUser(@RequestBody DomainUser user);
}

@Component
public class ConsumerFeignFallBack implements ConsumerFeign{
    @Override
    public Result<DomainUser> getUser(int id) {
        Result<DomainUser> result = new Result<>(ResultCode.Unavailable);
        return result;
    }

    @Override
    public Result<String> addUser(DomainUser user) {
        Result<String> result = new Result<>(ResultCode.Unavailable);
        return result;
    }
}
```

通过上面的代码，可以在 Feign 调用时就实现断路器逻辑，而且这种写法要比对单个方法逐个实现 fallback 简洁得多。

（3）监控页面展示

查看监控页面，如图 9-12 所示。发现除了在上一小节添加的 getUserFromProvider 方法被监控以外，Feign 中的 getUser 方法也受到了监控。但 Feign 中的 addUser 方法没有受到监控，因为还没有调用此方法，没有相关的监控数据，所以监控页面没有显示。在监控页面的下方，会发现又出现了一个线程池，线程池的名字默认使用所在类名或者 Feign 的注解中的名字。

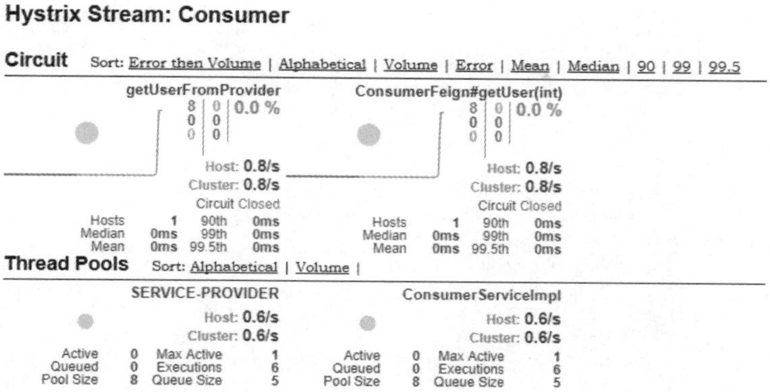

第 9 章 Spring Cloud

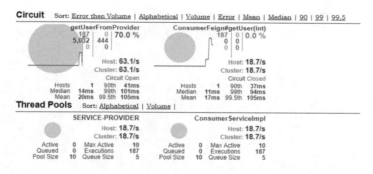

图 9-12　Hystrix 监控

（4）接口压力测试

下面对获取信息接口进行压力测试，观察其业务承载能力和监控数据的变化○，如图 9-13 所示。

图 9-13　服务压力监控

可见当程序压力较大时，大部分的请求都被熔断和线程拒绝了，这不是希望得到的结果。熔断的添加是保证程序达到上限后不会集体瘫痪，而不是限制程序的执行能力。所以需要了解 Hystrix 的一些配置。

9.3.3　Hystrix 相关配置

Hystrix 的配置比较多，包含设定熔断规则、能力开启、选择隔离方式以及线程数、并发数等设置，下面了解一些常用的 Hystrix 配置项。

（1）线程数配置

在配置文件中，添加如下配置，然后再次对服务进行压力测试，如图 9-14 所示。

```
hystrix:
  threadpool:
    default:
      coreSize: 100
```

○ 这里为了演示的连贯性，没有删除 getUserFromProvider 方法的熔断器代码，但是此方法和 Feign 接口类监控的是同一个调用，所以在实际工作中应该避免这种浪费性能的重复。

215

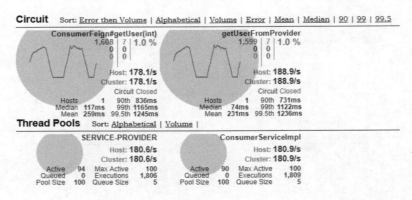

图 9-14　服务压力监控

从图中的显示结果可见，服务现在基本处于稳定运行的状态，仅有几个调用是超时的。还有一点应该注意，在 Thread Pools 部分，Pool Size 从 10 变成了 100。可见使用配置扩大线程数后，程序的执行能力确实提高了[一]。

（2）设置超时时间

上图中存在一些请求的超时情况，如果被调用服务的业务逻辑比较复杂，或者压力较大时执行速度较慢，但又不希望 Hystrix 很快就判定服务调用超时，这时可设置一个期望的值，作为熔断器判定超时的依据。

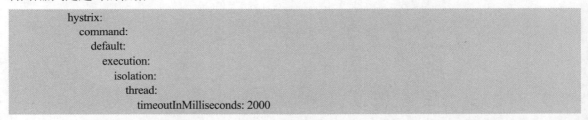

通过配置把超时时间设置为 2s，服务再次执行压力测试，没有出现超时情况，整个服务运行正常，如图 9-15 所示。注意超时时间不要设置过大，如果设置为几十秒就失去了意义。

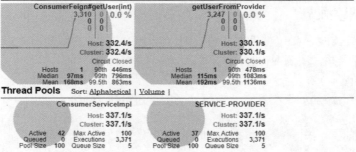

图 9-15　服务压力监控

[一] 在实际项目中，不要盲目扩大线程数，线程数的配置要和服务器每秒承载的请求数、单个请求的执行时间、服务器的资源情况等因素良好地配合。

第 9 章 Spring Cloud

（3）信号量隔离与并发数

另一种隔离方式是信号量隔离，它没有单独开辟出一个线程池，而是用并发数进行控制。使用信号量隔离也能承载相当的业务压力，只是它没有线程池监控⊖，具体配置如下。效果如图 9-16 所示。

```
hystrix:
  command:
    default:
      execution:
        isolation:
          strategy: SEMAPHORE
          semaphore:
            maxConcurrentRequests: 2000
          thread:
            timeoutInMilliseconds: 2000
```

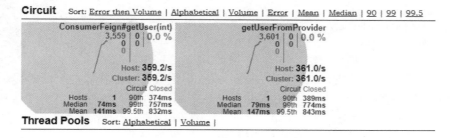

图 9-16　服务压力监控

（4）熔断触发次数

可以设置熔断执行的条件，例如下面的设定，是设置当滚动时间窗口内，达到几次失败则触发熔断机制，默认为 20，这里设定为 1，此设置仅作为演示，毕竟这个值实在是太小了。

```
hystrix:
  command:
    default:
      circuitBreaker:
        requestVolumeThreshold: 1
```

Hystrix 还有一些其他设置，例如设置线程池的队列大小，设置滚动时间窗口的长度等，如果没有特殊需求，使用默认值即可。当然如果对其非常感兴趣可以翻阅相关文档。

9.3.4　Hystrix 作为限流工具

上一小节中，把断路器添加在了 Consumer 服务的 Feign 调用中。如果在一个大的集群中，有很多个业务服务需要调用 Provider 服务的接口，而这些服务中有的没有添加熔断逻辑，那么会在 Provider 服务达到性能上限时被拖垮。是否可以在 Provider 服务的接口本身做流量限制和

⊖ 本处演示完后，会删除 getUserFromProvider 方法的熔断器配置。

快速失败呢？[一]

按照之前的介绍，对 Provider 服务先完成 Hystrix 的依赖引入和启动类、配置文件配置。之后修改 Controller 的代码如下：

```
@HystrixCommand(commandKey="provider-getuser",groupKey="provider-usercontroller",
fallbackMethod= "getUserFallBack")
@RequestMapping(value="/{Id}",method=RequestMethod.GET)
public Result<DomainUser> getUser(@PathVariable("Id") int id) {
    DomainUser user = service.getUser(id);
    Result<DomainUser> result = null;
    if(user!=null) {
        result = new Result<>(ResultCode.OK, user);
    }else {
        result = new Result<>(ResultCode.Not_Found);
    }
    return result;
}

public Result<DomainUser> getUserFallBack(int id){
    Result<DomainUser> result = new Result<>(ResultCode.Unavailable);
    return result;
}
```

代码中对接口添加了熔断，并且用 commandKey 配置了方法在监控页面的显示名称；用 groupKey 配置了线程池名字，同一 groupKey 的方法使用同一线程池；配置了一个快速失败的数据结果返回，返回错误码 503。

下面打开此服务的监控页面，观察其监控显示，如图 9-17 所示。

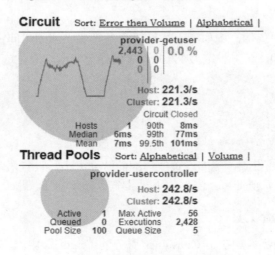

图 9-17　服务压力监控

[一] 这里作者翻阅了较多的资料，大部分资料是介绍 Hystrix 原理和使用方法的，但是使用在何处却论述较少。通常的用法是在服务调用时使用，但是作者认为在 Controller 的接口上使用 Hystrix 限流也是个不错的办法。这样可以在调用之初就避免接口向后的服务压力。毕竟 Hystrix 要避免程序执行过程中的故障，程序调用的源头也是属于执行流程之内的。

可见，代码中设置的方法名和线程组已经生效，设置这些属性对后面的聚合监控比较有好处，用户可以更加方便地看到某个服务的运行情况。

9.3.5　Turbine 聚合展示

Turbine 是 HystrixDashboard 的聚合展示，它可以通过配置，实现多个服务的监控以及同一服务多台实例的聚合报告。下面使用 Turbine 监控 Provider 服务和 Consumer 服务的运行情况。

（1）创建工程

新建一个工程，工程名为 SpringCloudTurbine，其他配置遵循前面介绍的 Spring Boot 工程的配置方法即可。

（2）添加依赖

由于 Turbine 需要监控各个服务，所以它需要连接到 Eureka 上，同时还要引入 Turbine 的功能组件；为了有显示页面，还要有 hystrix-dashboard 依赖。

```xml
<dependency>
    <groupId>org.springframework.cloud</groupId>
    <artifactId>spring-cloud-starter-eureka</artifactId>
</dependency>
<dependency>
    <groupId>org.springframework.cloud</groupId>
    <artifactId>spring-cloud-starter-turbine</artifactId>
</dependency>
<dependency>
    <groupId>org.springframework.cloud</groupId>
    <artifactId>
        spring-cloud-starter-hystrix-dashboard
    </artifactId>
</dependency>
```

（3）启动类添加注解

在启动类中开启 Eureka、HystrixDashborad 和 Turbine。

```
@EnableEurekaClient
@EnableTurbine
@EnableHystrixDashboard
```

（4）服务监控分组

在 Consumer 和 Provider 服务的配置文件中，eureka.instance 路径下，添加 matadata-map.cluster 配置，把应用服务加入一个监控组，具体如下：

```yaml
eureka:
  instance:
    metadata-map:
      cluster: main
    …
```

（5）Turbine 配置

```yaml
server:
  port: 18030
spring:
  application:
```

```yaml
      name: turbine

eureka:
  instance:
    hostname: turbine
    prefer-ip-address: true
    instance-id: ${spring.cloud.client.ipAddress}:${server.port}
  client:
    region: javadevmap
    availability-zones:
      javadevmap: map_eureka
    service-url:
      map_eureka: http://172.17.238.237:18001/eureka/,http://172.17.238.239:18001/eureka/

turbine:
  app-config: SERVICE-CONSUMER,SERVICE-PROVIDER
  aggregator:
    clusterConfig: main
  clusterNameExpression: metadata['cluster']
  combine-host-port: true
```

在 Turbine 服务中，除了配置常规的端口、名字，还要配置 Eureka 和 Turbine 特有的配置。其中 turbine.app-config 配置要监控的服务名；turbine.aggregator.clusterConfig 配置可选分组；turbine.clusterNameExpression 会根据页面参数传入的分组进行不同的分组监控；turbine.combine-host-port 为 true 表示根据 host 和 port 进行实例区分。

（6）页面展示

打开 turbine 服务部署地址的 Hystrix Dashboard 页面，即 47.95.113.117:18030/hystrix；在页面中输入 turbine 数据流地址和监控的分组，如图 9-18 所示。

图 9-18　Turbine 配置

由于使用浏览器打开的是 Turbine 服务自己的链接地址，所以在输入数据流时，使用的是本机的 IP 地址，并且设定分组为 main，输入框具体输入的地址为 http://127.0.0.1:18030/turbine.stream?cluster=main，进入后可以看到监控页面，如图 9-19 所示。

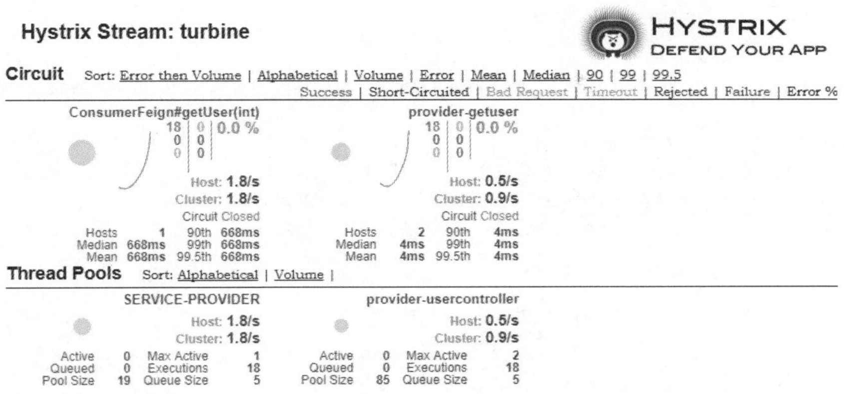

图 9-19　服务聚合监控

可见，两个服务的监控数据已经聚合，并且由于 provider 启动的是两个实例，所以在图中 provider-getuser 接口显示的 Hosts 数为 2。

（7）多分组监控

Turbine 的多分组设置相对简单，只要在被监控的不同实例中配置不同的分组名，然后在 Turbine 服务的 clusterConfig 配置项中配置要监控的分组列表，中间通过逗号隔开；在 Dashboard 页面中，用 cluster 选择不同的分组即可。

9.3.6　Turbine 通过总线聚合信息

Turbine 可以通过消息总线搜集服务的监控信息，也就是说被监控服务器的数据不是直接到达 Turbine 服务，而是发送到一个消息队列，然后 Turbine 服务从消息队列[①]中获取监控数据。

（1）Turbine 服务依赖

Turbine 服务依赖需要更换，不再使用之前的 Turbine 依赖，而使用 Turbine-amqp 依赖组件。

```
<dependency>
    <groupId>org.springframework.cloud</groupId>
    <artifactId>spring-cloud-starter-turbine-amqp</artifactId>
</dependency>
```

（2）启动项配置

更换启动项为@EnableTurbineStream。

（3）添加消息队列地址

配置 rabbitmq 的地址及账号。

```
spring:
  rabbitmq:
    host: 172.17.238.238
    port: 5673
    username: guest
    password: guest
```

---

① 这里消息队列使用 RabbitMQ，后面的章节会包含一个简单的 RabbitMQ 的安装方法，这里就不再介绍其安装。

通过以上这几步，Turbine 服务完成了新的配置。

（4）被监控服务依赖

对于需要监控的服务，添加如下依赖组件，使其能够把监控信息发送至消息队列。

```
<dependency>
    <groupId>org.springframework.cloud</groupId>
    <artifactId>spring-cloud-netflix-hystrix-amqp</artifactId>
</dependency>
```

（5）被监控服务添加消息队列地址

```
spring:
  rabbitmq:
    host: 172.17.238.238
    port: 5673
    username: guest
    password: guest
```

（6）RabbitMQ 队列情况

观察消息队列的通道和队列情况，如图 9-20、图 9-21 所示，发现以上程序正在通过消息队列进行监控数据的传输。

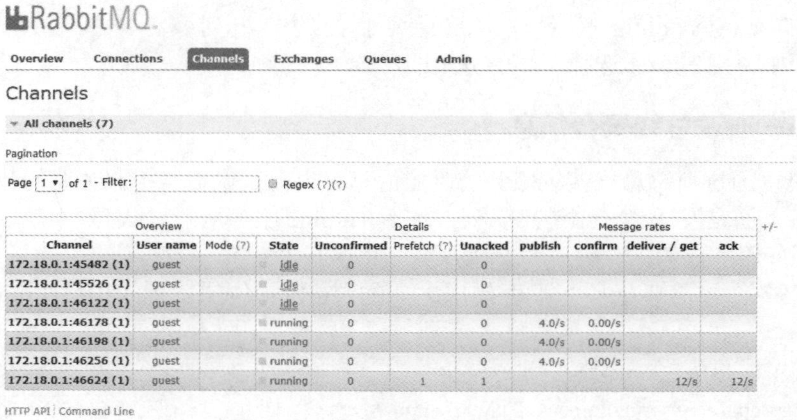

图 9-20　RabbitMQ 页面（一）

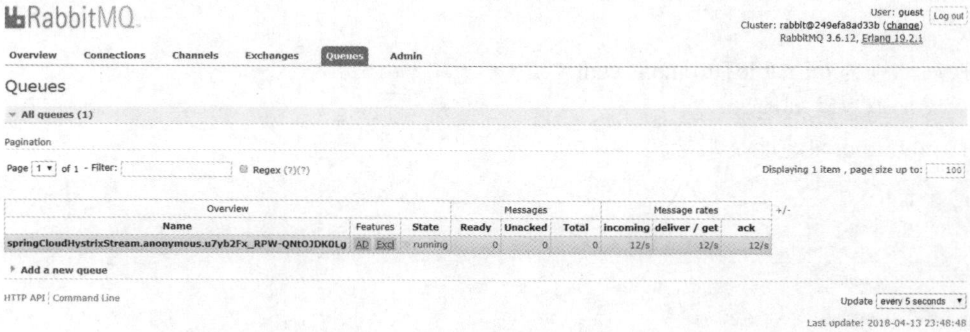

图 9-21　RabbitMQ 页面（二）

第 9 章 Spring Cloud

（7）监控页面展示

打开 Consumer 服务的 Hystrix 页面，输入 Turbine Stream 的地址，可以看到如图 9-22 所示监控数据。

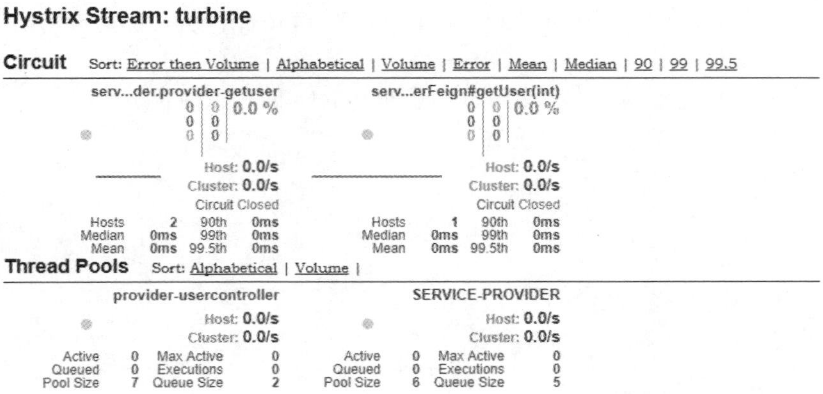

图 9-22　服务监控页面

9.4 Zuul

一个前端发起的 Web 请求，如果是通过域名访问的，那么首先会进行 DNS 解析，解析出来的是一个或者几个 IP 地址，这个 IP 地址一般是反向代理和负载的地址。负载会把请求转到微服务集群中进行业务逻辑的处理。如果服务集群中存在多个服务并且每个服务存在多个实例，那么负载要根据请求路径进行服务区分，并且对这一服务的多个实例地址进行负载策略，这样对于负载来讲工作量就有些大了，而且把这个规则配置到负载上，当多个服务的多个实例出现变化时，负载的配置工作量很大，所以微服务集群需要一个统一的入口，这就是 Zuul。

Zuul 的能力不仅仅是服务路由的能力，还可以在 Zuul 中对请求进行统一的身份认证、参数校验等其他逻辑。下面逐一介绍 Zuul 的这些常用方法。

9.4.1　Zuul 的基本使用

Zuul 服务也是一个 Spring Boot 工程，通过引入不同的依赖和配置，实现不同的能力。首先创建一个 SpringCloudZuul 工程，在工程中进行如下配置。

（1）添加依赖

由于本例使用 Zuul 根据服务名进行路由⊖，所以需要添加 Eureka Client 的服务发现依赖，同时需要添加 Zuul 的组件依赖。

```
<dependency>
    <groupId>org.springframework.cloud</groupId>
    <artifactId>spring-cloud-starter-eureka</artifactId>
</dependency>
<dependency>
```

⊖ 对于不针对服务，仅对单独 IP 端口的路由，可以不配置 Eureka，仅配置接口路径和地址映射即可。具体格式后面会有简单介绍。

```xml
        <groupId>org.springframework.cloud</groupId>
        <artifactId>spring-cloud-starter-zuul</artifactId>
    </dependency>
```

（2）启动项配置

在启动类中添加@EnableEurekaClient 和@EnableZuulProxy 注解，即完成启动类配置。

（3）添加配置文件

配置文件中，主要配置服务本身端口、名字和 Eureka 的相关内容。

```yaml
server:
  port: 18040
spring:
  application:
    name: zuul

eureka:
  instance:
    hostname: zuul
    instance-id: ${spring.cloud.client.ipAddress}:${server.port}
    prefer-ip-address: true
    metadata-map:
      cluster: main
  client:
    region: javadevmap
    availability-zones:
      javadevmap: map_eureka
    service-url:
      map_eureka: http://172.17.238.237:18001/eureka/,http://172.17.238.239:18001/eureka/
```

（4）服务调用路由情况

上面已经完成了整个 Zuul 服务的最基本使用的配置，但是没有配置集群内服务的相关内容，全部使用默认项。即使这样集群内所有服务也会被 Zuul 自动路由，路由的路径是服务名加本身接口路径。如 Consumer 服务的/consumer/{id}接口被路由为/service-consumer/consumer/{id}。下面使用 Postman 来访问此接口，效果如图 9-23 所示。

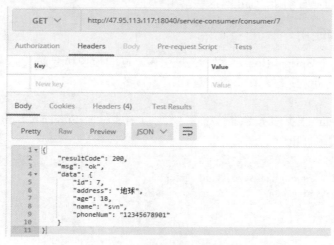

图 9-23　带路由的服务请求

9.4.2 Zuul 的配置

在上面的例子中，Consumer 和 Provider 都通过 Zuul 默认进行了路由，但服务路由会有一些限制，例如不希望某些服务被开放出去，或者希望对服务路由时的路径进行手动设置。下面看几种 Zuul 的配置。

（1）单实例路由

Zuul 不注册到 Eureka 上，仅添加 Zuul 依赖，对请求来的路径进行服务分发。

```
zuul:
  routes:
    consumer:
      path: /consumer/**
      strip-prefix: false
      url: http://localhost:18020/
```

上面的代码填写在 yml 文件中，其中 zuul.routes 标签下用于配置不同路径的路由规则，consumer 标签表示路由配置段，每段都有自己的名字；path 表示路径；url 表示转发到的服务地址；strip-prefix 比较特殊，表示在转发时不去掉前缀（本例中如果不把它置为 false，那么 consumer 服务收到的请求路径则会缺失/consumer，而仅剩后面的路径）。这段配置的目的就是把请求路径为/consumer/**的请求转发到http://localhost:18020/服务上。

请求路径中通配符的规则见表 9-1。

表 9-1 通配符规则

通配符	规则
?	匹配任意单个字符，例如/consumer/?可匹配/consumer/a
*	匹配任意数量的字符，例如/consumer/*可匹配/consumer/abc
**	匹配任意数量的字符，支持多级目录，例如/consumer/**可匹配/consumer/a/b/c

（2）根据服务名路由

一般情况下，集群内的某一服务不会仅仅布置一个实例，那么在多实例的情况下，使用服务名进行自带负载的路由策略是个明智的选择。根据服务名路由只要配置 Eureka 并且针对每一个服务在配置文件中设置映射关系。

```
zuul:
  routes:
    consumer:
      path: /consumer/**
      strip-prefix: false
      service-id: SERVICE-CONSUMER
    provider:
      path: /user/**
      strip-prefix: false
      service-id: SERVICE-PROVIDER
```

在配置路由规则时，和设置 IP 路由区别不大，仅把 url 配置改为正确的 service-id 即可完成根据服务名路由。例如通过 Zuul 访问 Provider 服务，可以在 Provider 多台实例的日志中看到轮询的日志打印。

（3）前缀的用处

如前面所讲，Zuul 请求的前面是反向代理和负载，那么在反向代理中识别请求路径时，如果只有这一个服务集群需要映射,那么肯定是能正确分发到 Zuul 服务的;如果反向代理负责映射多个服务集群,而多个服务集群中难免会有某些服务的请求路径是相同的,这种情况就需要给同一集群内的服务分配一个统一的前缀。

```yaml
zuul:
  prefix: /springcloud
  strip-prefix: true
  routes:
    consumer:
      path: /consumer/**
      strip-prefix: false
      service-id: SERVICE-CONSUMER
    provider:
      path: /user/**
      strip-prefix: false
      service-id: SERVICE-PROVIDER
```

上面代码中,对此 Zuul 服务通过 prefix 属性添加一个统一的前缀/springcloud,并且在进行具体服务映射之前删掉此前缀。从而在反向代理中通过识别此前缀,即可映射相应集群,实现通过请求路径标明集群的目的。如图 9-24 所示。

图 9-24 自定义的路由请求

(4)隐藏服务端口

Zuul 会自动为集群内的服务提供负载,供外网访问,但有时不希望某些服务的接口暴露给外网,所以可以通过 ignored-services 配置来设置某些服务不需要 Zuul 提供负载。

```yaml
zuul:
  prefix: /springcloud
  strip-prefix: true
  ignored-services: SERVICE-PROVIDER
  routes:
    consumer:
      path: /consumer/**
      strip-prefix: false
      service-id: SERVICE-CONSUMER
```

第 9 章 Spring Cloud

配置中去掉了 Provider 服务配置的映射规则，并且在 Zuul 的 ignored-services 配置项中添加 Provider 服务的 ServiceId，这样 Zuul 就不会把 Provider 服务的接口暴露出去。使用 Postman 无法通过 Zuul 访问 Provider[⊖]服务的接口，如图 9-25 所示。

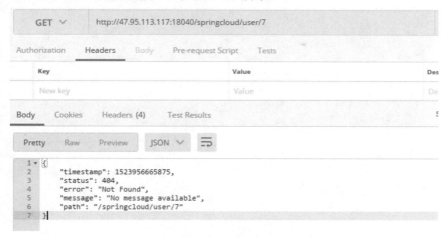

图 9-25 服务屏蔽后的访问情况

（5）其他配置

Zuul 标签下还有一些其他配置，例如 host 中包含连接数和超时时间等配置；ignored-*标签可以对不同类型数据进行过滤等，这些配置的使用要在实际项目中根据具体需要进行选择。

9.4.3 Filter 基本使用

可以通过自定义 Filter 来实现在 Zuul 网关上的请求拦截与过滤。在同一个 Zuul 服务中，可以定义多个 Filter，多个 Filter 根据分类和优先级的定义有序执行，这种有序的关系也便于把一些通用并且简单的过滤器放到前面，如果这个过滤器验证不通过，后面的过滤器通过检测上一个过滤器的状态选择性执行，从而节省服务器开销。下面介绍 Filter 的使用方法。

（1）Filter 简单实现

在 Zuul 工程中创建一个 CommonFilter 类，这个类实现如下。

```
@Component
public class CommonFilter extends ZuulFilter{
    ObjectMapper mapper = new ObjectMapper();
    @Override
    public Object run() {
        try {
            RequestContext ctx = RequestContext.getCurrentContext();
            HttpServletRequest request = ctx.getRequest();
            String versionString = request.getHeader("version");
            String typeString = request.getHeader("clienttype");
            if(versionString!=null && typeString!=null) {
                ctx.setSendZuulResponse(true);
                ctx.set("isFilterSuccess", true);
            }else {
                ctx.setSendZuulResponse(false);
```

⊖ 为了演示方便，后面的例子中放开了对 Provider 服务的屏蔽。

```java
            ctx.set("isFilterSuccess", false);
            ctx.setResponseStatusCode(400);
            Result<String> result = new Result<>(ResultCode.Bad_Request);
            ctx.setResponseBody(mapper.writeValueAsString(result));
        }
    } catch (Exception e) {
        e.printStackTrace();
        RequestContext ctx = RequestContext.getCurrentContext();
        ctx.setSendZuulResponse(false);
        ctx.set("isFilterSuccess", false);
        ctx.setResponseStatusCode(400);
    }
    return null;
}

@Override
public boolean shouldFilter() {
    return true;
}

@Override
public int filterOrder() {
    return 0;
}

@Override
public String filterType() {
    return "pre";
}
}
```

（2）方法含义

CommonFilter 继承自 ZuulFilter 抽象类，并且实现了 4 个抽象方法，其含义见表 9-2。

表 9-2　ZuulFilter 抽象方法

方法名	含　　义	说　　明
filterType	过滤器类型	pre：请求路由前被调用 route：请求路由时调用 error：处理请求发生错误时被调用 post：在 route 和 error 之后被调用
filterOrder	过滤器优先级	数字越小，越先执行
shouldFilter	过滤器是否执行的判断	true 表示执行
run	过滤器具体逻辑	可以通过此方法对数据进行检测，并且可以提前结束路由

在上面的例子中，在 run 方法中验证了请求头中是否包含 version 和 clienttype 这两个数据项，如果不包含则验证不通过；方法中使用 ctx.set("isFilterSuccess", boolean)在上下文中存放了一个数据，可以使用这种方法传递 Filter 之间的数据。

（3）实际效果演示

如果不添加 version 和 clienttype 这两个数据项，则不能验证通过，效果如图 9-26 所示。

第 9 章　Spring Cloud

图 9-26　不带 Header 参数的请求

添加 Header 信息后，可以正确返回数据，如图 9-27 所示。

图 9-27　带 Header 参数的请求

9.4.4　简单的鉴权服务

用户的登录鉴权可以放到 Zuul 的 Filter 中进行，也可以新建一个服务专门作为用户登录鉴权使用。这里新建一个 Spring Boot 工程，叫作 SpringCloudAuth[一]，专门对外提供鉴权能力，Filter 可以调用此服务从而实现鉴权验证。

在此工程中，对外提供两个接口，一个是获取用户 token（令牌）的接口，另一个是验证用户 token 的接口。集群中的服务在被访问前，都会先通过 Zuul 去此服务验证 token，验证通过才能继续访问集群服务，否则不允许访问[二]。

○ 在此演示的服务实现仅仅是一个鉴权服务的主要思路，并非真正的鉴权服务，这个示例还需扩充能力，例如生成的 token 应该缓存起来，token 应该具备时效性，不同的账号会有不同的权限等，大家可以自己丰富鉴权服务的实现或者使用一些开源的鉴权组件。

○ 此 Auth 服务可以加入微服务集群中，这样就需要在获取 token 时，在用户鉴权 Filter 中对路径进行特殊处理，不进行 token 验证，否则用户在没有 token 的情况下永远无法获取 token。可以使用判断 HttpServletRequest 中请求路径的方法，若发现是获取 token 的请求，则不进行 token 验证。

（1）添加 Controller 类接口。

```java
@RestController
@RequestMapping("/auth")
public class AuthController {
    class Admin{
        public int id = 1;
        public String name = "admin";
        public String password = "password";
    }
    private Admin admin = new Admin();

    @RequestMapping(value="/gettoken",method=RequestMethod.GET)
    public Result<String> getToken(@RequestParam("name") String name,@RequestParam("password") String password) {
        if(name!=null && password!=null && name.equals(admin.name) &&
                                            password.equals(admin.password)) {
            String token = generateToken();
            Result<String> result = new Result<>(ResultCode.OK,token);
            return result;
        }else {
            Result<String> result = new Result<>(ResultCode.Bad_Request);
            return result;
        }
    }

    @RequestMapping(value="/validtoken",method=RequestMethod.GET)
    public Result<String> validToken(@RequestParam("token") String token) {
        if(token != null && token.equals(generateToken())) {
            Result<String> result = new Result<>(ResultCode.OK);
            return result;
        }
        else {
            Result<String> result = new Result<>(ResultCode.Bad_Request);
            return result;
        }
    }

    private String generateToken() {
        StringBuilder stringBuilder = new StringBuilder();
        stringBuilder.append(admin.id).append(admin.name).append(admin.password);
        try {
            return md5Encode(stringBuilder.toString());
        } catch (Exception e) {
            e.printStackTrace();
        }
        return null;
    }

    private String md5Encode(String inStr) throws Exception {
        MessageDigest md5 = null;
        try {
            md5 = MessageDigest.getInstance("MD5");
```

```
            } catch (Exception e) {
                System.out.println(e.toString());
                e.printStackTrace();
                return "";
            }

            byte[] byteArray = inStr.getBytes("UTF-8");
            byte[] md5Bytes = md5.digest(byteArray);
            StringBuffer hexValue = new StringBuffer();
            for (int i = 0; i < md5Bytes.length; i++) {
                int val = ((int) md5Bytes[i]) & 0xff;
                if (val < 16) {
                    hexValue.append("0");
                }
                hexValue.append(Integer.toHexString(val));
            }
            return hexValue.toString();
        }
    }
```

在 Controller 类中提供了一个默认的账号密码，当 getToken 接口使用此账号密码获取 token 时，如果都验证正确，可以返回一个简易的 token，此 token 是用字符串拼接的并且进行了一次 MD5[⊖]；当 validToken 接口进行 token 验证时，使用接口获取的值和系统的 token 进行验证，二者相同则验证通过。

（2）获取 token

使用 Postman 先获取 token 如下：c8a02ee0a25f0ec1dd6c5df642c304da。如图 9-28 所示。

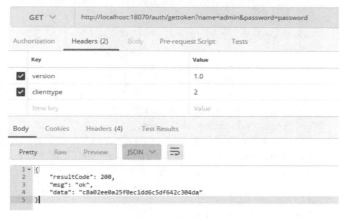

图 9-28　获取 token 信息

9.4.5　Filter 使用其他服务进行鉴权

新建一个 AuthFilter，此 Filter 调用上一小节的鉴权服务进行 token 的验证，使用 Feign 进行服务间调用，所以要添加 Feign 的相关配置，具体配置方法参照之前的章节，如果验证通过则对服务进行路由。下面列出 Feign 和 Filter 的代码。

⊖ Message Digest Algorithm MD5（中文名为消息摘要算法第 5 版）为计算机安全领域广泛使用的一种散列函数，用以提供消息的完整性保护。

```java
@FeignClient(name="auth",url="http://172.17.238.237:18070")
public interface AuthFeign {
    @RequestMapping(value="/auth/gettoken",method=RequestMethod.GET)
    public Result<String> getToken(@RequestParam("name") String name,@RequestParam("password") String password);

    @RequestMapping(value="/auth/validtoken",method=RequestMethod.GET)
    public Result<String> validToken(@RequestParam("token") String token);
}

@Component
public class AuthFilter extends ZuulFilter{
    ObjectMapper mapper = new ObjectMapper();
    @Autowired
    private AuthFeign feign;

    @Override
    public boolean shouldFilter() {
        RequestContext ctx = RequestContext.getCurrentContext();
        return (boolean) ctx.get("isFilterSuccess");
    }

    @Override
    public Object run() {
        try {
            RequestContext ctx = RequestContext.getCurrentContext();
            HttpServletRequest request = ctx.getRequest();
            String token = request.getHeader("token");
            if(token!=null) {
                Result<String> authResult = feign.validToken(token);
                if(authResult.getResultCode() == 200) {
                    ctx.setSendZuulResponse(true);
                    ctx.set("isFilterSuccess", true);
                }else {
                    ctx.setSendZuulResponse(false);
                    ctx.set("isFilterSuccess", false);
                    ctx.setResponseStatusCode(401);
                    Result<String> result = new Result<>(ResultCode.Unauthorized);
                    ctx.setResponseBody(mapper.writeValueAsString(result));
                }
            }else {
                ctx.setSendZuulResponse(false);
                ctx.set("isFilterSuccess", false);
                ctx.setResponseStatusCode(400);
                Result<String> result = new Result<>(ResultCode.Bad_Request);
                ctx.setResponseBody(mapper.writeValueAsString(result));
            }
        } catch (Exception e) {
            e.printStackTrace();
            RequestContext ctx = RequestContext.getCurrentContext();
            ctx.setSendZuulResponse(false);
            ctx.set("isFilterSuccess", false);
            ctx.setResponseStatusCode(400);
```

第 9 章 Spring Cloud

```
        }
        return null;
    }

    @Override
    public String filterType() {
        return "pre";
    }

    @Override
    public int filterOrder() {
        return 1;
    }
}
```

AuthFilter 类中的 filterOrder 方法返回的数字是 1，所以这个 filter 会晚于 CommonFilter 执行；shouldFilter 方法中，使用从上一个 Filter 获取的数据，如果请求没有通过上一个过滤器，那么本过滤器不执行。

通过 Postman 调用结果如图 9-29 所示。

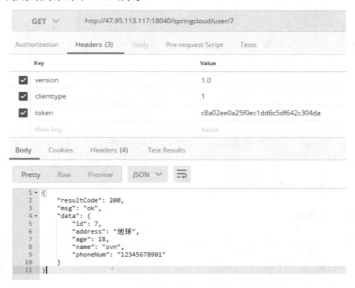

图 9-29　带 token 的请求

9.4.6　Zuul 的其他使用方法

（1）监控

Zuul 依赖默认集成了 Hystrix，所以添加 Zuul 的监控非常简单，只要在配置文件中添加服务分组信息 eureka.instance.metadata-map.cluster=main[○]，即可把 Zuul 服务的监控聚合到 Turbine[○]上，如图 9-30 所示。

　㊀ 这里使用此风格是为了便于横向书写，项目代码中采用的是 yaml 形式的书写方式。
　㊀ 这里需要保证 Zuul 与 Turbine 的数据采集方式统一，默认情况下 Zuul 不是通过消息队列采集的，如果 Turbine 想通过消息队列采集，那么 Zuul 上也要配置消息队列；如果 Zuul 不想使用消息队列，那么两者都要改为非消息队列的采集方式。

233

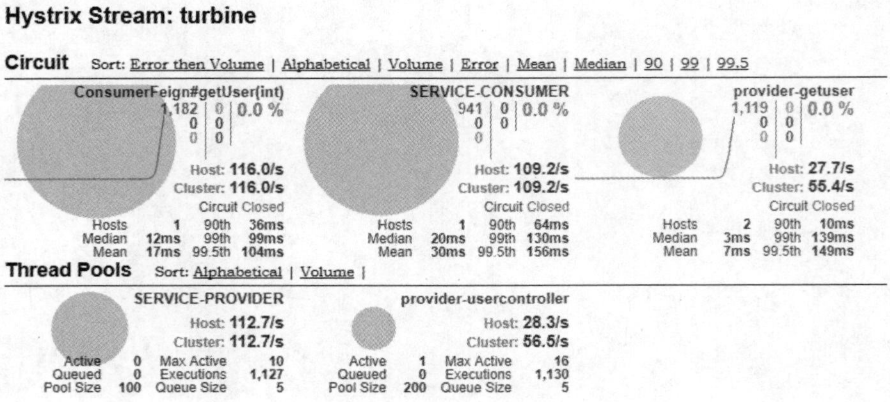

图 9-30　带 Zuul 服务的监控页面

由上图可见，Zuul 的监控级别是服务级的，SERVICE-CONSUMER 就是从 Zuul 获取的监控数据。

（2）高可用

Zuul 的高可用相对简单，在微服务集群中不用进行什么配置，启动多台 Zuul 服务的实例，在前面的负载中完成 Zuul 服务的负载策略即可。

9.5　Config

Config 是 Spring Cloud 微服务集群的配置中心，可以把各个服务的配置通过 Config 服务进行统一获取，这样可以保证配置文件的集中管理，同时可以针对某些配置进行动态的更新，这样也解决了业务服务更新配置时重新部署的问题。

9.5.1　配置 Config 服务端

Config 服务也是个 Spring Boot 程序。基本的 Config 服务主要配置三个地方：pom 文件、yml 文件和启动类，但是有一个特殊的地方是 Config 服务需要配置一个版本管理服务器，用作真实的配置文件的存放组件。Config 对多种版本管理工具进行了支持，本节使用 Git 作为配置文件存放的工具。

（1）建立 Git 工程存放配置

在 Git 中创建一个工程，工程路径如下：https://gitee.com/hwhe/SpringCloudConfig.git。在此工程中，创建两个分支，一个是 master 默认分支，一个是 dev 分支。

两个分支中的目录结构见表 9-3。

两个分支的文件结构相同，在不同的分支中，可以根据服务需要设置不同的配置内容。后面会在 Config 服务中，根据文件夹、程序名、profile 和分支获取不同的配置项。

表 9-3　目录结构

分支	文件夹	文件
master	configs	service-consumer-dev.yml
		service-consumer-prod.yml
		service-provider-dev.yml
		service-provider-prod.yml
dev	configs	service-consumer-dev.yml
		service-consumer-prod.yml
		service-provider-dev.yml
		service-provider-prod.yml

第 9 章　Spring Cloud

（2）Config 服务配置

在 pom 文件中，添加如下依赖：

```
<dependency>
    <groupId>org.springframework.cloud</groupId>
    <artifactId>spring-cloud-config-server</artifactId>
</dependency>
```

在启动类中，添加@EnableConfigServer 注解。

在 yml 文件中，添加如下配置：

```yaml
server:
  port: 18050

spring:
  application:
    name: config-server
  cloud:
    config:
      server:
        git:
          uri: https://gitee.com/hwhe/SpringCloudConfig.git
          search-paths: configs
          username: your@mail.com
          password: yourpassword
```

上面的配置中，uri 是 Git 服务的地址；search-paths 是 Git 工程的文件夹，这里对应 Git 工程的 configs 文件夹；username 和 password 是 Git 的账号密码。

（3）通过 Config 获取配置

可以通过请求 Config 服务来获取 Git 上的配置属性，具体请求可以使用 Config 服务地址加"/{applicationname}/{profile}/{label}"路径的形式来获取，获取的内容如下。使用 Postman 请求效果如图 9-31 所示。

```
{
    "name": "SpringCloudServiceProvider",
    "profiles": [
        "dev"
    ],
    "label": null,
    "version": "f30aa5b053cb613dd9ce503967592326928fba5b",
    "state": null,
    "propertySources": [
        {
            "name": "https://gitee.com/hwhe/SpringCloudConfig.git/configs/service-provider-dev.yml",
            "source": {
                "custom.foo": "springcloudprovider-dev    label is master"
            }
        }
    ]
}
```

图 9-31　获取配置信息

由图可见，使用 Postman 工具获取信息，仅使用/{applicationname}/{profile}路径就得到了 Git 上的配置项，而没有使用 label，这是因为在不设置 label 的情况下，默认使用 master 分支。在获取的数据中，propertySources 属性下的内容是 Git 的文件地址信息和配置项。下面尝试在路径中添加 label 来获取 dev 分支下的数据，获取的内容如下。使用 Postman 请求效果如图 9-32 所示。

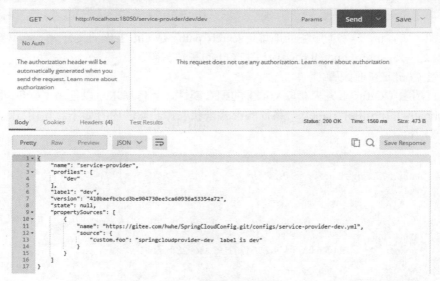

图 9-32　获取配置信息

```
{
    "name": "SpringCloudServiceProvider",
    "profiles": [
        "dev"
    ],
    "label": "dev",
    "version": "6a26fc099b163fae362df3872425c3d5db12c734",
    "state": null,
```

```
            "propertySources": [
                {
                    "name": "https://gitee.com/hwhe/SpringCloudConfig.git/configs/ SpringCloudServiceProvider- dev.yml",
                    "source": {
                        "custom.foo": "springcloudprovider-dev    label is dev"
                    }
                }
            ]
        }
```

可以使用 spring.cloud.config.server.default-label 属性来设置默认的分支，这样这个分支就成为 Config 服务的默认数据源。

9.5.2 服务通过 Config 获取配置

本节先演示业务服务 Consumer 在程序中获取一个本服务自定义的配置，然后再加入 Config 服务的客户端依赖，使用 Config 服务获取配置，并且观察配置的优先级别。

（1）服务中获取自定义配置

在 yml 文件中，添加如下配置，自定义一个 custom.foo 属性，里面包含了一些标记信息。

```yml
custom:
    foo: application config version 1
```

更改 Consumer 服务的 Controller，添加如下代码，可以通过接口获取配置信息。

```java
@Value("${custom.foo}")
String fooValue;

@RequestMapping(value="/foo",method=RequestMethod.GET)
public Result<String> getFoo() {
    Result<String> result = null;
    if(fooValue!=null) {
        result = new Result<>(ResultCode.OK, fooValue);
    }else {
        result = new Result<>(ResultCode.Not_Found);
    }
    return result;
}
```

通过 Postman 直接访问服务的接口可以得到配置信息，如图 9-33 所示。

（2）从 Config 服务获取配置

业务服务可以通过 Config 服务获取配置信息，只要在存储配置信息的 Git 工程中添加服务对应的配置项，在业务服务中添加 Config 客户端依赖，并且新建一个 bootstrap.yml 文件，配置 Config 服务的信息即可。

添加 Git 配置信息：

在管理配置文件的 Git 工程中，选择 master 分支的 service-consumer-dev.yml 文件，在其中添加如下配置：

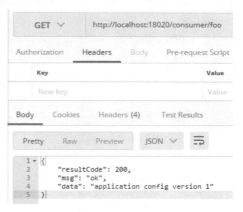

图 9-33　通过接口获取配置

在 Consumer 工程中添加依赖：

```
<dependency>
    <groupId>org.springframework.cloud</groupId>
    <artifactId>spring-cloud-starter-config</artifactId>
</dependency>
```

在 Consumer 工程中添加 bootstrap.yml 文件，并且进行如下配置：

```
spring:
  application:
    name: service-consumer
  cloud:
    config:
      profile: dev
      uri: http://localhost:18050/
```

custom:
    foo: configserver config version 1 from master

在 bootstrap.yml 文件中设置业务服务名，Config 服务的 uri 地址和选择的 profile。先后启动 Config 服务和 Consumer 服务，通过 Postman 获取 foo 信息，如图 9-34 所示。

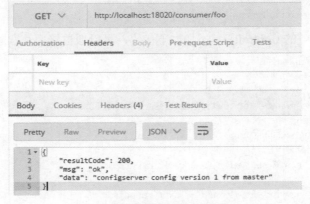

图 9-34　通过接口获取配置

由请求输出结果可见，可以通过 Config 服务获取配置，并且从 Config 服务获取的配置覆盖了业务服务本身的配置，这是 Spring Boot 的配置优先级导致的。

如果业务服务想从其他分支获取配置，只要在 bootstrap.yml 文件中添加 spring.cloud.config.label={branch 具体分支}配置即可。

9.5.3　添加加密

如果想对配置进行保护，可以在 Config 服务中设定账号密码，这样只有配置了此账号密码的服务才能获取配置。

（1）Config 服务添加安全组件

在 pom 文件中添加如下依赖：

```
<dependency>
    <groupId>org.springframework.boot</groupId>
    <artifactId>spring-boot-starter-security</artifactId>
```

```
</dependency>
```

在 yml 文件中添加如下配置：

```yaml
security:
  user:
    name: user
    password: password
```

（2）应用服务添加账号密码

把 Consumer 工程的 bootstrap.yml 文件修改为如下内容：

```yaml
spring:
  application:
    name: service-consumer
  cloud:
    config:
      profile: dev
      uri: http://localhost:18050/
      username: user
      password: password
```

这样就完成了最简单的账号密码的验证。

9.5.4 通过 Config 服务名读取配置

前面介绍的业务服务直接配置了 Config 服务的 IP 地址进行配置的获取，其实 Config 服务也可以通过 Eureka 的注册发现能力，让其他服务通过服务名去访问，这样 Config 服务只要配置多台实例即可实现服务的高可用。

要实现此能力只要完成两件事，一是 Config 服务完成 Eureka 的配置，把 Config 服务注册到 Eureka 上，关于此配置前面已经介绍很多，这里就不再重复介绍。二是在业务服务的 bootstrap 文件中，修改为如下配置：

```yaml
spring:
  application:
    name: service-consumer
  cloud:
    config:
      profile: dev
      discovery:
        enabled: true
        service-id: CONFIG-SERVER
      username: user
      password: password
eureka:
  instance:
    hostname: service-consumer
    instance-id: ${spring.cloud.client.ipAddress}:${server.port}
    prefer-ip-address: true
    metadata-map:
      cluster: main
  client:
    region: javadevmap
```

```
            availability-zones:
                javadevmap: map_eureka
            service-url:
                map_eureka: http://172.17.238.237:18001/eureka/,http://172.17.238.239:18001/eureka/
```

这里做的事情主要是把 Eureka 的相关配置从 application.yml 文件中移至 bootstrap.yml 文件。另外一个不同的地方就是去掉了 spring.cloud.config.uri 配置，添加了 spring.cloud.config.discovery 配置，添加此配置后，业务服务将通过服务名查找 Config 服务获取配置。

9.5.5 配置动态刷新

如果想更新一条业务配置，但是由于某些原因又不能重新启动业务服务，这个时候就用到了动态更新。

要实现此能力，只要基于之前的工作，并且在业务服务中保证/refresh 端点可访问，在需要刷新属性值的类上方添加@RefreshScope 注解即可。

（1）配置业务服务

在 Consumer 工程的 yml 文件中，添加如下配置，此配置添加了/refresh 端点，并且屏蔽了安全认证。

```
endpoints:
    refresh:
        enabled: true
        sensitive: false
```

在 Controller 类的上方添加@RefreshScope 注解。

（2）动态配置演示

首先通过接口直接获取配置项，可以得到如下返回：

```
{
    "resultCode": 200,
    "msg": "ok",
    "data": "configserver config version 1 from master"
}
```

修改 Git 配置工程中对应的文件信息，改为：

```
custom:
    foo: configserver config version 2 from master
```

再次通过接口获取服务的配置项，发现配置项的返回没有变化。这时需要调用此服务的/refresh 端点。用 post 请求调用 http://47.95.113.117:18020/refresh 后，再通过/consumer/foo 接口获取信息，可以得到如下返回数据：

```
{
    "resultCode": 200,
    "msg": "ok",
    "data": "configserver config version 2 from master"
}
```

9.5.6 批量刷新配置

上一节实现了动态刷新，但是这种使用方法在服务实例众多的时候明显不太合适，因为每个服务都要调用/refresh 进行刷新。有没有一种办法，能够实现批量的服务刷新？这就需要总线来进行配合了。

在 Config 服务上配置总线的地址，并且在业务服务中配置总线的地址。这样只要调用 Config 服务的/bus/refresh 端口，总线就会通知所有服务进行刷新。

（1）服务配置

在 Config 服务和业务服务的 pom 文件中添加如下依赖：

```
<dependency>
    <groupId>org.springframework.cloud</groupId>
    <artifactId>spring-cloud-starter-bus-amqp</artifactId>
</dependency>
```

在 Config 服务和业务服务的 yml 文件中添加 RabbitMQ 的配置：

```
spring:
  rabbitmq:
    host: 172.17.238.238
    port: 5673
    username: guest
    password: guest
```

这样就完成了批量刷新的全部配置工作。

（2）全量服务刷新

所有服务启动后，修改 Git 文件的配置。例如把 master 分支上的 service-consumer-dev.yml 文件修改为：

```
custom:
  foo: configserver config version 3 from master
```

调用 Config 服务刷新，由于对 Config 服务添加了安全组件，所以在 Postman 中提交刷新时，需要配置账号密码，如图 9-35 所示。

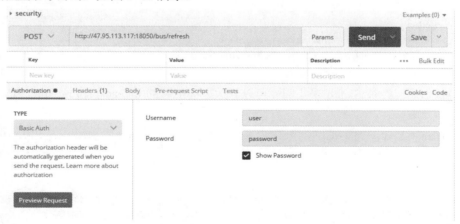

图 9-35　带安全组件的配置刷新

需要注意的是，刷新是 post 请求，并且路径是/bus/refresh。

（3）部分服务刷新

如果只想刷新某个服务，可以在请求路径中添加 destination=applicationname。例如只刷新 Consumer 服务，可以使用如下请求，如图 9-36 所示。

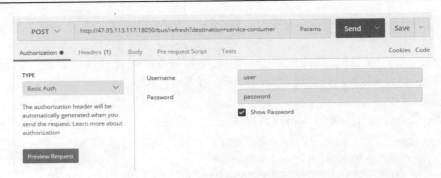

图 9-36 特定服务刷新

9.6 Sleuth 与 Zipkin

现在集群中主要包含两个业务服务,一个是 Consumer,另一个是 Provider,调用关系非常清晰。但是如果集群内业务不断健全,新功能不断增加,而微服务的数量也不断增加,想搞明白一个业务具体调用了哪些服务以及整个调用链中性能瓶颈出现在哪里,就会变得越来越困难。

在这种情况下,可以添加 Sleuth 和 Zipkin 来跟踪服务链路情况。Sleuth 提供了信息采集的能力,它可以对一个请求分配唯一的 id,整个链路中所有调用路径都会记录此 id,同时 Sleuth 还对每一次操作进行了更加详细的记录,本节会在后面讲解 Sleuth 的记录内容。Zipkin 可以从多台服务器中采集这些记录信息,进行统一的整理、展示和数据存储。本节详细介绍这两个组件的使用方法。

9.6.1 Sleuth 信息采集

Sleuth 具备信息采集能力,只要在业务服务中添加如下依赖即可。

```
<dependency>
    <groupId>org.springframework.cloud</groupId>
    <artifactId>spring-cloud-starter-sleuth</artifactId>
</dependency>
```

如何确认信息已经被 Sleuth 采集呢?可以修改之前配置的 logback-spring.xml 日志文件,把信息的输出部分由

```
<pattern>%d{yyyy-MM-dd}    %d{HH:mm:ss.SSSZ}    %-5level    %logger{36}[${APP_Name}],[%15.15t] : %m%n%wEx</pattern>
```

改为

```
<pattern>%d{yyyy-MM-dd}  %d{HH:mm:ss.SSSZ} %-5level %logger{36}  [${APP_Name}], %16X{X-B3-TraceId}, %16X{X-B3-SpanId}, %5X{X-Span-Export}], [%15.15t] : %m%n%wEx</pattern>
```

这样,就可以对比修改前后的服务调用,可以看到输入的日志中多出如下内容:[, 3bd7d8737aaca993, 3bd7d8737aaca993, false],这些就是通过 Sleuth 采集到的数据,TraceId 就是服务调用的唯一标识,Sleuth 会保证此 id 在同一请求中保持唯一。SpanId 是各个任务的 id。

现在虽然在日志中记录了调用链路的唯一标识,但是逐个服务实例查看一个请求的 TraceId 明显是不现实的,所以需要一个聚合的展示工具,下面就用到了 Zipkin。

9.6.2　Zipkin 数据聚合展示

Sleuth 记录的信息分散在各个节点中，可以通过 Zipkin 进行统一的收集和分析。使用 Spring Boot 建立一个工程，工程名叫 SpringCloudZipkin，然后完成 Zipkin 组件的引用和配置，即可搭建一个 Zipkin 服务。

（1）搭建 Zipkin 服务

添加 Zipkin 服务的依赖组件。

```
<dependency>
    <groupId>io.zipkin.java</groupId>
    <artifactId>zipkin-server</artifactId>
</dependency>
<dependency>
    <groupId>io.zipkin.java</groupId>
    <artifactId>zipkin-autoconfigure-ui</artifactId>
</dependency>
```

在启动类中添加注解@EnableZipkinServer，在 yml 文件中设置服务的端口为 18060，服务名为 zipkin-server。

（2）配置业务服务

在 Consumer 和 Provider 业务服务中，添加如下依赖，可以把 Sleuth 收集到的数据发送至 Zipkin 服务。

```
<dependency>
    <groupId>org.springframework.cloud</groupId>
    <artifactId>spring-cloud-starter-sleuth</artifactId>
</dependency>
<dependency>
    <groupId>org.springframework.cloud</groupId>
    <artifactId>spring-cloud-sleuth-zipkin</artifactId>
</dependency>
```

在 yml 配置文件中添加如下配置：

```yaml
spring:
  zipkin:
    enabled: true
    sender:
      type: web
    base-url: http://172.17.238.237:18060
  sleuth:
    enabled: true
    rxjava:
      schedulers:
        hook:
          enabled: false
    sampler:
      percentage: 1.0
```

此配置中，通过 spring.zipkin.base-url 设置了 Zipkin 服务器的地址；通过 spring.zipkin.sender.type 指定了发送类型；在 spring.sleuth 中关闭了 rxjava；并且设置了 spring.sleuth.sampler.percentage 为 1.0，表示百分之百采样，这个配置项的默认值是 0.1，表示百分之十采

样，可以根据需要设置不同的值，实现不同比例的采样率。

（3）页面展示

访问 Zipkin 服务的地址，可以看到如图 9-37 所示页面。

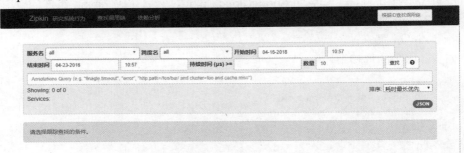

图 9-37　Zipkin 页面

在页面中，可以根据提示输入不同的搜索条件，进行链路跟踪的查看。服务名是指链路记录的服务；跨度名是指不同的请求记录；可以设置时间区间进行搜索；Annotations Query 可以设置更详细的过滤条件；持续时间可以设置请求耗时时间，注意这里使用的时间单位是微秒而不是毫秒；数量设置查询输出结果的个数；排序可以选择按最耗时排序或者按照时间排序等。

请求几次 Consumer 服务，点击查找可以查询 Zipkin 记录的链路跟踪情况。如图 9-38 所示。

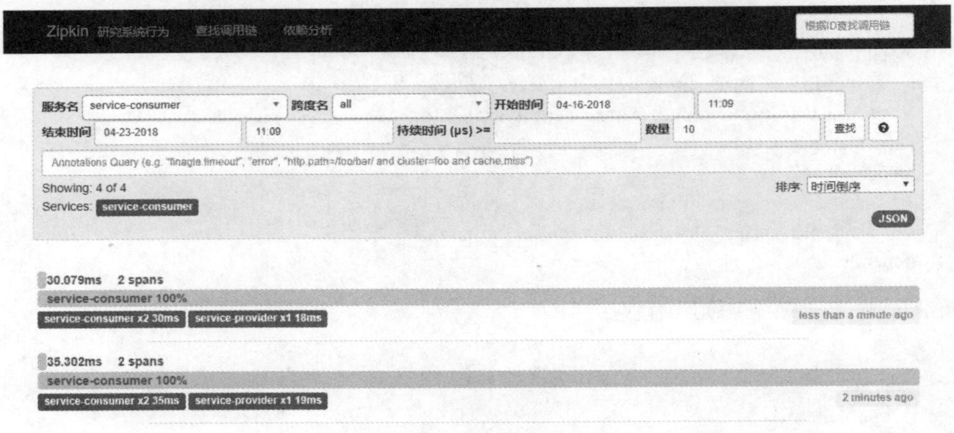

图 9-38　Zipkin 服务链路记录

点击其中一条数据，进入这条数据的详情，可以看到链路中服务的耗时等信息，如图 9-39 所示。

图 9-39　Zipkin 服务链路详情

点击右上角的 JSON 按钮，可以看到如下数据内容。

```
[
  {
    "traceId": "dd7c3c184a764f2b",
    "id": "dd7c3c184a764f2b",
    "name": "http:/consumer/7",
    "timestamp": 1524452937415000,
    "duration": 30079,
    "annotations": [
      {
        "timestamp": 1524452937415000,
        "value": "sr",
        "endpoint": {
          "serviceName": "service-consumer",
          "ipv4": "172.17.238.237",
          "port": 18020
        }
      },
      {
        "timestamp": 1524452937445079,
        "value": "ss",
        "endpoint": {
          "serviceName": "service-consumer",
          "ipv4": "172.17.238.237",
          "port": 18020
        }
      }
    ],
    "binaryAnnotations": [
      {
        "key": "http.host",
        "value": "47.95.113.117",
        "endpoint": {
          "serviceName": "service-consumer",
          "ipv4": "172.17.238.237",
          "port": 18020
        }
      }
      …
    ]
    …
  }
]
```

由于数据过长，所以仅节选了其中的部分数据进行展示。还是希望大家能够亲自搭建一个 Zipkin 看一看实际的数据内容。那么数据中记录的内容主要在描述什么呢？下一节会讲解这个问题。

点击界面上方的"依赖分析"，可以看到服务间的依赖关系如图 9-40 所示。

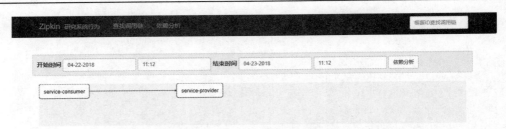

图 9-40　Zipkin 服务链路依赖

9.6.3　数据解读

简单来说，Zipkin 的数据记录在一个请求链路中，某个服务收到请求，发出请求，接收应答等全量信息。图 9-41 可表示链路中发生的事件。

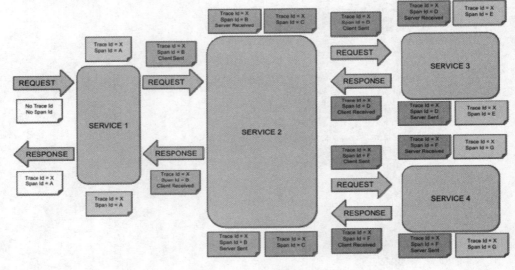

图 9-41　Zipkin 链路事件

图中的 Span 表示一个业务单元的请求发送与收到应答的完整事务，在上一节的 Json 数据中，annotations 数据项标识了 Span 记录的信息，一个 Span 中最多记录 4 组数据内容。每组内容中 value 数据项标识不同的数据类型：cs 标识客户端请求发出，sr 标识服务端收到，ss 标识服务端应答，cr 标识客户端收到应答，这样就形成了最基础的业务单元。

一个或者多个 Span 组成一个 trace，traceId 是这一次业务记录的唯一标识。上一节的数据中，请求调用了 Consumer 服务，Consumer 服务又调用了 Provider 服务，所以 Consumer 服务收到请求并向 Web 返回数据这一流程是一个 Span，Consumer 服务向 Provider 请求并且收到 Provider 服务的返回这一流程也是一个 Span，这两个 Span 组成了一个 trace。

在上一节的 Json 数据中，binaryAnnotations 数据项记录了一些其他辅助信息，例如服务名、IP、端口、请求路径、处理请求的类名、方法等等。

9.6.4　通过消息中间件收集信息

链路跟踪信息也可以通过消息中间件进行收集，这样业务服务的跟踪信息在 Zipkin 没有启动的时候也可以通过消息中间件记录下来。

第 9 章　Spring Cloud

（1）业务服务改造

在业务服务中，保留 spring-cloud-starter-sleuth 依赖，删掉 spring-cloud-sleuth-zipkin 依赖，添加另外两个依赖用来发送跟踪数据到消息中间件。

```
<dependency>
    <groupId>org.springframework.cloud</groupId>
    <artifactId>spring-cloud-starter-sleuth</artifactId>
</dependency>
<dependency>
    <groupId>org.springframework.cloud</groupId>
    <artifactId>spring-cloud-sleuth-zipkin-stream</artifactId>
</dependency>
<dependency>
    <groupId>org.springframework.cloud</groupId>
    <artifactId>spring-cloud-starter-stream-rabbit</artifactId>
</dependency>
```

修改业务服务配置文件：

```
spring:
  rabbitmq:
    host: 172.17.238.238
    port: 5673
    username: guest
    password: guest
sleuth:
  stream:
    enabled: true
  enabled: true
rxjava:
  schedulers:
    hook:
      enabled: false
sampler:
  percentage: 1.0
```

配置中添加了消息中间件的地址，并且设置了 sleuth 通过 stream 发送。把本章使用的两个业务服务部署上线，对 Consumer 请求几次，可以在 RabbitMQ 中看到 sleuth.sleuth 队列有数据等待处理，如图 9-42 所示。这是由于 Zipkin 服务还未改造为消息中间件模式从而无法消费消息。下面修改 Zipkin 服务。

图 9-42　消息中间件收集链路信息

（2）Zipkin 服务改造

修改 Zipkin 服务的依赖，去掉 zipkin-server 依赖，改为如下依赖：

```xml
<dependency>
    <groupId>org.springframework.cloud</groupId>
    <artifactId>spring-cloud-sleuth-zipkin-stream</artifactId>
</dependency>
<dependency>
    <groupId>org.springframework.cloud</groupId>
    <artifactId>spring-cloud-starter-stream-rabbit</artifactId>
</dependency>
<dependency>
    <groupId>io.zipkin.java</groupId>
    <artifactId>zipkin-autoconfigure-ui</artifactId>
</dependency>
```

修改启动类注解，去掉@EnableZipkinServer 注解，添加@EnableZipkinStreamServer 注解。
在配置文件中添加消息中间件配置项：

```
spring:
  rabbitmq:
    host: 172.17.238.238
    port: 5673
    username: guest
    password: guest
  sleuth:
    enabled: false
```

之后，把 Zipkin 服务重新部署，观察消息中间件的变化，如图 9-43 所示。可见 Zipkin 服务启动后，即处理了消息中间件中的数据。

图 9-43　消息中间件数据被消费

9.6.5 数据保存

虽然 Zipkin 的数据不像业务服务数据一样需要永久保存，但是有时确实希望能够保留一段时间内的跟踪数据，而不是 Zipkin 服务重启后数据就全部丢失了。在这个前提下，可以使用 MySQL 或 ElasticSearch[一]两种存储组件。Zipkin 的存储配置相对简单，只要准备好相应存储组件并且进行如下配置即可完成 Zipkin 数据的存储。

（1）MySQL 存储

在 Zipkin 服务中添加如下依赖：

```
<dependency>
    <groupId>mysql</groupId>
    <artifactId>mysql-connector-java</artifactId>
    <scope>runtime</scope>
</dependency>
<dependency>
    <groupId>org.springframework.boot</groupId>
    <artifactId>spring-boot-starter-jdbc</artifactId>
</dependency>
```

在配置文件中添加如下配置：

```
spring:
  ...
  datasource:
    schema: classpath:/mysql.sql
    driver-class-name: com.mysql.jdbc.Driver
    url: jdbc:mysql://172.17.238.238:3306/springcloudzipkin
    username: root
    password: mypass
    initialize: true
    continue-on-error: true
zipkin:
  storage:
    type: mysql
```

在对应的 MySQL 中建立 springcloudzipkin 数据库。

经过以上几步，就完成了 Zipkin 的 MySQL 数据库存储。

（2）ElasticSearch 存储

在 Zipkin 服务中添加如下依赖，这里需要注意的是引入的依赖版本号要能适配 ElasticSearch，否则 Zipkin 无法收集链路信息。

```
<dependency>
    <groupId>io.zipkin.java</groupId>
    <artifactId>zipkin-autoconfigure-storage-elasticsearch-http</artifactId>
    <version>1.24.0</version>
    <optional>true</optional>
</dependency>
```

在配置文件中添加如下配置：

```
zipkin:
```

[一] ElasticSearch 是一个分布式的、基于 RESTful 接口的搜索和分析引擎。

```
    storage:
        type: elasticsearch
        elasticsearch:
            cluster: elasticsearch
            hosts: http://172.17.238.238:9200
            index: zipkin
            index-shards: 5
            index-replicas: 1
```

经过如上配置，即完成了 Zipkin 通过 ElasticSearch 存储。

至此，本章已经完成了 Spring Cloud 大部分组件的讲解。下面查看 Eureka 页面，来看看整个服务集群的运行情况，如图 9-44 所示，一个健全的集群系统已经搭建好了。

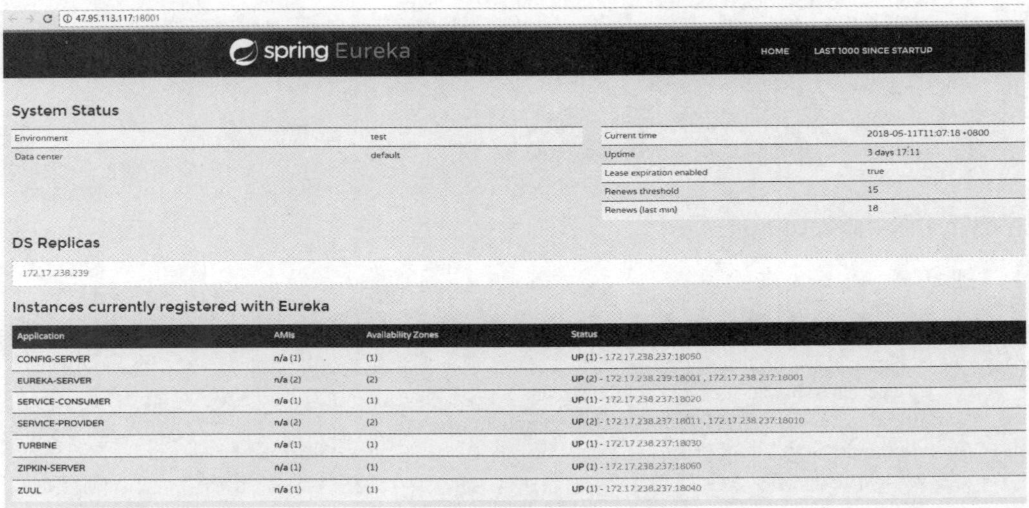

图 9-44　Spring Cloud 服务集群

第三篇 组 件 篇

前面的章节已经接触了部分系统组件[一]，例如在项目中整合了 MyBatis，继而使用了 MySQL 数据库；整合了 Logback，从而使服务可以打印日志。这些系统组件的使用在项目中是必不可少的。不同的系统组件提供了不同的能力，所以在程序设计阶段，需要根据项目需求选择相应的系统组件。

本篇将首先介绍 MySQL 数据库，包含操作数据库的语言、MyBatis 的整合及自定义的语句、事务、数据库的优化等内容。希望通过以上内容的讲解，能让读者对数据库有更深层的认识。

继 MySQL 之后，将介绍两种不同的 NoSQL 存储，分别是 MongoDB 和 Redis。这两种 NoSQL 的存储方式主要是针对 MySQL 在某些场景下无法充分满足业务需求而设计的，例如快速存取、大量文本保存等。这两种 NoSQL 存储方式虽然不能替代 MySQL，但是在自己擅长的领域表现良好。

在业务中，除了存储，还有一些需求需要其他的组件来实现，例如保存系统配置和实现注册中心，这里就会用到 Zookeeper；为了满足文件存储的能力，就会用到 FastDFS；为了在平台内实现快速的搜索，会用到 ElasticSearch；为了让平台能够定时执行某些任务，会用到分布式定时任务 ElasticJob；为了在平台内实现不同服务的解耦或者降低服务压力的目的，对某些实时性要求不高的业务可以使用消息队列，本篇会介绍 RabbitMQ 作为消息队列的用法。本篇的最后，会介绍一种日志管理方式，可以方便地汇总、查询和统计日志情况，这就是 ELK。

希望读者通过本篇的学习，能够了解相关组件的特性和基本用法，并且可以在实际业务中选择性地使用合适的业务组件。

[一] 由于本篇介绍的内容较多且较难归类，并且本篇所介绍的内容都是为了满足或者扩充系统程序的某些能力，所以把本篇介绍的内容统称为系统组件。

第 10 章 MySQL

常见的数据库种类很多，例如 Oracle、MySQL、SQL Server 等，各有特点和应用范围，其中 MySQL 由于支持多语言开发、成本低、可定制、社区活跃度高、开放源码等特点，成为许多项目的首选。MySQL 是一个多用户、多线程的关系型数据库管理系统，本章将讲解 MySQL 的特性、命令及用法。

10.1 MySQL 基本介绍和使用场景

业务开发中，常用数据库存储业务数据。那么什么是数据库呢？数据库（Database）是按照数据结构来组织、存储和管理数据的软件，每个数据库都有一个或多个不同的 API 接口用于创建、访问、管理、搜索和复制所保存的数据。而 MySQL 是常用的数据库产品之一。

10.1.1 MySQL 概述

MySQL 是一种关系型数据库管理系统，将数据保存在不同的表中，而不是将所有数据放在一个大表内，这样就加快了存取速度并提高了灵活性。MySQL 的特点有：
- 开源免费；采用了 GPL 协议，可以修改源码适应自己的系统。
- 可以处理有上千万条记录的大型数据。
- 可移植性高，安装简单方便。
- 支持常见的 SQL 语句规范，使用标准的 SQL 数据语言形式。
- 良好的运行效率。
- 调试、管理、优化简单。

10.1.2 MySQL 常用存储引擎

数据库存储引擎是数据库底层软件组织。不同的存储引擎提供不同的存储机制、索引技巧、锁定水平等功能，可针对不同的业务场景使用不同的存储引擎。这里介绍常用的两种存储引擎 MyISAM 与 InnoDB。

InnoDB 和 MyISAM 是使用 MySQL 时最常用的两个存储引擎，这两个引擎各有优劣，视具体应用场景而定。

MyISAM 类型的表强调的是性能，其执行速度比 InnoDB 类型快，但是不提供事务支持，而 InnoDB 提供事务支持以及外部键等高级数据库功能。

从上面的差别可以看出，InnoDB 更适合作为生产环境中的事务处理，而 MyISAM 更适合作为 ROLAP（关系型联机分析处理）型数据仓库。

10.1.3 MySQL 使用场景

MySQL 应用场景有以下几种：

（1）Web 应用系统

MySQL 可应用在站点数据管理，因为 MySQL 数据库安装配置简单，而且性能出色。还有一个非常重要的原因，MySQL 是开放源代码的，完全可以免费使用，常用于电商系统、博客系统、企业管理系统等。

（2）日志系统

对需要大量插入和查询日志记录的系统来说，MySQL 是非常不错的选择。在使用 MyISAM 存储引擎的时候，MySQL 数据库的插入和查询性能都非常高效，如果设计较好，两者可以做到互不锁定，达到很高的并发性能。例如处理用户相关的操作日志等。

（3）数据仓库系统

随着数据仓库数据量的飞速增长，需要的存储空间越来越大。有几个主要的解决思路。一个是采用昂贵的高性能主机以提高计算性能，用高端存储设备提高 I/O 性能，效果理想，但是成本非常高，例如使用 Oracle。另一个是通过将数据水平拆分，使用多台廉价的计算机安装 MySQL 存放数据，每台计算机上面只存放一部分数据，解决了数据量的问题，所有计算机并行计算，解决了计算能力问题，通过中间代理程序调配各台计算机的运算任务，既可以解决计算性能问题又可以解决 I/O 性能问题。所以 MySQL 也是一个不错的选择。

（4）嵌入式系统

嵌入式环境对软件系统最大的限制是硬件资源非常有限，在嵌入式环境下，软件系统必须是轻量级、低消耗的软件。MySQL 在资源使用方面的伸缩性非常大，对嵌入式环境来说，是一种非常合适的数据库系统，而且 MySQL 有专门针对嵌入式环境的版本。

10.2 MySQL 基本操作

本书重点不在 MySQL 环境的安装。MySQL 软件环境采用 Docker 容器部署，具体部署命令参照第 19 章的内容。

SQL 语言共分为四大类：数据定义语言 DDL，数据查询语言 DQL，数据操纵语言 DML，数据控制语言 DCL。接下来介绍这四部分语言如何在 MySQL 环境下进行使用。

10.2.1 MySQL 创建和删除数据库

MySQL 需使用 create 命令创建数据库，语法如下：

```
CREATE DATABASE <数据库名>;
```

用 drop 命令删除数据库：

```
DROP DATABASE <数据库名>;
```

在删除数据库过程中，务必谨慎，因为在执行删除命令后，所有数据将会清除。

10.2.2 DDL 基本操作

数据定义语言(DDL)用来创建数据库中的各种对象（表、视图等）。

在介绍表之前，先了解一下 MySQL 的数据类型：数字、日期/时间、字符串、空间坐标等。这几个类型又更细致地划分了许多子类型，见表 10-1。

表 10-1 数据类型

数据类型	具体子类型
数字类型	整数：tinyint、smallint、mediumint、int、bigint 浮点数：float、double、real、decimal
日期和时间	date、time、datetime、timestamp、year
字符串类型	字符串：char、varchar 文本：tinytext、text、mediumtext、longtext
二进制	（可用来存储图片、音乐等）tinyblob、blob、mediumblob、longblob
空间类型	Geometry、Point、Curve、LineString、MultiPoint 等

（1）MySQL 表创建

数据库表的信息包括表名、表字段以及各个字段的类型。

创建 MySQL 数据表的 SQL 通用格式：

```
CREATE TABLE table_name (column_name column_type);
```

例如创建商品表 product 的 SQL 语句如下：

```
CREATE TABLE `product` (
`id`   int(11) NOT NULL AUTO_INCREMENT ,
`product_name`   varchar(150) CHARACTER ,
`price`   int(11) NULL DEFAULT NULL ,
`product_desc`   varchar(500),
`product_pic`   varchar(255),
PRIMARY KEY (`id`)
) ENGINE=InnoDB DEFAULT CHARSET=utf8;
```

- AUTO_INCREMENT 是定义此列为自增的属性，一般用于主键，数值会自动加 1。
- PRIMARY KEY 关键字用于定义列为主键。也可使用多列来定义主键，列间以逗号分隔。
- 表中某些字段可以为 NULL，也可以设置字段不能为空（如 NOT NULL），对于设置不能为空的字段，若在操作数据库时赋值为 NULL，就会报错。
- ENGINE 设置存储引擎，本节使用 InnoDB；CHARSET 设置编码，常用 utf8。

（2）MySQL 表的修改删除

当需要修改数据表名或表字段时，就需要使用 MySQL 的 ALTER 命令。例如要将商品表 product 的 product_desc 字段大小改成 varchar(400)，使用 MODIFY 关键字，具体操作如下：

```
ALTER TABLE product MODIFY product_desc varchar(400);
```

修改字段名称使用 CHANGE 关键字，例如将商品表中的 product_desc 字段修改名称为 product_desc01，字段类型长度不变：

```
ALTER TABLE product CHANGE product_desc product_desc01 varchar(400);
```

（3）MySQL 修改表名

如需修改表的名称，可以在 ALTER TABLE 语句中使用 RENAME 子句：

```
ALTER TABLE <原表名> RENAME TO <新表名>;
```

10.2.3　DQL 基本操作

数据查询语言(DQL)基本结构由 SELECT 子句、FROM 子句、WHERE 子句组成。
（1）MySQL 数据库中查询数据通用的 SELECT 语法：

```
SELECT column_name,column_name
FROM table_name
[WHERE Clause]
[LIMIT N][ OFFSET M]
```

- SELECT：可以读取一条或者多条记录。不指定列名而使用星号（*），SELECT 语句会返回表的所有字段数据。
- From：可使用一个或多个表，用逗号（,）分隔。
- WHERE：设置查询条件。
- LIMIT：设定返回的记录数，例如 limit m,n 其中 m 是指记录开始的 index，n 是指从第 m+1 条开始，取 n 条。
- OFFSET：指定 SELECT 语句开始查询的数据偏移量。默认情况下偏移量为 0。

（2）WHERE 子句常用语法如下：

```
SELECT field1, field2...fieldN FROM table_name1, table_name2...
[WHERE condition1 [AND [OR]] condition2... ]
```

WHERE 可以使用 AND 或者 OR 指定一个或多个条件。WHERE 子句类似于程序语言中的 if 条件，根据 MySQL 表中的字段值来读取指定的数据。

例如查询商品价格在[100,300]区间的数据：

```
select * from product where price >=100 and price <=300
```

10.2.4　DML 基本操作

数据操纵语言（DML）主要有三种形式：
（1）插入：INSERT
MySQL 数据表插入数据使用 INSERT INTO 语法：

```
INSERT INTO table_name ( field1, field2,...fieldN ) VALUES ( value1, value2,...valueN );
```

例如在商品表中插入一条数据：

```
INSERT INTO `product` (id,product_name,price,product_desc,product_pic) VALUES ('10', 'java dev map Book', '212', '商品描述 desc', 'download?filename=javamapdev.png');
```

（2）更新：UPDATE
UPDATE 命令是修改 MySQL 数据表数据的常用语法：

```
UPDATE table_name SET field1=newvalue1, field2=newvalue2 [WHERE Clause]
```

可以同时更新多个字段并且在 WHERE 子句中指定更新条件。例如更新 id 为 107 商品的商品名称，更改为 java dev Version02：

```
update product set product_name = 'java dev Version02' where id =107
```

（3）删除：DELETE
MySQL 数据表中删除数据的通用语法：

```
DELETE FROM table_name [WHERE Clause]
```

如果未指定 WHERE 子句，MySQL 表中的所有记录都将被删除。可以在 WHERE 子句中指定删除条件。例如删除商品表中 id 为 107 的数据：

```
delete from product where id =107
```

10.2.5　DCL 基本操作

数据控制语言（DCL）用来授予或回收访问数据库的某种特权，可以限制用户访问哪些库、哪些表；限制用户对哪些表执行 SELECT、CREATE、DELETE、UPDATE、ALTER 等操作；限制用户登录的 IP 或域名；限制用户自己的权限是否可以授权给别的用户。

（1）GRANT：授权。

MySQL 赋予用户权限命令的简单格式：

```
grant 权限 on 数据库对象 to 用户
```

例如授权用户 user01 在数据库（javadevmap）中表 product 的所有权限，并指定 user01 的登录密码为 123456。具体如下：

```
grant all privileges on javadevmap.product to 'user01'@'%' identified by '123456' with grant option;
```

上述命令具体含义如下：

- all privileges：表示将所有权限授予用户；也可指定具体的权限，例如 SELECT、CREATE、DROP 等。
- on：表示这些权限对哪些数据库和表生效。格式：数据库名.表名，如果写 "*" 表示所有数据库、所有表。例子中指定为 javadevmap 数据库的 product 表。
- to：将权限授予哪个用户。格式：'用户名'@'登录 IP 或者域名'。%表示没有限制，在任何主机都可以登录。例如：'user01'@'192.168.3.%'，表示 user01 这个用户只能在 192.168.3 的 IP 段登录。
- identified by：指定用户的登录密码。
- with grant option：表示允许用户将自己的权限授权给其他用户。

对用户进行权限变更之后，需重新加载权限，将权限信息从内存中写入数据库。具体操作命令：

```
flush privileges;
```

查看当前用户权限：

```
show grants;
```

回收权限：删除 user01 这个用户的 create 权限，该用户将不能创建数据库和表。

```
revoke create on *.* from user01@%;
flush privileges;
```

（2）回滚：ROLLBACK

ROLLBACK [WORK] TO [SAVEPOINT]：回退到某一点。回滚命令使数据库状态回到上次最后提交的状态。

MySQL 默认是打开了自动提交的，关闭自动提交有以下方法：

- Session 级别：使用 START TRANSACTION 或者 BEGIN 来开始一个事务，使用 ROLLBACK/COMMIT 来结束一个事务；或使用 SET autocommit=0 关闭当前 session 的自动提交，每次提交需要手动 COMMIT。
- 全局级别：SET GLOBAL autocommit=0 关闭全局的自动提交。

用商品表来演示 ROLLBACK 的使用。即删除 id 为 107 的商品，然后执行 ROLLBACK，查看数据是否回滚。具体操作命令如下：

```
select * from product;
start transaction;
delete from product where id=107;
select * from product;
rollback;
select * from product;
```

操作结果如图 10-1 所示。

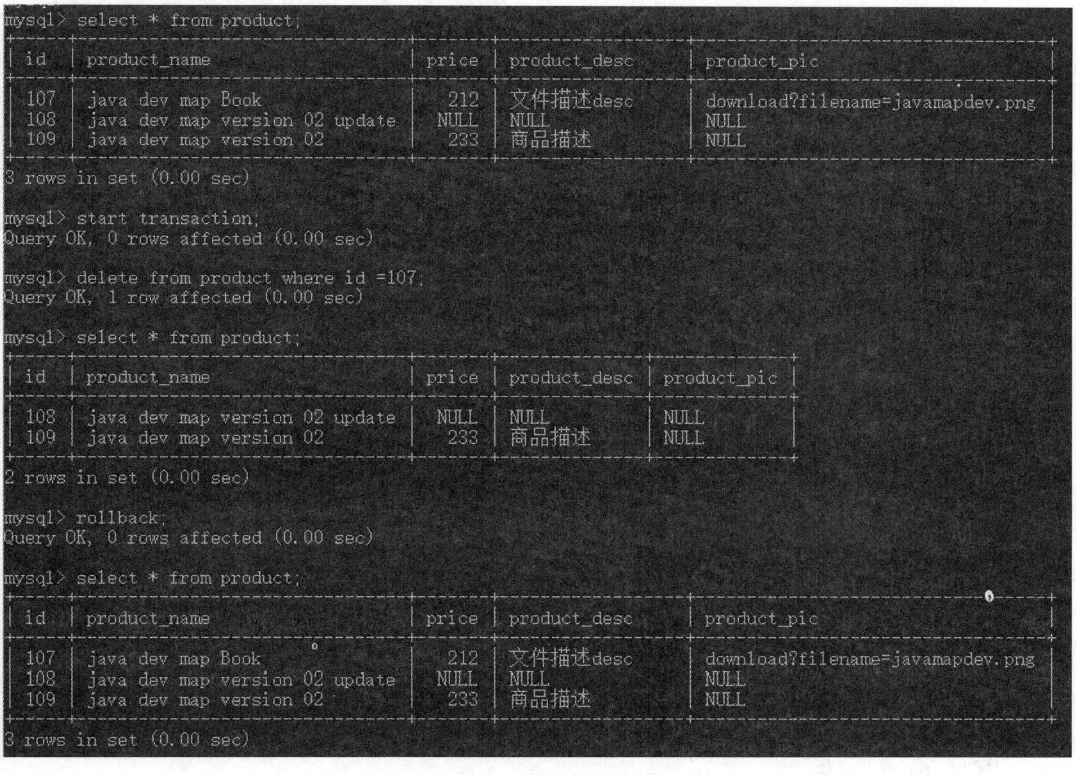

```
mysql> select * from product;
+-----+--------------------------------+-------+---------------+--------------------------------+
| id  | product_name                   | price | product_desc  | product_pic                    |
+-----+--------------------------------+-------+---------------+--------------------------------+
| 107 | java dev map Book              |   212 | 文件描述desc   | download?filename=javamapdev.png|
| 108 | java dev map version 02 update |  NULL | NULL          | NULL                           |
| 109 | java dev map version 02        |   233 | 商品描述       | NULL                           |
+-----+--------------------------------+-------+---------------+--------------------------------+
3 rows in set (0.00 sec)

mysql> start transaction;
Query OK, 0 rows affected (0.00 sec)

mysql> delete from product where id =107;
Query OK, 1 row affected (0.00 sec)

mysql> select * from product;
+-----+--------------------------------+-------+---------------+-------------+
| id  | product_name                   | price | product_desc  | product_pic |
+-----+--------------------------------+-------+---------------+-------------+
| 108 | java dev map version 02 update |  NULL | NULL          | NULL        |
| 109 | java dev map version 02        |   233 | 商品描述       | NULL        |
+-----+--------------------------------+-------+---------------+-------------+
2 rows in set (0.00 sec)

mysql> rollback;
Query OK, 0 rows affected (0.00 sec)

mysql> select * from product;
+-----+--------------------------------+-------+---------------+--------------------------------+
| id  | product_name                   | price | product_desc  | product_pic                    |
+-----+--------------------------------+-------+---------------+--------------------------------+
| 107 | java dev map Book              |   212 | 文件描述desc   | download?filename=javamapdev.png|
| 108 | java dev map version 02 update |  NULL | NULL          | NULL                           |
| 109 | java dev map version 02        |   233 | 商品描述       | NULL                           |
+-----+--------------------------------+-------+---------------+--------------------------------+
3 rows in set (0.00 sec)
```

图 10-1　事务命令结果

ROLLBACK 只能在一个事务处理内使用（在执行一条 START TRANSACTION 命令之后）。分析上面的例子，从查询 product 表开始，首先执行一条 SELECT 语句查看表中数据。然后开始一个事务处理，用一条 DELETE 语句删除 id 为 107 的商品数据。接下来执行 SELECT 语句验证 id 为 107 的记录被删除。此时执行 ROLLBACK 命令，回退 START TRANSACTION 之后的所有语句，最后一条 SELECT 语句显示 id 为 107 的商品记录被恢复了。

（3）提交：COMMIT

MySQL 语句都是直接针对数据库表执行和编写的。这就是所谓的隐含提交（implicit

commit），即提交操作是自动进行的。但是，在事务处理块中，提交不会隐含地进行。为进行明确的提交，需要使用 COMMIT 语句。这里用商品表和用户举例，实现删除 id 为 1 的用户表记录和商品 id 为 107 的商品表记录；如下所示：

```
start transaction;
delete from user where id =1;
delete from product where id =107;
commit;
```

上面例子涉及删除用户表 user 和商品表 product 中的数据，所以使用事务处理块来保证业务不被部分删除（一方删除成功，一方未删除成功）。最后的 COMMIT 语句仅在不出错时生效。如果第一条 DELETE 语句起作用，但第二条失败，则第一条 DELETE 语句会被自动撤销。

10.3 事务处理

上一节多少涉及了事务，本节学习事务处理。首先了解什么是数据库事务：数据库事务（database transaction）是指作为单个逻辑工作单元执行的一系列操作，要么完全执行，要么完全不执行。事务处理可以确保事务性单元内的所有操作都成功完成时才更新面向数据的资源。

通俗理解，在关系数据库中，一个事务可以是一条 SQL 语句，一组 SQL 语句或整个程序或者理解为事务是用来管理 insert、update、delete 语句单个或组合使用的。例如银行转账场景：从一个账号扣款并在另一个账户增款，要么都执行，要么都不执行。

10.3.1 事务概述

一般来说，事务必须满足 4 个条件（ACID）：

- 原子性（Atomicity，或称不可分割性）：一个事务（transaction）中的所有操作，要么全部完成，要么全部不完成，不会结束在中间某个环节。事务在执行过程中发生错误，会回滚（rollback）到事务开始前的状态，就像这个事务从来没有执行过一样。
- 一致性（Consistency）：在事务开始之前和事务结束以后，数据库的完整性不会被破坏。这表示写入的内容必须完全符合所有的预设规则。
- 隔离性（Isolation，又称独立性）：数据库允许多个并发事务同时对其数据进行读写和修改的能力，隔离性可以防止多个事务并发执行时由于交叉执行而导致数据的不一致。事务隔离分为不同级别，包括读未提交（read uncommitted）、读提交（read committed）、可重复读（repeatable read）和串行化（serializable）。
- 持久性（Durability）：事务处理结束后，对数据的修改是永久的，会被持久化到本地。

10.3.2 事务处理方法

本节演示的例子基于 Spring Boot 整合 MyBatis，MyBatis 整合进 Spring Boot 工程的方法已经在前面章节有详细的讲解，不再赘述。按照之前的方法新建一个工程 MysqlExample。

Spring Boot 项目事务配置步骤：

1）注解依赖

需要的注解为 @EnableTransactionManagement 和 @Transactional，它们来自 spring-tx-4.3.14.RELEASE.jar 包，该包在配置 MyBatis 依赖时，通过起步依赖 mybatis-spring-boot-starter 已自动引入。

2）业务类添加@Transactional 注解

@Transactional 注解如果加在类上，则该类所有的方法都会被事务管理，如果加在方法上，则仅对该方法进行事务管理。一般都是加在方法上，因为只有涉及增、删、改才会需要事务。

在工程中，添加业务逻辑处理类 ProductServiceImpl，此类中包含 addProduct 方法，用于添加商品，在此方法上添加事务注解，并指定 REQUIRED 事务传播行为，事务隔离级别为底层数据库的默认隔离级别，事务超时时间为 30s，针对 Exception 进行回滚：

```
@Service
public class ProductServiceImpl implements ProductService {
    @Autowired
    ProductDao productDao;

    @Transactional(propagation = Propagation.REQUIRED,
            isolation = Isolation.DEFAULT,
            timeout = 30,
            rollbackFor = Exception.class)
    @Override
    public void addProduct(String name, int price, String desc) {
        Product product = new Product();
        product.setPrice(price);
        product.setProductName(name);
        product.setProductDesc(desc);
        productDao.save(product);
    }
}
```

上面配置@Transactional 注解时使用了相关属性，属性含义见表 10-2。

表 10-2　事务属性

属　　性	类　　型	描　　述
value	String	可选的限定描述符，指定使用的事务管理器
propagation	enum:Propagation	可选的事务传播行为设置
isolation	enum:Isolation	可选的事务隔离级别设置
readOnly	boolean	读写或只读事务，默认读写
timeout	int	事务超时时间设置（秒）
rollbackFor	Class 对象数组，必须继承自 Throwable	导致事务回滚的异常类数组
rollbackForClassName	类名数组，必须继承自 Throwable	导致事务回滚的异常类名字数组
noRollbackFor	Class 对象数组，必须继承自 Throwable	该属性用于设置不需要进行回滚的异常类数组，当方法中抛出指定异常数组中的异常时，不进行事务回滚
noRollbackForClassName	类名数组，必须继承自 Throwable	该属性用于设置不需要进行回滚的异常类名称数组，当方法中抛出指定异常名称数组中的异常时，不进行事务回滚

3）开启事务

其实目前的事务已经是默认开启的，但是为了标记此服务中包含事务处理，可以在工程的启动类中添加注解@EnableTransactionManagement。

通过上面的三步，即完成了事务的配置。那么配置事务与未配置事务有什么区别呢？这里编写了两个测试方法进行验证。即在 ProductServiceImpl 类中添加两个方法，其执行内容一致，不同的地方在于方法 modifyProductsByTransaction() 添加了 @Transactional 注解，而方法 modifyProducts()未添加，在方法体里面通过 "int i = 4 / 0;" 语句来抛出异常。

```java
@Autowired
ProductDao productDao;
@Override
public void modifyProducts() {

    Product product = new Product();
    product.setPrice(233);
    product.setProductName("java dev map version 02");
    product.setProductDesc("商品描述");
    productDao.save(product);

    int i = 4 / 0;
    product = new Product();
    product.setPrice(800);
    product.setProductName("java dev map version 03");
    product.setProductDesc("商品描述 03");
    productDao.save(product);
}

@Transactional(propagation = Propagation.REQUIRED,
        isolation = Isolation.DEFAULT,
        timeout = 30,
        rollbackFor = Exception.class)
@Override
public void modifyProductsByTransaction() {
    Product product = new Product();
    product.setPrice(233);
    product.setProductName("java dev map version 02");
    product.setProductDesc("商品描述");
    productDao.save(product);

    int i = 4 / 0;
    product = new Product();
    product.setPrice(800);
    product.setProductName("java dev map version 03");
    product.setProductDesc("商品描述 03");
    productDao.save(product);
}
```

添加测试类 ProductMapperTest，分别执行两个方法。

```java
@RunWith(SpringRunner.class)
@SpringBootTest
public class ProductMapperTest {
    @Autowired
    ProductService productServiceImpl;
    @Test
    public void testTransaction(){
        productServiceImpl.modifyProducts();
```

```
        }
        @Test
        public void testTransaction02(){
            productServiceImpl.modifyProductsByTransaction();
        }
    }
```

执行两个测试方法时，均抛出了异常：

```
        java.ang.ArithmeticException: / by zero
```

但是方法 modifyProducts()的第一条数据仍然被持久化到数据库中。而添加了@Transactional 的方法 modifyProductsByTransaction()执行后，两条记录都没有被持久化到数据库中。可见 @Transactional 作用可以保证事务的原子性。

10.4　MyBatis 插入获取主键

业务开发时，有时候插入一条数据，需要立刻得到插入数据的 id，例如插入一条商品数据，然后将插入成功的 id 返回给前端。但是自动生成的 mapper 中的 insert 方法，默认是不会返回主键的。这里使用如下方法来演示如何实现插入数据后返回主键 id。

新建自定义 Mapper 类 ProductManualMapper，添加如下内容：

```
    public interface ProductManualMapper {
        Integer insertProduct(@Param("pro") Product product);
    }
```

在项目的 resources/mybatis/manual 文件夹下面新建 ProductManualMapper.xml[⊖]文件，在此文件中添加如下内容：

```
        <?xml version="1.0" encoding="UTF-8"?>
        <!DOCTYPE mapper
                PUBLIC "-//mybatis.org//DTD Mapper 3.0//EN"
                "http://mybatis.org/dtd/mybatis-3-mapper.dtd">
        <mapper namespace="com.javadevmap.mysqlexample.mapper.ProductManualMapper">
            …
            <insert id="insertProduct" parameterType="Product">
                <selectKey resultType="java.lang.Integer" keyProperty="pro.id" order="AFTER" >
                    SELECT LAST_INSERT_ID()
                </selectKey>
                INSERT INTO
                product(product_name, price, product_desc)
                VALUES (
                #{pro.productName},
                #{pro.price},
                #{pro.productDesc}
                )
            </insert>
        </mapper>
```

<insert></insert> 标签中没有 resultType 属性，但<selectKey></selectKey>标签是有的。通过

[⊖] 生成此文件后，记得检查工程的 application.yml 文件，查看 mybatis 属性项是否添加了对此文件的扫描。

设置<selectKey>中 order 的属性值，即 order="AFTER"，使其先执行插入语句，再执行查询语句。keyProperty="pro.id"表示将自增长后的 Id 赋值给实体类中用@Param 注解标注为 'pro' 的类实例字段。如果没有添加@Param 注解指定类实例名，那么<insert>标签中的 pro 名字要去掉。

selectKey 标签属性和含义见表 10-3。

表 10-3 selectKey 标签属性和含义

属性名	作用
resultType	执行 SQL 语句后的返回数据类型
order	执行 SQL 的顺序；AFTER 表示先执行插入语句，之后再执行查询语句
keyProperty	keyProperty 是 Java 对象的属性名

在 DAO 中注入自定义 Mapper，并且添加保存方法：

```
public interface ProductDao {
    int save(Product product);
}

@Repository
public class ProductDaoImpl implements ProductDao{
    @Autowired
    ProductManualMapper productManualMapper;

    @Override
    public int saveProduct(Product product) {
        return productManualMapper.insertProduct(product);
    }
}
```

编写测试代码，具体如下：

```
@Autowired
private ProductDao productDao;
@Test
public void testGetInsertDataId() {
    Product product = new Product();
    product.setPrice(2332);
    product.setProductName("java dev map " + System.currentTimeMillis());
    product.setProductDesc("商品描述 dao save ");
    Integer num = productDao.save(product);
    System.out.println(product);
    System.out.println("insertProductId: " + product.getId());
}
```

运行结果如下：

```
Product [Hash = 1029790510, id=146, productName=java dev map 1526027394815, price=2332, productDesc=商品描述 dao save, productPic=null, serialVersionUID=1]
    insertProductId: 146
```

通过上面的运行结果可以看到程序返回了插入商品的 id。

当然如果使用的是 Maven 的 mybatis-generator-maven-plugin 插件，可以配置 generateKey

属性，让插件在自动生成 Insert Mapper 语句时，生成一条正确的 selectKey 元素。在 generatorConfig.xml 配置 table 标签时添加< generatedKey >元素，具体配置如下：

```xml
<generatorConfiguration>
    ….
    <table tableName="product">
        <property name="useActualColumnNames" value="false" />
        <!-- 数据库表主键 -->
        <generatedKey column="id" sqlStatement="Mysql" identity="true" />
    </table>
</context>
</generatorConfiguration>
```

当执行生成命令 mybatis-generator:generate 后，就会发现 Product 对应的 Mapper 文件已经在<insert>标签内生成了返回主键的<selectKey>标签元素。建议使用此方法，因为手写<insert>毕竟很麻烦，而且容易出错。

10.5　MyBatis 多表查询

MyBatis 自动生成单表的增删改查功能给开发带来了很多的便利，但是有时候需求会涉及多张表的操作，就需要手动编写 mapper 文件以及对应的 DAO 类了，本节实现订单与产品信息关联查询，这里演示一对一的关系，即每个订单中含有一个产品 id，需要连接 product 表和 t_order 表。

（1）定义返回实体 Bean

新建一个 OrderAndProductModel 类，具体如下：

```java
package com.javadevmap.mysqlexample.model;
import java.io.Serializable;
/**
 * 订单与商品 bean
 */
public class OrderAndProductModel implements Serializable{
    private String orderName;
    private Integer id;
    private String productName;
    private Integer price;
    private String productDesc;
    private String productPic;

    //… 省略 get 与 set 方法
    @Override
    public String toString() {
        return "OrderAndProductModel{" +
                "orderName='" + orderName + '\'' +
                ", id=" + id +
                ", productName='" + productName + '\'' +
                ", price=" + price +
                ", productDesc='" + productDesc + '\'' +
                ", productPic='" + productPic + '\'' +
                '}';
```

        }
    }

（2）定义 Mapper 文件

在项目的 ProductManualMapper.xml 文件中，添加内容，定义两个表的查询语句，具体如下：

```xml
<mapper namespace="com.javadevmap.mysqlexample.mapper.ProductManualMapper">
    …
    <resultMap id="BaseResultMap"
            type="com.javadevmap.mysqlexample.model.OrderAndProductModel" >
        <id column="id" property="id" jdbcType="INTEGER" />
        <result column="name" property="orderName" jdbcType="VARCHAR" />
        <result column="product_name" property="productName" jdbcType="VARCHAR" />
        <result column="price" property="price" jdbcType="INTEGER" />
        <result column="product_desc" property="productDesc" jdbcType="VARCHAR" />
        <result column="product_pic" property="productPic" jdbcType="VARCHAR" />
    </resultMap>

    <select id="getOrderProductList" resultMap="BaseResultMap">
        select pro.id, pro.product_name,pro.price,pro.product_desc,ord.name
                from product pro,t_order ord where pro.id=ord.product_id;
    </select>
</mapper>
```

Mapper 文件中定义了与实体 OrderAndProductModel 映射的 resultMap 结果集，同时增加了一个 select 的查询语句。

（3）定义数据操作

在 ProductManualMapper 类中增加一个方法 getOrderProductList()，方法名与 mapper 文件中的 select id 一致，具体如下：

```java
public interface ProductManualMapper {
    …
    List<OrderAndProductModel> getOrderProductList();
}
```

编写测试类 ProductMapperTest，测试定义的 getOrderProductList 方法，具体如下：

```java
@RunWith(SpringRunner.class)
@SpringBootTest
public class ProductMapperTest {
    @Autowired
    private ProductManualMapper productManualMapper;
    List<OrderAndProductModel> orderProductList=null;
    @Test
    public void testSelfMapper() {
        orderProductList = productManualMapper.getOrderProductList();
        System.out.println("result size() = " + orderProductList.size());
        if (orderProductList.size() > 0) {
            for (OrderAndProductModel item : orderProductList) {
                System.out.println(item);
            }
        }
```

                }
            }

运行结果如下：

        result size() = 2
        OrderAndProductModel{orderName='order01', id=108, productName='java dev map version 02 update', price=34, productDesc='null', productPic='null'}
        OrderAndProductModel{orderName='order02', id=109, productName='java dev map version 02', price=233, productDesc='商品描述', productPic='null'}

10.6　查询优化

业务应用的访问性能由多方面因素决定。数据库 MySQL 是业务应用的组成部分，也是决定其性能的重要部分。所以提升 MySQL 的性能至关重要。MySQL 性能的提升可分为三部分，包括硬件、网络、软件。其中硬件、网络取决于公司的硬件设备以及网络带宽。软件部分又分很多种，本节着手于 MySQL 软件层面的优化，从查询优化入手进行性能的提升。

10.6.1　优化查询的方向

首先需要了解 MySQL 服务器状态信息，例如 MySQL 启动后的运行时间，MySQL 的客户端会话连接数，服务器执行的慢查询数，执行了多少 SELECT/UPDATE/DELETE/INSERT 语句等统计信息，以便根据当前 MySQL 服务器的运行状态进行相应的调整或优化。

使用 show status 指令查看 MySQL 服务器的状态信息，如图 10-2 所示。

执行 show status 语句后，MySQL 将会列出多达 340 条的状态信息记录，为了快速实现优化需要特别关注 Slow_queries（慢查询次数）、Com_(CRUD) 操作的次数、Uptime(上线时间)等几个重要信息。

图 10-2　MySQL 状态信息

- 查看查询时间超过 long_query_time 秒的查询的个数：

    show global status like 'slow_queries';

- 查看 select 语句的执行数：

    show global status like 'com_select';

- 查看 insert 语句的执行数：

    show global status like 'com_insert';

- 查看 update 语句的执行数：

    show global status like 'com_update';

- 查看 delete 语句的执行数：

    show global status like 'com_delete';

- 查询当前 MySQL 本次启动后的运行统计时间（秒）：

    show global status like 'uptime';

通过上面的常用 MySQL 指令，能够清楚地了解当前 MySQL 是否存在问题，以及问题分布在哪里。接下来针对出现的问题着手优化。

10.6.2 EXPLAIN 分析

在日常工作中，可以通过开启慢查询记录一些执行时间比较久的 SQL 语句，设置的具体步骤如下：

1）将 slow_query_log 全局变量设置为 "ON" 状态。

    set global slow_query_log='ON';

2）设置慢查询日志存放的位置。

    set global slow_query_log_file='/var/log/mysql/slow.log'⊖

3）设置查询超过 4s 就记录。

    set global long_query_time=4;

通过上面设置，在 log 文件中记录了超过 4s 的 SQL 语句，可以用 explain 命令来查看这些 SQL 语句的执行计划，查看 SQL 语句有没有使用索引，有没有全表扫描。例如查看查询商品语句的执行计划：

    EXPLAIN SELECT * FROM product

执行结果如图 10-3 所示。

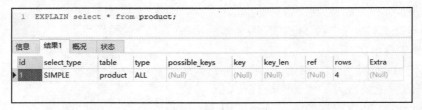

图 10-3　全表扫描 EXPLAIN 执行结果

执行 explain 命令后，展现信息有 10 列，分别是 id、select_type、table、type、possible_keys、key、key_len、ref、rows、Extra，下面对这些字段进行讲解：

- id：SELECT 识别符。这是 SELECT 查询序列号。
- select_type：表示查询中每个 select 子句的类型。
- table：表示查询的表。
- type：表示 MySQL 在表中找到所需行的方式，又称"访问类型"。

常用的类型有：NULL、system、const、eq_ref、ref、range、index、ALL（从左到右，性能从好到差）。

⊖ 设置此值时需要注意 MySQL 的文件操作权限。

1）NULL：MySQL 在优化过程中分解语句，执行时甚至不用访问表或索引，例如从一个索引列里选取最小值可以通过单独索引查找完成。

2）system：表仅有一行，这是 const 类型，一般不会出现，可以忽略不计。

3）const：数据表最多只有一个匹配行，因为只匹配一行数据，所以很快，常用于 PRIMARY KEY 或者 UNIQUE 索引的查询，可以认为 const 是最优化的。如图 10-4 所示。

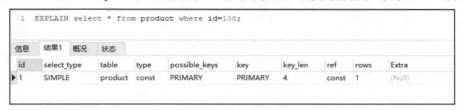

图 10-4　主键查询 EXPLAIN 执行结果

4）eq_ref：使用的索引是唯一索引，对于每个索引键值，表中只有一条记录匹配，简单来说，就是多表连接中使用 primary key 或者 unique key 作为关联条件。这里演示使用方式，添加一个订单表 t_order。表结构如下：

```
CREATE TABLE `t_order` (
`id`    int(10) UNSIGNED NOT NULL AUTO_INCREMENT ,
`name`    varchar(255) NULL DEFAULT NULL ,
`product_id`    int(11) NULL DEFAULT NULL ,
`num`    int(11) NULL DEFAULT NULL ,
`user_id`    int(11) NULL DEFAULT NULL ,
PRIMARY KEY (`id`)
)
ENGINE=InnoDB DEFAULT CHARACTER SET=utf8 COLLATE=utf8_unicode_ci
```

注意 t_order 表的 product_id 与商品表 product 的 id 字段进行关联。如图 10-5 所示。

图 10-5　多表关联查询 EXPLAIN 执行结果

5）ref：查询条件索引既不是 UNIQUE 也不是 PRIMARY KEY 的情况。ref 可用于"="操作符的带索引的列。这里以 product 表的 price 字段进行演示，此字段需建立索引。如图 10-6 所示。

图 10-6　索引关联查询 EXPLAIN 执行结果

6）Range：只检索给定范围的行，使用一个索引来选择行。注意 product 表的 price 字段建立了索引。如图 10-7 所示。

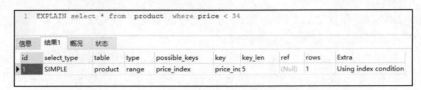

图 10-7　范围查询 EXPLAIN 执行结果

7）Index：Full Index Scan，index 与 ALL 的区别是 index 类型只遍历索引树。

8）ALL：MySQL 将遍历全表以找到匹配的行（性能最差）。

- possible_keys: 指出 MySQL 能使用哪个索引在该表中找到行。如果该列为 NULL，说明没有使用索引，可以对该列创建索引来提高性能。
- key：显示 MySQL 实际决定使用的键（索引）。如果没有选择索引，键是 NULL。可以强制使用索引或者忽略索引。
- key_len：表示索引中使用的字节数，可通过该列计算查询中使用的索引的长度。
- ref：表示上述表的连接匹配条件，即哪些列或常量被用于查找索引列上的值。
- rows：显示 MySQL 认为它执行查询时必须检查的行数。
- Extra：该列包含 MySQL 解决查询的详细信息。

10.6.3　小结

对于数据库的优化要注意以下原则：

- 合理的索引能够加速数据读取效率，不合理的索引会拖慢数据库的响应速度。
- 索引越多，更新数据的速度越慢。
- 当程序和数据库结构或 SQL 语句已经优化到一定的程度，而程序瓶颈并不能顺利解决，则应该考虑使用诸如 Redis 这样的分布式缓存。
- 用 EXPLAIN 来分析 SQL 语句的性能。
- 索引字段上进行运算会使索引失效。尽量避免在 WHERE 子句中对字段进行函数或表达式操作，这将导致引擎放弃使用索引而进行全表扫描。
- 避免使用!=或<>、IS NULL 或 IS NOT NULL、IN、NOT IN 等这样的操作符。因为这会使系统无法使用索引，而只能直接搜索表中的数据。

当然优化的准则远远不止这些，需要多在实际工作中研究探索。

10.7　数据库主从复制原理

MySQL 的主从体系中，多个从服务器采用异步的方式更新主数据库的变化，业务服务器执行写或修改操作是在主服务器上进行的，读操作则在各从服务器上进行。

MySQL 集群之间复制的基础是二进制日志文件（binary log file）。一台 MySQL 数据库一旦启用二进制日志，其作为 master，以"事件"的方式把数据库中相应操作记录到二进制日志中；从数据库 slave 通过一个 I/O 线程与主服务器 master 保持通信，并监控 master 的二进制日志文件的变化，如果发现 master 二进制日志文件发生变化，则会把变化复制到自己的中继日志中，然后 slave 的一个 SQL 线程会把相关的"事件"执行到自己的数据库中，以此实现从数据库和主数据库的一致性，也就实现了主从复制。

第 11 章 MongoDB

随着互联网业务的发展,传统的关系型数据库 RDBMS[⊖](MySQL 等)在一些场合遇到挑战。首先,对数据库存储的容量要求越来越高,单机无法满足需求,很多时候需要用集群来解决问题,而 RDBMS 由于要支持 join、union 等操作,一般不支持分布式集群。其次,在大数据流行的今天,很多数据都"频繁读和增加,不频繁修改",而 RDBMS 对所有操作一视同仁,这就带来了空间浪费以及查询性能问题。另外,互联网业务的不确定性导致数据库的存储模式也需要频繁变更,不自由的存储模式增大了实现的复杂性和扩展的难度。而非关系型数据库 NoSQL[⊖]正好填补了这块空白,MongoDB 正是非关系型数据库的代表产品之一。

11.1 MongoDB 基本介绍和使用场景

MongoDB 用于超大规模数据的存储。例如文章信息、页面缓存、地理位置、用户生成的数据和用户操作日志,如果要对这些数据进行存储,那么关系型数据库相对于 MongoDB 就逊色不少,MongoDB 数据库的使用能很好地处理这些数据。

11.1.1 MongoDB 概述

MongoDB 是用 C++语言编写的,是一个基于分布式文件存储的开源数据库系统。在高负载的情况下,可添加更多的节点,可以保证服务器性能;可为应用提供可扩展的高性能数据存储解决方案;可将数据存储为一个文档,保存为键值对形式。

MongoDB 有以下优势:
- 文档结构的存储方式,获取数据方便快捷。
- 高效存储二进制大对象(例如文件、照片、视频)。
- 内置 GridFS,支持大容量的存储。
- 类似 Json 的存储格式。
- 动态查询,全索引支持,扩展到内部对象和内嵌数组。
- 复制(复制集)和支持自动故障恢复。
- MapReduce 支持复杂聚合。

MongoDB 也有不足:
- 不支持事务操作。
- MongoDB 占用空间很大。
- 无法进行关联表查询,不适用于关系多的数据。

⊖ RDBMS 即关系数据库管理系统(Relational Database Management System),是将数据组织为相关的行和列的系统,而管理关系数据库的计算机软件就是关系数据库管理系统。

⊖ NoSQL,指的是非关系型的数据库。NoSQL 有时也称作 Not Only SQL 的缩写,是对不同于传统的关系型数据库的数据库管理系统的统称。

- 复杂聚合操作通过 MapReduce 创建，速度慢。

11.1.2 MongoDB 使用场景

基于 MongoDB 的特性，这里列举 MongoDB 的使用场景：

（1）日志/内容/图片/视频等业务

MongoDB 更侧重数据写入性能，而非事务安全，MongoDB 很适合业务系统中有大量"低价值"数据的场景。但是应当避免在需要事务安全性的业务中使用 MongoDB，除非能在架构设计上保证事务安全。

（2）高可用性业务

MongoDB 的复制集（Master-Slave）配置方便，可快速处理单节点故障，自动、安全地完成故障转移。

（3）业务数据量很大

关系型数据库的弱点是完成数据的扩展较为困难，例如 MySQL，需要通过数据库和表的拆分完成扩展。而 MongoDB 内建了多种数据分片的特性，可以很好地适应大数据量的需求。

（4）地理坐标数据查询

MongoDB 支持二维空间索引，因此可以快速、精确地从指定位置获取数据。

（5）存储不同结构数据

MongoDB 是文档型数据库，为非结构化文档数据增加一个新字段操作简单，并且不会影响到已有数据。另外当业务数据发生变化时，不需要修改表结构。MongoDB 可以使用非标准的关系型思想（结构化）来处理数据，也可以把数据直接序列化成 Json 存储到 MongoDB 中。

11.2 MongoDB 基本操作

MongoDB 的安装相对简单，访问官网 https://www.mongodb.com，然后下载合适的版本即可。例如 Windows 系统，只需在 MongoDB 官网下载 msi 文件，双击运行即可。

本书的 MongoDB 软件环境采用 Docker 容器部署。具体部署命令见第 19 章。

11.2.1 MongoDB 基本命令

在 MongoDB 中默认数据库是 test。如果没有创建过任何数据库，则集合/文档将存储在 test 数据库中。

创建数据库：

```
> use db_name
```

要检查当前选择的数据库，使用命令 db：

```
> db
```

检查数据库列表：

```
> show dbs
```

为单个数据库添加管理用户，例如给 admin 数据库创建一个用户名为 root、密码为 root 的用户。

```
> use admin
```

第 11 章 MongoDB

> db.createUser({user:"root",pwd:"root",roles:[{role:'root',db:'admin'}]})

创建完 root 用户后，再次操作 admin 数据库时，需要进行身份认证，具体如下：

> mongo 宿主机 ip/数据库名 -u 用户名 -p 密码

例如用 root 用户登录本机 admin 数据库：

> mongo localhost/admin –uroot –proot

下面介绍 MongoDB 集合操作。集合的创建方式分两种：隐式创建集合和显式创建集合。

- 隐式创建集合

当向集合中插入文档时，如果集合不存在，系统会自动创建，所以向一个不存在的集合中插入数据也就是创建了集合。

```
> db
test
> show tables
> db.products.insert({"name": "javadevmap", "level": 6})
WriteResult({ "nInserted" : 1 })
> show tables
products
```

- 显式创建集合

db.createCollection("集合名"，配置参数)

显示创建集合可以通过一些配置参数创建一些特殊的集合，如固定集合。

```
> db.createCollection("orders")
{ "ok" : 1 }
```

删除集合，格式为：db.集合名字.drop()。

```
> db.orders.drop()
true
```

集合写入数据。

```
> var product = {"name": "java dev map","price":199}
> db.products.insert(product)
WriteResult({ "nInserted" : 1 })
```

到此，本节已经介绍了常用的 MongoDB 命令，如果想了解更多 MongoDB 命令，可以登录 MongoDB 官网进行学习。

11.2.2 MongoDB 图形化工具

MongoDB 自带的 Shell 是一个很好的工具，但是它在操纵大数据集时就没那么直观了。使用 MongoDB 客户端管理工具，可以大大提高 MongoDB 应用的开发效率。这里介绍一个简单的 MongoDB 可视化查看工具 Robomongo[⊖]。只要通过如图 11-1 所示页面，正确配置 MongoDB 的地址和密码，即可登录 MongoDB，然后查看 MongoDB 中的数据，如图 11-2 所示。

⊖ Robomongo 是一个基于 Shell 的跨平台开源 MongoDB 可视化管理工具。官网是 https://robomongo.org/。

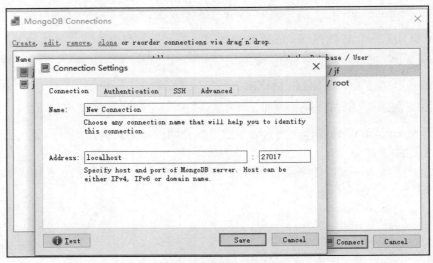

图 11-1　Robomongo 配置

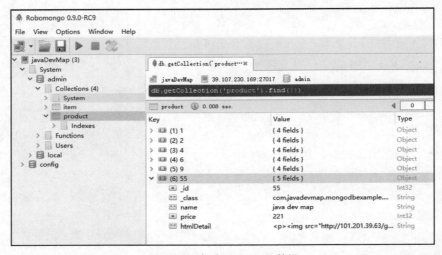

图 11-2　查看 MongoDB 数据

因为这是个可视化工具，所以使用起来非常简单，使用此工具基本可以完成对 MongoDB 所支持的增删改查等任何操作。此工具的具体使用这里就不再演示，只要熟悉使用方法即可。

11.3　SpringBoot 集成 MongoDB

创建一个 Spring Boot 工程，工程名使用 MongodbExample，具体创建方法见第 7 章。本节使用 Spring Boot 工程整合 MongoDB 进行数据操作。

11.3.1　整合 MongoDB

（1）添加依赖

为了整合 MongoDB，需要添加起步依赖 spring-boot-starter-data-mongodb；后面的章节需要使用页面进行演示，所以还要添加起步依赖 spring-boot-starter-thymeleaf，具体如下：

```xml
<dependencies>
    <dependency>
        <groupId>org.springframework.boot</groupId>
        <artifactId>spring-boot-starter-data-mongodb</artifactId>
    </dependency>
    <dependency>
        <groupId>org.springframework.boot</groupId>
        <artifactId>spring-boot-starter-thymeleaf</artifactId>
    </dependency>
</dependencies>
```

（2）修改配置文件

在配置文件 application.yml 中添加如下配置，使用 root 用户操作数据库 admin：

```yml
spring:
  application:
    name: mongodb-example
  data:
    mongodb:
      host: 39.107.230.169
      port: 27017
      database: admin
      username: root
      password: root
```

只要以上两步，即可完成 MongoDB 的 Spring Boot 工程整合。

11.3.2 操作数据

Mongodb 实现商品的增删改查功能，首先要定义商品的数据结构。创建包 com.javadevmap.mongodbexample.model，然后新建 Product 类，具体如下：

```java
package com.javadevmap.mongodbexample.model;
import org.springframework.data.annotation.Id;
public class Product {
    @Id
    private Integer id;
    private String name;
    private int price;
    /** 产品页面对应的商品详情 */
    private String htmlDetail;

    public Product() {
    }
    public Product(Integer id, String name, int price) {
        this.id = id;
        this.name = name;
        this.price = price;
    }
    //...省略 get 与 set 方法
    @Override
    public String toString() {
        return "Product{" +
                "id=" + id +
```

```
                ", name='" + name + "\"" +
                ", price=" + price +
                '}';
    }
}
```

注意类中的 id 字段上面，增加了一个 @Id 注解，让 id 字段作为 Product 文档的唯一标识，MongoDB 常用注解见表 11-1。

表 11-1 MongoDB 常用注解

标签名	作 用
@Id	文档的唯一标识，在 MongoDB 中为 ObjectId，它是唯一的
@Document	把一个 Java 类声明为 MongoDB 的文档，可以通过 collection 参数指定这个类对应的文档
@Indexed	声明该字段需要索引，建索引可提高查询效率
@CompoundIndex	复合索引的声明，建复合索引可以有效地提高多字段的查询效率
@GeoSpatialIndexed	声明该字段为地理信息的索引
@Transient	映射忽略的字段，该字段不会保存到 MongoDB

创建完上面的实体类 Product，接下来创建一个操作类 ProductMongoRepository 来进行 MongoDB 的操作，此类提供了针对 MongoDB 的增删改查功能，只要继承 MongoRepository，就继承了一些默认的方法，主要包含如下几类：delete、find、save、count、exists。此类实现具体如下：

```
@Component
public interface ProductMongoRepository extends MongoRepository<Product, Integer> {
}
```

类中不需要实现任何方法，当前 ProductMongoRepository 已经具备了基本的增删改查能力了。编写一个测试类进行验证：

```
@RunWith(SpringRunner.class)
@SpringBootTest
public class ProductMapperTest {
    @Autowired
    ProductMongoRepository productMongoRepository;

    @Test
    public void testMongoOrigin() {
        productMongoRepository.deleteAll();
        productMongoRepository.insert(new Product(101, "product01", 18));
        System.out.println("mongo data find by id");
        System.out.println(productMongoRepository.findOne(101));
        productMongoRepository.delete(101);
    }
}
```

运行结果如下：

```
mongo data find by id
Product{id=101, name='product01', price=18}
```

当这些默认方法不能覆盖所有业务的数据操作需求时，就需要扩展方法。扩展使用方法时，就好像用英文直译要做的事情一样来定义方法名，例如想通过 id 来查找某条数据，只要定义一个方法 findById，这样就具备了通过 id 查找的能力。下面在 ProductMongoRepository 中定义几个方法来扩充基础的数据库操作能力。

```
@Component
public interface ProductMongoRepository extends MongoRepository<Product, Integer> {
    Product findByName(String name);
    List<Product> findByPrice(int id);
    List<Product> findByPriceLessThan(int price);
    Product findOneByPrice(Integer price);
    Product findOneByPriceAndName(Integer price,String name);
    List<Product> findByPrice(Integer price,Pageable page);
}
```

如上面所写，只要把想做的事按照 JPA 的规则命名一个方法，即完成了 MongoDB 操作语句的建立，使用以上规则（参见 7.5.2 节），就可以自由组装接口方法来扩展数据库操作能力。

11.3.3 缓存商品详情页面功能

本节实现用 MongoDB 存储商品详情 html 页面的功能。将部分页面信息存储到 MongoDB 中，当用户访问某个商品详情时，根据商品 id 从 MongoDB 中取出，返回给前台。

（1）准备数据

准备商品的详情页面主要内容作为要缓存的数据，具体如下：

```
<p>< img src="http://t.cn/R3cquJp" alt=""/>< img src="http://t.cn/R3c5hfe" alt=""/></p>
```

把此内容通过 test 类保存到 Mongodb 中。

```
@RunWith(SpringRunner.class)
@SpringBootTest
public class ProductMapperTest {
    @Autowired
    ProductMongoRepository productMongoRepository;
    @Test
    public void addProductOne(){
        productMongoRepository.delete(55);
        Product product = new Product();
        product.setId(55);
        product.setName("java dev map");
        product.setPrice(221);
        product.setHtmlDetail("<p>< img src="http://t.cn/R3cquJp" alt=""/>< img src="http://t.cn/R3c5hfe" alt=""/></p>");
        Product result = productMongoRepository.insert(product);
        System.out.println(result);
    }
}
```

执行以上程序，页面数据已经保存到 MongoDB 中。当然实际业务中，上面的商品详情代码需后台管理人员编辑商品时进行生成。

（2）准备展示页面

接下来定义一个展示页面，在 resources 文件夹下新建一个 template 文件夹，新建页面

detail.html 文件，此页面用来展示从 MongoDB 中获取的商品详情代码，具体如下：

```html
<html xmlns:th="http://www.thymeleaf.org">
<head>
    <meta charset="UTF-8"/>
    <title>商品详情页面</title>
    <link rel="stylesheet" href="/product.css"/>
</head>
<body>
<div class="switchable-panel">
    <div>
        <label style="font-size:48px;display:block;width:100%;text-align:center"
                          th:text="${tips}"></label>
    </div>
    <div></div>
    <div class="panel-intro" th:utext="${desc}"></div>
</div>
</body>
</html>
```

detail 文件中使用了一个 CSS 样式文件 product.css，此文件为静态资源文件，和图片等静态资源一起存放在/resources/static 文件夹下，主要是为了美化页面样式，这里就不再展示代码。

detail 展示页面需要将一段 html 代码放到页面中，需要使用 thymeleaf 进行非转义文本处理，即使用 th:utext 标签进行处理。

（3）处理 HTTP 请求

创建包 com.javadevmap.mongodbexample.controllers，定义 ProductDetailController 类，将 MongoDB 中的数据与页面关联起来，具体如下：

```java
@Controller
@RequestMapping(value="/product")
public class ProductDetailController {
    @Autowired
    private ProductMongoRepository productRepository;

    @RequestMapping(value="/{productId}/detail")
    public ModelAndView getProductDesc(@PathVariable("productId") Integer productId) {

        Product product = productRepository.findOne(productId);
        ModelAndView modelAndView = new ModelAndView("detail");
        if(null == product){
            modelAndView.addObject("tips",String.format("商品 id 为 %s 不存在",productId));
        }else{
            modelAndView.addObject("tips",String.format("商品%s 的商品详情页面",product.getName()));
            modelAndView.addObject("desc",product.getHtmlDetail());
        }
        return modelAndView;
    }
}
```

在 ProductDetailController 中定义一个接口方法，用于接收前端传递的 id，然后去 MongoDB 中查询商品，如果有该商品，就使用 detail 页面进行展示。

启动服务后访问http://localhost:18082/product/55/detail，运行效果如图 11-3所示。

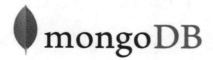

图 11-3　商品详情页面

对于一个电商平台来讲，可以使用如上方法把商品详情保存在 MongoDB 中，并且根据请求返回页面信息。

第 12 章　Redis

在实际业务中，如果仅使用关系型数据库，当然可以对数据进行增删改查等操作，但随着业务量的上升，就会遇到数据库的性能瓶颈，虽然可以通过优化数据库或者使用读写分离等技术使数据库的承载能力提高，但还有一种可以显著提高业务性能的方式：添加缓存。

缓存凭借着超强的数据读写能力，能够承担非常大的业务请求压力，并且一些不适合存入数据库的数据放入缓存中也是一种不错的选择。缓存分为本地缓存和分布式缓存，本章仅介绍分布式缓存 Redis。

Redis 支持数据的持久化，可以在重启 Redis 时把持久化的数据再加载进缓存；Redis 提供了 5 种数据格式的存储，分别是 String、List、Hash、Set、ZSet；Redis 支持数据的备份，也就是可以建立多节点。单台 Redis 在不考虑网络的情况下，可以达到每秒 10 万次左右的读能力。所以对于一名研发者来讲，使用好 Redis 会让工作变得简单且流畅。

12.1　基本的 Redis 操作

本章使用 Spring Boot 工程添加 Redis 组件来操作 Redis 缓存。新建一个工程 RedisExample，然后进行如下操作。

（1）添加组件依赖

```
<dependency>
    <groupId>org.springframework.boot</groupId>
    <artifactId>spring-boot-starter-data-redis</artifactId>
</dependency>
```

（2）添加 Redis 的配置信息至 yml 文件

```
server:
  port: 18081
spring:
  application:
    name: redis-example
  redis:
    database: 0
    host: 39.106.10.196
    port: 6379
    password: mypass
    pool:
      max-active: 1000
      max-wait: 1000
      max-idle: 300
      min-idle: 100
    timeout: 1200
```

（3）添加 Redis 基本操作

在工程内,新建 RedisDao 接口和它的实现类,本节先简单地向 Redis 中添加一个 String 类型的数据。

```java
public interface RedisDao {
    public void testRedisString();
}

@Repository
public class RedisDaoImpl implements RedisDao{
    @Autowired
    private RedisTemplate<String, String> redisTemplate;

    @Override
    public void testRedisString() {
        redisTemplate.opsForValue().set("String:stringredis", "string redis value");
        String redisString = redisTemplate.opsForValue().get("String:stringredis");
        System.out.println(redisString);
    }
}
```

上面的代码中,使用 RedisTemplate 对 Redis 进行操作,对于 Redis 中的 String 类型,使用 redisTemplate.opsForValue()作为数据操作的方式。由于 Redis 主要是通过键值对进行存储的,所以 set 方法把"String:stringredis"作为 key,"string redis value"作为 value 存入 Redis 中。在 key 字符串中的":"起到了命名空间的作用,可以标识 Redis 中某一类数据的分组,并且在查看工具中会根据命名空间自动分类聚合。

(4)使用测试类测试效果

在 test 类中,添加如下测试代码,检验 Redis 的使用情况。

```java
@RunWith(SpringRunner.class)
@SpringBootTest
public class RedisExampleApplicationTests {
    @Autowired
    private RedisDao redisDao;

    @Test
    public void testRedisString() {
        redisDao.testRedisString();
    }
}
```

运行结果如下:

```
string redis value
```

在测试类中注入 RedisDao 类的实例,然后调用 Dao 中的方法,由输出可见可以获取 Redis 中的数据。

12.2 Redis 常用命令和可视化工具

上一节通过代码把一个 String 类型的数据存入 Redis 中。那么如何查看 Redis 中的数据呢?这里介绍两种办法。一种是直接登录 Redis 缓存,通过命令查看 Redis 数据;另一种是通过可视化工具查看 Redis 中的数据。

12.2.1 Redis 命令

Redis 的命令较多，本节简单介绍几个常用命令，如图 12-1 所示。

图中，使用 redis-cli 命令连接到了 Redis；通过 AUTH 命令输入密码登录；通过 keys 命令查看符合正则规则的 key 列表，本例匹配了全部的 key；通过 del 命令删除某个 key 值。Redis 的命令还有很多，可以通过 Redis 命令直接操作 Redis，实现各种类型的数据操作，但是对于研发者来讲，这些命令了解即可，使用可视化的操作工具会更加方便。

12.2.2 可视化工具

本节介绍一个简单的 Redis 可视化查看工具 Redis Desktop Manager。只要通过如图 12-2 所示界面，正确配置 Redis 的地址和密码，即可登录到 Redis 上。然后就可以查看 Redis 中的数据，如图 12-3 所示。

图 12-1 Redis 命令　　　　图 12-2 Redis Desktop Manager 配置

图 12-3 Redis Desktop Manager 操作数据

因为这是个可视化工具，所以使用起来非常简单，使用此工具基本可以完成对 Redis 所支持的增删改查等任何操作，并且还可以设置数据展示的格式。此工具的具体使用这里就不再演示，大家自行熟悉使用方法即可。

12.3 Redis 的五种数据格式的操作

Redis 支持五种数据类型的操作，所以可以使用 Redis 做很多事情，例如简单的信息放入 String[一]类型中；自带键值对类型的数据放入 Hash 中；队列可以放入 List 中；一些集合数据可以放入 Set 和 ZSet 中。五种类型的数据操作方式见表 12-1。

表 12-1 Redis 操作数据类型

类　　型	数据操作方式
String	redisTemplate.opsForValue()
List	redisTemplate.opsForList()
Hash	redisTemplate.opsForHash()
Set	redisTemplate.opsForSet()
ZSet	redisTemplate.opsForZSet()

12.3.1 String 操作

本节使用的方法如下。读者应首先了解各个方法的作用，然后阅读代码和输出，了解 Redis 是如何操作 String 类型数据的。

- set：向 Redis 中设置键和值，重载方法中包含一个带有过期时间的方法，可以设置超时删除。
- get：获取 Redis 中的键和值。
- getAndSet：获取旧的值，并且设置新的值。
- size：获取字符串长度。
- append：如果 Redis 中不包含此键，则添加键和值，如果已经包含此键，则在值的后面追加数据。
- setIfAbsent：如果没有此键，则向 Redis 中添加键和值。
- multiSet：同时添加多个 String 的键和值。
- multiGet：同时获取多个键的值。
- increment：对某一值进行数值操作。

```
@Override
public void testRedisString() {
    redisTemplate.opsForValue().set("String:stringredis", "string redis value");
    String redisString = redisTemplate.opsForValue().get("String:stringredis");
    System.out.println(redisString);
    redisString = redisTemplate.opsForValue().getAndSet("String:stringredis", "string redis value 2");
    System.out.println(redisString);
    redisString = redisTemplate.opsForValue().get("String:stringredis");
    System.out.println(redisString);
    Long size = redisTemplate.opsForValue().size("String:stringredis");
    System.out.println("size is " + size);

    redisTemplate.opsForValue().append("String:stringappend", "stringappend%");
```

[一] Java 对象的保存也可以转成 Json 格式后，存入 String 类型中。

```java
            redisString = redisTemplate.opsForValue().get("String:stringappend");
            System.out.println(redisString);
            redisTemplate.opsForValue().append("String:stringappend", "stringappend");
            redisString = redisTemplate.opsForValue().get("String:stringappend");
            System.out.println(redisString);

            redisTemplate.opsForValue().set("String:stringredistimeout", "stringredis timeout 10 seconds",10, TimeUnit.SECONDS);
            redisString = redisTemplate.opsForValue().get("String:stringredistimeout");
            System.out.println(redisString);

            boolean settype = redisTemplate.opsForValue().setIfAbsent("String:stringredisifabsent", "string redis ifabsent");
            System.out.println("set type is " + settype);
            settype = redisTemplate.opsForValue().setIfAbsent("String:stringredis", "string redis ifabsent");
            System.out.println("set type is " + settype);
            redisTemplate.delete("String:stringredis");
            settype = redisTemplate.opsForValue().setIfAbsent("String:stringredis", "string redis ifabsent");
            System.out.println("set type is " + settype);

            Map<String, String> map = new HashMap<String, String>();
            map.put("String:strings1", "strings1");
            map.put("String:strings2", "strings2");
            map.put("String:strings3", "strings3");
            redisTemplate.opsForValue().multiSet(map);
            List<String> keys = new ArrayList<String>();
            keys.add("String:strings1");
            keys.add("String:strings2");
            keys.add("String:strings3");
            List<String> values = redisTemplate.opsForValue().multiGet(keys);
            System.out.println(values.toString());

            redisTemplate.opsForValue().increment("String:num", 1);
            redisString = redisTemplate.opsForValue().get("String:num");
            System.out.println("num is " + redisString);
            redisTemplate.opsForValue().increment("String:num", 1);
            redisString = redisTemplate.opsForValue().get("String:num");
            System.out.println("num is " + redisString);

            Set<String> stringkeys = redisTemplate.keys("String*");
            redisTemplate.delete(stringkeys);
    }
```

运行结果如下：

```
string redis value
string redis value
string redis value 2
size is 20
stringappend%
stringappend%stringappend
stringredis timeout 10 seconds
set type is true
```

```
set type is false
set type is true
[strings1, strings2, strings3]
num is 1
num is 2
```

上面的代码并不复杂，没有包含业务逻辑，仅仅是 Redis 操作的 API 效果演示。其中注意几点：getAndSet 方法取出来的是旧值；append 方法在没有键时，效果和 set 相同；TimeUnit 的超时类型有很多种，可以根据实际需要选择；setIfAbsent 在有键时会设置失败；increment 方法在没有键和值时默认从 0 开始；最后，通过 redisTemplate.delete()方法删除了所有的符合正则规则的键和值。

12.3.2 List 操作

对于 List 类型的操作，可以直接把 Redis 中的键设想为 List 的名字，Redis 中的值就是一个 List，可以像普通 List 一样进行 Push、Pop 等操作，List 操作的具体方法如下：

- rightPushAll：从 List 的末尾，插入全部的 List 参数中的内容。
- range：获取 List 某区间内的内容。
- leftpush：从 List 的前端插入，重载方法中包含从 List 的某个值的前端插入。
- size：获取 List 的数据个数。
- leftPop：从前端弹出。
- rightPopAndLeftPush：从一个 List 的后端弹出，并插入一个 List 的前端。
- set：修改 List 中的某个值。
- trim：截取 List。

代码如下：

```
@Override
public void testRedisList() {
    List<String> list = new ArrayList<String>();
    list.add("listString1");
    list.add("listString2");
    list.add("listString3");
    list.add("listString4");
    redisTemplate.opsForList().rightPushAll("List:list", list);
    List<String> retlist = redisTemplate.opsForList().range("List:list", 0, 2);
    System.out.println(retlist.toString());
    retlist = redisTemplate.opsForList().range("List:list", 0, -1);
    System.out.println(retlist.toString());

    redisTemplate.opsForList().leftPush("List:list", "listString0");
    redisTemplate.opsForList().rightPushAll("List:list", "listString5","listString6","listString8");
    redisTemplate.opsForList().leftPush("List:list", "listString8", "listString7");
    retlist = redisTemplate.opsForList().range("List:list", 0, -1);
    System.out.println(retlist.toString());
    long size = redisTemplate.opsForList().size("List:list");
    System.out.println("size is " + size);

    redisTemplate.opsForList().leftPop("List:list");
    redisTemplate.opsForList().rightPopAndLeftPush("List:list", "List:list");
```

```
            retlist = redisTemplate.opsForList().range("List:list", 0, -1);
            System.out.println(retlist.toString());

            redisTemplate.opsForList().set("List:list", 0, "listString0");
            retlist = redisTemplate.opsForList().range("List:list", 0, -1);
            System.out.println(retlist.toString());
            redisTemplate.opsForList().trim("List:list", 1, 3);
            retlist = redisTemplate.opsForList().range("List:list", 0, -1);
            System.out.println(retlist.toString());

            Set<String> listkeys = redisTemplate.keys("List*");
            redisTemplate.delete(listkeys);
        }
```

运行结果如下：

```
[listString1, listString2, listString3]
[listString1, listString2, listString3, listString4]
[listString0, listString1, listString2, listString3, listString4, listString5, listString6, listString7, listString8]
size is 9
[listString8, listString1, listString2, listString3, listString4, listString5, listString6, listString7]
[listString0, listString1, listString2, listString3, listString4, listString5, listString6, listString7]
[listString1, listString2, listString3]
```

上面的代码中，先通过 rightPushAll 插入一个 List 数据；用 range 节选输出后，在 List 前端和后端分别向 List 中添加数据，并且还在特定位置添加了一个数据；之后做了前端弹出的操作，然后把最后一个数据放到 List 的前面；用 set 方法修改 List 中的值；用 trim 方法对 List 进行节选。

12.3.3　Hash 操作

Hash 类型的 Redis 比较特殊，因为此类型的 Redis 值中保存的还是一组键值对。首先看看 Hash 类型的常用方法。

- putAll：把 hashmap 全部加入 Redis 某值中。
- size：获取 Redis 值中的 hashmap 的键值对个数。
- keys：获取 Redis 值中的 hashmap 中的键的集合。
- values：获取 Redis 值中的 hashmap 中的值的列表。
- entries：获取 Redis 值中的 hashmap。
- hasKey：判断 Redis 值中的 hashmap 是否包含此 key。
- get：获取 Redis 值中的 hashmap 某个键对应的值数据。
- delete：删除 Redis 值中的 hashmap 的某个键值对。
- putIfAbsent：如果 Redis 值中的 Hashmap 不包含此键值对，则添加。
- put：向 Redis 值中的 hashmap 添加键值对。

代码如下：

```
        @Override
        public void testRedisHash() {
            Map<String, String> map = new HashMap<>();
            map.put("key1", "hash value 1");
            map.put("key2", "hash value 2");
```

```java
        map.put("key3", "hash value 3");
        map.put("key4", "hash value 4");
        redisTemplate.opsForHash().putAll("Hash:map", map);
        System.out.println("size is " + redisTemplate.opsForHash().size("Hash:map"));
        Set<Object> keys = redisTemplate.opsForHash().keys("Hash:map");
        System.out.println(keys.toString());
        List<Object> values = redisTemplate.opsForHash().values("Hash:map");
        System.out.println(values.toString());
        Map<Object, Object> retmap = redisTemplate.opsForHash().entries("Hash:map");
        System.out.println(retmap.toString());

        System.out.println("has key4 = " + redisTemplate.opsForHash().hasKey("Hash:map", "key4"));
        String value = (String) redisTemplate.opsForHash().get("Hash:map", "key4");
        System.out.println("key4 value is " + value);
        redisTemplate.opsForHash().delete("Hash:map", "key4");
        retmap = redisTemplate.opsForHash().entries("Hash:map");
        System.out.println(retmap.toString());
        redisTemplate.opsForHash().putIfAbsent("Hash:map", "key4", "hash value 4");
        redisTemplate.opsForHash().put("Hash:map", "key5", "hash value 5");
        retmap = redisTemplate.opsForHash().entries("Hash:map");
        System.out.println(retmap.toString());

        Set<String> mapkeys = redisTemplate.keys("Hash*");
        redisTemplate.delete(mapkeys);
    }
```

运行结果如下：

```
size is 4
[key1, key2, key3, key4]
[hash value 1, hash value 2, hash value 3, hash value 4]
{key4=hash value 4, key1=hash value 1, key3=hash value 3, key2=hash value 2}
has key4 = true
key4 value is hash value 4
{key3=hash value 3, key2=hash value 2, key1=hash value 1}
{key2=hash value 2, key4=hash value 4, key5=hash value 5, key3=hash value 3, key1=hash value 1}
```

上面的代码较为简单，仅按照代码的顺序和输出对应阅读即可理解。Redis 对 Hash 类型的支持让 Redis 的使用场景更广了，但是在工作中一定要确定必须使用此种类型的情况下再使用，毕竟 Redis 是键值对模式，Hash 又在 Redis 的值中保存了键值对，这种方式如果不能熟练使用的话，是很容易用错的。

12.3.4 Set 操作

使用 Redis 的 Set 可以保存集合数据，集合数据的唯一性在 Redis 中也是支持的。如果理解集合的话，那么使用 Redis 的 Set 也非常简单，只要熟悉其写法即可。下面列出 Redis 的 Set 常用方法：

- members：获取集合中的元素。
- add：向集合中添加数据。
- remove：移除集合中的数据。
- size：获取集合中元素的个数。
- isMember：判断集合中是否包含某数据。

- intersect:集合间求交集。
- union:集合间求并集。
- unionAndStore:集合间求并集,并且保存到另一个集合。
- difference:计算两个集合的差集。
- randomMember:获取集合中的随机元素。

代码如下:

```java
@Override
public void testRedisSet() {
    redisTemplate.opsForSet().add("Set:set1", "set value 1","set value 2","set value 3");
    Set<String> set = redisTemplate.opsForSet().members("Set:set1");
    System.out.println(set.toString());
    redisTemplate.opsForSet().add("Set:set1", "set value 1");
    redisTemplate.opsForSet().remove("Set:set1", "set value 3");
    set = redisTemplate.opsForSet().members("Set:set1");
    System.out.println(set.toString());

    redisTemplate.opsForSet().add("Set:set2", "set value 2","set value 3","set value 4");
    System.out.println("set 2 size is " + redisTemplate.opsForSet().size("Set:set2"));
    System.out.println("set value 4 is member " +
                    redisTemplate.opsForSet().isMember("Set:set2", "set value 4"));

    set = redisTemplate.opsForSet().intersect("Set:set1", "Set:set2");
    System.out.println(set.toString());

    set = redisTemplate.opsForSet().union("Set:set1", "Set:set2");
    System.out.println(set.toString());

    redisTemplate.opsForSet().unionAndStore("Set:set1", "Set:set2","Set:set3");
    set = redisTemplate.opsForSet().members("Set:set3");
    System.out.println(set.toString());

    set = redisTemplate.opsForSet().difference("Set:set1", "Set:set2");
    System.out.println(set.toString());

    for(int i = 0; i<3; i++) {
        String member = redisTemplate.opsForSet().randomMember("Set:set2");
        System.out.println("random member is " + member);
    }

    Set<String> setkeys = redisTemplate.keys("Set*");
    redisTemplate.delete(setkeys);
}
```

运行结果如下:

```
[set value 2, set value 3, set value 1]
[set value 2, set value 1]
set 2 size is 3
set value 4 is member true
[set value 2]
[set value 2, set value 3, set value 1, set value 4]
[set value 2, set value 3, set value 1, set value 4]
```

```
        [set value 1]
        random member is set value 3
        random member is set value 3
        random member is set value 4
```

阅读上面的代码，可以发现，重复向集合中添加数据是无用的。这里使用了两个 Redis 集合，并且通过这两个集合进行交集、并集、差集的操作，并集操作通过 unionAndStore 方法保存到了另一个集合，交集等操作也有类似方法，这里就不再演示。

12.3.5 ZSet 操作

ZSet 和 Set 的区别在于 ZSet 是有序的，ZSet 中的数据通过一个排序值进行排序，并且可以通过排序值来获取 ZSet 中的元素。ZSet 主要方法如下：

- add：向集合中添加数据。
- scan：获取游标。
- range：获取某区间的数据。
- incrementScore：对集合中的排序值进行操作。
- rank：获取集合中某个值的位置。
- rangeByScore：根据排序值获取集合中的数据。
- count：获取某个区间的集合数据的个数。
- removeRangeByScore：根据排序值移除集合中的数据。

代码如下：

```java
        @Override
        public void testRedisZSet() {
            redisTemplate.opsForZSet().add("ZSet:set1", "set value 1", 1.0);
            redisTemplate.opsForZSet().add("ZSet:set1", "set value 2", 2.0);
            redisTemplate.opsForZSet().add("ZSet:set1", "set value 4", 4.0);
            Cursor<TypedTuple<String>> cursor = redisTemplate.opsForZSet().scan("ZSet:set1", ScanOptions.NONE);
            while (cursor.hasNext()) {
                TypedTuple<String> item = cursor.next();
                System.out.println(item.getValue());
            }
            redisTemplate.opsForZSet().add("ZSet:set1", "set value 3", 3.0);
            Set<String> retSet = redisTemplate.opsForZSet().range("ZSet:set1", 0, -1);
            System.out.println(retSet.toString());

            redisTemplate.opsForZSet().incrementScore("ZSet:set1", "set value 1", 1.0);
            redisTemplate.opsForZSet().incrementScore("ZSet:set1", "set value 2", -1.0);
            retSet = redisTemplate.opsForZSet().range("ZSet:set1", 0, -1);
            System.out.println(retSet.toString());

            Long index = redisTemplate.opsForZSet().rank("ZSet:set1", "set value 1");
            System.out.println("set value 1 index is " + index);

            retSet = redisTemplate.opsForZSet().rangeByScore("ZSet:set1", 2.0, 4.0);
            System.out.println(retSet.toString());

            Long count = redisTemplate.opsForZSet().count("ZSet:set1", 2.0, 4.0);
```

```
            System.out.println("count is " + count);

            redisTemplate.opsForZSet().removeRangeByScore("ZSet:set1", 3.0, 4.0);
            retSet = redisTemplate.opsForZSet().range("ZSet:set1", 0, -1);
            System.out.println(retSet.toString());

            Set<String> zsetkeys = redisTemplate.keys("ZSet*");
            redisTemplate.delete(zsetkeys);
    }
```

运行结果如下：

```
set value 1
set value 2
set value 4
[set value 1, set value 2, set value 3, set value 4]
[set value 2, set value 1, set value 3, set value 4]
set value 1 index is 1
[set value 1, set value 3, set value 4]
count is 3
[set value 2, set value 1]
```

上面的代码中，通过游标的方式遍历集合和通过区间的方式获取集合，都能得到一组有序的数据；incrementScore 方法可以对排序值进行增减，增减后集合自动排序；rank 方法获取某个值在集合中的位置，默认顺序从小到大，返回 0 是表示第一个数据；注意 rangeByScore 的取值区间是封闭的。

12.4 Redis 事务处理

当要一次批量地对 Redis 进行很多操作的时候，或者在操作 Redis 某个值的时候不希望其他程序对这个值进行改变，就需要用到 Redis 的事务处理。

12.4.1 批量操作

请先阅读下面的代码，然后根据代码输出的耗时时间思考 Redis 的性能问题。

```
    @Override
    public void testRedisMulti() {
        long starttime = System.currentTimeMillis();
        for(int i=0;i<1000;i++) {
            redisTemplate.opsForValue().set("String:Strings A:" + i, "strings " + i);
        }
        long endtime = System.currentTimeMillis();
        System.out.println("duration is " + (endtime-starttime));

        starttime = System.currentTimeMillis();
        Map<String, String> map = new HashMap<String, String>();
        for(int i=0;i<1000;i++) {
            map.put("String:Strings B:" + i, "strings " + i);
        }
        redisTemplate.opsForValue().multiSet(map);
```

```
            endtime = System.currentTimeMillis();
            System.out.println("duration is " + (endtime-starttime));

            starttime = System.currentTimeMillis();
            redisTemplate.setEnableTransactionSupport(true);
            redisTemplate.multi();
            for(int i=0;i<1000;i++) {
                redisTemplate.opsForValue().set("String:Strings C:" + i, "strings " + i);
            }
            redisTemplate.exec();
            endtime = System.currentTimeMillis();
            System.out.println("duration is " + (endtime-starttime));

            Set<String> stringkeys = redisTemplate.keys("String*");
            redisTemplate.delete(stringkeys);
        }
```

运行结果如下：

```
    duration is 5443
    duration is 16
    duration is 56
```

第一种方法最耗时，因为用 for 循环的方式对 redis 操作了 1000 次，而且演示环境访问的是外网的 Redis，大部分时间浪费在了网络上；第二种方法最省时，因为虽然向 Redis 中添加了 1000 条数据，但其实只是执行了一次 Redis 操作；第三种方法使用批量提交的方式，所以节省了大量的网络时间，但是 Redis 的操作次数没有变化，所以耗时比第二种方法要长一些。在使用 Redis 时一定要注意使用方法，完成同样的工作使用不同的方法耗时是有明显区别的。

12.4.2　对值进行监控

假设程序通过 Redis 的某个键值对记录了当前系统的某一数值，有多个服务实例都可以访问这个键值。这会出现一个问题，当其中一个实例通过一系列计算要设置这一键值时，其他程序可能在第一个实例计算期间已经修改了这个原始键值，这样第一个实例再设置这个键值可能会发生错误。这时就用到了 watch 方法，它可以监控一个值的变化。

```
        @Override
        public void testRedisTransaction() {
            redisTemplate.opsForValue().set("watchvalue", "1");
            redisTemplate.setEnableTransactionSupport(true);
            redisTemplate.watch("watchvalue");
            redisTemplate.multi();
            redisTemplate.opsForValue().getAndSet("watchvalue", "3");
            List<Object> list = redisTemplate.exec();
            System.out.println(list.toString());
        }
```

正常执行输出为：

    [1]

中途值被修改时输出为：

    []

在上面的方法中,如果在 watch 之后,其他程序修改了 watchvalue 中的值,那么 Redis 操作将执行失败。可以根据 exec 的返回值判断程序是否执行成功。如果此代码是属于不可失败的代码,那么可以放入 While 循环中监控返回值,直到执行成功为止。

12.5　Redis 分布式锁

一个程序内可以有很多个线程,同一程序内的众多线程访问公共资源时需要加锁。但是一个系统集群内的众多程序要访问同一公共资源时,应该怎么限制多个程序的共同访问呢?分布式锁就是解决此问题的。实现分布式锁的能力可以借用多种工具,这里介绍使用 Redis 实现分布式锁的办法。

代码如下:

```java
@Repository
public class RedisLock {
    @Autowired
    private RedisTemplate<String, String> redisTemplate;
    private String lockKey = "redislock";
    private volatile boolean locked = false;
    private int expireMsecs = 10 * 1000;
    private int timeoutMsecs = 2 * 1000;
    private static final int DEFAULT_ACQUIRY_RESOLUTION_MILLIS = 50;

    public boolean Lock() throws InterruptedException {
        int timeout = timeoutMsecs;
        int index = 0;
        Random random = new Random();
        while (timeout > 0) {
            System.out.println(Thread.currentThread().getName() +
                    " index value = " + (++index));
            long expires = System.currentTimeMillis() + expireMsecs + 1;
            String expiresStr = String.valueOf(expires);
            if (redisTemplate.opsForValue().setIfAbsent(lockKey, expiresStr)) {
                System.out.println(Thread.currentThread().getName() + " locked setIfAbsent");
                locked = true;
                return true;
            }

            String currentValueStr = redisTemplate.opsForValue().get(lockKey);
            if (currentValueStr != null
                    && Long.parseLong(currentValueStr) < System.currentTimeMillis()) {
                String oldValueStr = redisTemplate.opsForValue().getAndSet(lockKey, expiresStr);
                if (oldValueStr != null && oldValueStr.equals(currentValueStr)) {
                    System.out.println(Thread.currentThread().getName() +
                            " locked getAndSet");
                    locked = true;
                    return true;
                }
            }
            int temp = random.nextInt(DEFAULT_ACQUIRY_RESOLUTION_MILLIS);
```

```java
                    timeout -= temp;
                    Thread.sleep(temp);
                }
                return false;
            }

            public void unLock() {
                if (locked) {
                    System.out.println(Thread.currentThread().getName() + " unlock");
                    locked = false;
                    redisTemplate.delete(lockKey);
                }
            }

            public void success() {
                redisTemplate.opsForList().rightPush("redislocksuccesslist",
                                    Thread.currentThread().getName());
            }

            public void fail() {
                redisTemplate.opsForList().rightPush("redislockfaillist",
                                    Thread.currentThread().getName());
            }
        }
```

上面的代码中，主要包含 4 个方法。Lock 方法是获取分布式锁，如果获取成功会返回 true。这里分布式锁的实现主要依靠 Redis 的 setIfAbsent 方法，由于此方法只能在 Redis 中没有此键值时才能写入，所以可以写入的程序就获得了锁。当然方法中添加了特殊情况的补救逻辑和超时逻辑，并且重新尝试获取锁的时间间隔使用了随机数；unLock 方法是释放锁，当某个程序已经执行完既定的操作，使用此方法释放，即删除 Redis 锁使用的键值。success 和 fail 仅作为测试统计用，并无实际意义。

看一段测试代码：

```java
        @RunWith(SpringRunner.class)
        @SpringBootTest
        public class RedisExampleApplicationTests {
            @Autowired
            private RedisLock redisLock;

            private void redisDoSmth() {
                try {
                    if(redisLock.Lock()){
                        System.out.println(Thread.currentThread().getName() + " do some thing");
                        redisLock.unLock();
                        redisLock.success();
                    }else {
                        System.out.println(Thread.currentThread().getName() + " get lock fail");
                        redisLock.fail();
                    }
                } catch (Exception e) {
                    e.printStackTrace();
```

```java
            }
        }

        @Test
        public void testRedisLock() {
            ExecutorService eService = Executors.newFixedThreadPool(50);
            for (int i = 0; i < 50; i++) {
                eService.execute(new Runnable() {
                    @Override
                    public void run() {
                        redisDoSmth();
                    }
                });
            }
            eService.shutdown();

            try {
                Thread.sleep(20000);
            } catch (Exception e) {
                e.printStackTrace();
            }
        }
}
```

由于服务器资源有限，所以上面的测试方法中，使用多线程的方式模拟多个程序的抢占（实际项目中，同一程序内的锁不要用此方法），所以这里可以把线程想象为已经实现了同步逻辑的程序。每个程序抢占分布式锁，并且执行业务，如果执行成功则添加进成功队列，否则添加进失败队列。

12.6 Redis 实现秒杀

秒杀就是在众多请求同时发起时，尽量筛选出既定数目的头几个请求作为成功的请求。下面用 Redis 实现一段秒杀程序。

```java
@Repository
public class RedisSecKill {
    @Autowired
    private RedisTemplate<String, String> redisTemplate;

    private static final int TOTALCOUNT = 10;

    public void redisSecKill() {
        Long ranking = 0L;
        int index = 0;
        System.out.println(Thread.currentThread().getName() +
                " redisSecKill start************");
        boolean type = false;
        long starttime = System.currentTimeMillis();
        redisTemplate.setEnableTransactionSupport(true);
        List<Object> list = null;
        try {
```

```java
            while(list == null || list.size()==0 || (Long)list.get(0)==0) {
                System.out.println(Thread.currentThread().getName() +
                                " redisSecKill index value = " + (++index));
                long now = System.currentTimeMillis();
                long costTime = now - starttime;
                if(costTime > 1000) {
                    System.out.println(Thread.currentThread().getName() +
                                    " redisSecKill timeout break*********");
                    type = false;
                    break;
                }
                String temp = redisTemplate.opsForValue().get("redisSecKill");
                if(Long.valueOf(temp)>=TOTALCOUNT) {
                    System.out.println(Thread.currentThread().getName() +
                                    " redisSecKill out total break*********");
                    type = false;
                    break;
                }
                redisTemplate.watch("redisSecKill");
                redisTemplate.multi();
                redisTemplate.opsForValue().increment("redisSecKill", 1);
                redisTemplate.opsForValue().get("redisSecKill");
                //Thread.sleep(20);
                list = redisTemplate.exec();
                costTime = now - starttime;
                System.out.println(Thread.currentThread().getName() +
                                " costime = " + costTime);
            }
        } catch (Exception e) {
            System.out.println(e);
        }

        if(list!=null && list.size()>0) {
            String temp = (String)list.get(1);
            ranking = Long.valueOf(temp);
            if(ranking>TOTALCOUNT) {
                System.out.println(Thread.currentThread().getName() +
                                " not ok and value = " + ranking);
                type = false;
            }else {
                type = redisTemplate.opsForValue().setIfAbsent("redisSecKill success:" +
                                        ranking, Thread.currentThread().getName());
            }
        }

        if(type) {
            System.out.println(Thread.currentThread().getName() + " redisSecKill ok");
            redisTemplate.opsForList().
                        rightPush("redisSecKillList",Thread.currentThread().getName());
        }else {
            System.out.println(Thread.currentThread().getName() + " redisSecKill fail");
        }
}
```

        }
    }

在上面的代码中，只包含一个方法，就是秒杀的业务逻辑，由于用于演示，所以这个方法没有立刻返回秒杀结果，仅把秒杀成功的请求放到一个队列中。

此段代码的主要逻辑就是首先对一个值进行计数加 1 和获取当前计数的操作，然后把此事务得到的结果使用 setIfAbsent 方法写入一个值中，如果可以写入则确认此排名成功，并把成功的请求计入成功队列。当然方法中包含一些异常逻辑的处理，这里就不过多介绍。

以上程序的测试代码如下：

```java
@RunWith(SpringRunner.class)
@SpringBootTest
public class RedisExampleApplicationTests {
    @Autowired
    private RedisSecKill redisSecKill;

    @Test
    public void testSecKill() {
        ExecutorService eService = Executors.newFixedThreadPool(50);
        for (int i = 0; i < 50; i++) {
            eService.execute(new Runnable() {
                @Override
                public void run() {
                    redisSecKill.redisSecKill();
                }
            });
        }
        eService.shutdown();

        try {
            Thread.sleep(20000);
        } catch (Exception e) {
            e.printStackTrace();
        }
    }
}
```

在这段代码中，也是用一个线程指代一个服务，多线程则表示多服务实例抢占。在测试前先在 Redis 中添加键值对 redisSecKill 为 0。测试完成后，可以在秒杀成功队列中查看哪些线程秒杀成功。

第 13 章 Zookeeper

Zookeeper 从英文直译是"动物管理员"。各个系统就好比动物园里的动物，为了使各个系统能正常提供统一的服务，必须用一种机制来进行协调，这就是 ZooKeeper 的作用。

13.1 Zookeeper 介绍

Zookeeper[1]是一个开放源码的分布式应用程序协调服务，是为分布式应用提供一致性服务的软件，提供的功能包括：配置维护、命名服务、分布式同步、组服务等。Zookeeper 是用 Java 语言开发的，提供 Java 和 C 语言的客户端 API。

Zookeeper 的集群模式为 2n+1[2]个服务（奇数），只允许 n 个失效。Zookeeper 集群服务有 Leader、Follower、Observer 三个角色。

- Leader：提供写服务，针对 Zookeeper 进行数据更新相关操作。
- Follower：提供读服务，Leader 宕机后会在 Follower 中重新选举新的 Leader。
- Observer：是一种新型的 Zookeeper 节点，不参与投票，只是简单地接收投票结果，增加再多的 Observer，也不会影响集群的写性能。除了这个差别，其他方面和 Follower 基本上一样。

Zookeeper 结构是由 znode 节点组成的树形结构。格式类似分层的文件目录树形式，每个节点可以存放数据，也可以有子节点。节点的访问路径为绝对路径，不存在相对路径。

znode 节点根据存活时间，分为持久节点和临时节点。节点的类型在创建时就确定下来，并且不能改变。

- 持久节点的存活时间不依赖于客户端会话，只有客户端在显式执行删除节点操作时，节点才消失。
- 临时节点的存活时间依赖于客户端会话，当会话结束，临时节点将会被自动删除（当然也可以手动删除临时节点）。ZooKeeper 中临时节点不能拥有子节点。

Zookeeper 的应用场景是：

- 分布式命名服务：按名称标识集群中的节点。
- 数据发布与订阅：应用启动时主动获取配置信息，并在节点上注册一个观察者（watcher），每次配置更新都会通知到应用。
- 分布式通知/协调：不同的系统都监听同一个节点，一旦有了更新，另一个系统能够收到通知。
- 分布式锁：Zookeeper 能保证数据的强一致性，用户任何时候都可以相信集群中每个节

---

[1] Zookeeper 的官网为 http://zookeeper.apache.org/。
[2] Zookeeper 集群中只要有过半的机器正常工作，就是可用的。例如集群有 2 个 Zookeeper 节点，那么只要有 1 个宕机，Zookeeper 就不能用，因为 1 没有过半，2 个 Zookeeper 的宕机容忍度为 0；同理，集群有 3 个 Zookeeper 节点，一个宕机，剩下 2 个正常的，过半了，所以 3 个 Zookeeper 的容忍度为 1；多列举几个：2->0;3->1;4->1;5->2;6->2，会发现一个规律，2n 和 2n-1 的容忍度是一样的，都是 n-1。多出一台用处不大，所以集群总数为奇数。

点的数据都是相同的。
- 集群管理：服务加入集群时创建一个节点，写入当前服务的状态。监控父节点的应用会收到通知，进行相应的处理。离开时删除节点，监控节点的应用同样也会收到通知。

13.2 基本操作

下面介绍如何使用 Zookeeper 自带客户端以及 Java 客户端进行 Zookeeper 的相关数据操作。

登录 Zookeeper 官网 https://zookeeper.apache.org/releases.html 下载 ZooKeeper 压缩包，本书编写时，最新的稳定版本为 3.4.10。下载后解压到本机。Zookeeper 核心目录以及含义见表 13-1。

表 13-1 Zookeeper 核心目录

目录名称	内容说明
bin	Zookeeper 的可执行脚本目录，包括 zk 服务进程，Zookeeper 客户端等脚本。其中，.sh 是 Linux 环境下的脚本，.cmd 是 Windows 环境下的脚本
conf	配置文件目录。zoo_sample.cfg 为样例配置文件，需要修改为自己的名称，一般为 zoo.cfg
lib	Zookeeper 依赖的包
contrib	一些用于操作 Zookeeper 的工具包
docs	Zookeeper 的使用帮助手册
recipes	Zookeeper 某些用法的代码示例

13.2.1 Zookeeper 客户端操作

进入 Zookeeper 解压文件的 bin 目录，使用 zkCli 客户端来学习 Zookeeper 相关的命令操作。这里以 Linux 服务器环境为例进行讲解，如果当前为 Windows 环境，打开 cmd 命令行执行相应的.cmd 后缀命令即可。

- 启动 Zookeeper 服务：

```
$ ./zkServer.sh start
```

- 查看服务状态，包括节点类型：

```
$./zkServer.sh status
```

- 停止服务：

```
$./zkServer.sh stop
```

- 重启服务：

```
$ ./zkServer.sh restart
```

- 启动客户端，通过-server 指定连接的服务地址及端口：

```
$ ./zkCli.sh -server 127.0.0.1:2181
```

- 进入客户端后，用 ls 命令查看节点信息：

```
ls /path
```

第 13 章 Zookeeper

- 创建节点，指令格式为：

    create [-s] [-e] path data acl

创建节点/zkpath，并存放数据 zkDatas：

    create /zkpath "zkDatas"

创建临时节点,需要使用参数-e：

    create -e /temp datas

创建有序节点,需要使用参数-s：

    create -s /sequ- datas

- 获取节点内容。get 命令获取一个节点存储的数据内容，同时可获取该节点的 stat 信息。znode 的 stat 字段含义见表 13-2。

表 13-2　stat 信息

字 段 名	含 义
cZxid	创建节点事务的 zxid[一]
ctime	表示 znode 的创建时间
mZxid	znode 最近修改的 zxid
mtime	表示 znode 最近修改的时间
pZxid	该节点或子节点的最近一次创建或删除 zxid
cversion	znode 子节点修改次数
dataVersion	znode 节点数据修改次数
aclVersion	znode 的 ACL 修改次数
ephemeralOwner	如果 znode 是临时节点，则指示节点所有者的会话 ID；如果不是临时节点，则为零
dataLength	znode 数据长度
numChildren	当前节点包含的子节点个数

例如，获取/zkpath 数据信息：

    get /zkpath

输出结果为：

    "zkDatas"
    cZxid = 0x1c5
    ctime = Tue May 22 12:14:05 CST 2018
    mZxid = 0x1c5
    mtime = Tue May 22 12:14:05 CST 2018
    pZxid = 0x1c5
    cversion = 0

[一] zxid：ZooKeeper 状态的每一次改变，都对应着一个递增的 Transaction id，此 id 称为 zxid。由于 zxid 的递增性质，如果 zxid1 小于 zxid2，那么 zxid1 肯定先于 zxid2 发生。创建任意节点或者更新任意节点的数据或者删除任意节点，都会导致 Zookeeper 状态发生改变，从而导致 zxid 的值增加。

```
dataVersion = 0
aclVersion = 0
ephemeralOwner = 0x0
dataLength = 9
numChildren = 0
```

可以看出通过 get 方法返回了 znode 信息。

- 修改/zkpath 节点数据：

```
set /zkpath "zkDatasupdate"
```

- 删除节点。delete 命令可以用于删除一个节点，但它只能删除没有任何子节点的节点。例如，删除/zkpath 节点：

```
delete /zkpath
```

- quit 命令退出当前客户端：

```
quit
```

- 执行 help 命令，可查看更多命令：

```
[zk: 127.0.0.1:2181(CONNECTED) 2] help
ZooKeeper -server host:port cmd args
    stat path [watch]
    set path data [version]
    ls path [watch]
    delquota [-n|-b] path
    ls2 path [watch]
    setAcl path acl
    setquota -n|-b val path
    history
    redo cmdno
    printwatches on|off
    delete path [version]
    sync path
    listquota path
    rmr path
    get path [watch]
    create [-s] [-e] path data acl
    addauth scheme auth
    quit
    getAcl path
    close
    connect host:port
```

13.2.2　Java 客户端操作 Zookeeper

Zookeeper 客户端提供了基本的操作，但是有许多不足之处，例如其 Session 超时后没有重试机制，无法级联删除，一次性的 Watcher 机制等，因此平时业务开发常用 ZkClient[一]开源客户端。其在 Zookeeper 原有客户端基础上进行了封装，实现了 Watcher 反复注册、Session 超时重连、节点级联删除等功能。

一　网址是 http://mvnrepository.com/artifact/com.101tec/zkclient。

第 13 章　Zookeeper

本节使用 Spring Boot 工程来操作 Zookeeper。新建一个工程 ZookeeperExample，然后进行如下操作。

（1）添加组件依赖

```xml
<dependency>
    <groupId>com.101tec</groupId>
    <artifactId>zkclient</artifactId>
    <version>0.10</version>
</dependency>
```

（2）添加 Zookeeper 的配置信息至 yml 文件

```yaml
server:
  port: 18094
spring:
  application:
    name: zookeeper-example
zk:
  address: 127.0.0.1:2181
  connectionTimeout: 5000
```

（3）Zookeeper 配置注入

编写一个 ZookeeperModel 类，用于注入 yml 文件中的配置信息。

```java
@Component
@ConfigurationProperties(prefix="zk")

public class ZookeeperModel {
    /** 服务器地址列表*/
    private String address;
    /**毫秒*/
private int connectionTimeout;
//省略 getset 方法
}
```

（4）添加 Zookeeper 基本操作

在工程内，新建 ZookeeperDao 接口和它的实现类，简单实现 Zookeeper 的增删改查操作。

```java
public interface ZookeeperDao {
    public void testZkCRUD();
    public void testWathChildChange();
    public void testDataChanges();
}

@Component
public class ZookeeperDaoImpl implements ZookeeperDao{
    @Autowired
    private ZookeeperModel zookeeperModel;
    public void testZkCRUD(){
        String address = zookeeperModel.getAddress();
        ZkClient zkClient = new ZkClient(address, zookeeperModel.getConnectionTimeout());
```

```java
String pathParant="/persis";
String child01Path=pathParant+"/child01";
String child02Path=pathParant+"/child02";
String ephemeral="/ephemeral";
//1. 创建临时有序节点
String ephemeralSequential = zkClient.
        createEphemeralSequential("/ephemeralSequential-", "epheSequentialDatas");
String data4pathEph= zkClient.readData(ephemeralSequential);
System.out.println(String.format("临时有序节点路径为：%s,数据为：%s",ephemeralSequential,data4pathEph));
ephemeralSequential = zkClient.
        createEphemeralSequential("/ephemeralSequential-", "epheSequentialDatas");
data4pathEph= zkClient.readData(ephemeralSequential);
System.out.println(String.format("临时有序节点路径为：%s,数据为：%s",ephemeralSequential,data4pathEph));
zkClient.createPersistent(child01Path, true);
boolean exists= zkClient.exists(child01Path);
if(exists){
    System.out.println(String.format("节点:%s, 级联创建成功",child01Path));
}
zkClient.createEphemeral(ephemeral);
exists= zkClient.exists(ephemeral);
if(exists){
    System.out.println(String.format("临时节点:%s, 创建成功",ephemeral));
}
//delete node
zkClient.delete(ephemeral);
zkClient.deleteRecursive(pathParant);
exists= zkClient.exists(pathParant);
System.out.println(String.format("节点%s,是否存在：%s",pathParant,exists));

//2. 创建节点以及子节点
zkClient.createPersistent(pathParant, "rootDatas");
String data = zkClient.readData(pathParant);
System.out.println(String.format("数据节点%s,数据为：%s",pathParant,data));
zkClient.createPersistent(child01Path, "datas of child01");
zkClient.createPersistent(child02Path, "datas of child02");

List<String> list = zkClient.getChildren("/persis");
for (String p : list) {
    System.out.println("child path is "+p);
    String path = pathParant +"/"+ p;
    data = zkClient.readData(path);
    System.out.println(String.format("数据节点:%s,数据:%s",path,data));
}

//3. 更新节点数据
boolean isExists=zkClient.exists(child01Path);
System.out.println(String.format("数据节点%s,是否存在：%s",child01Path,isExists));
if(isExists){
    data=zkClient.readData(child01Path).toString();
    System.out.println(String.format("数据节点%s,数据为：%s",child01Path,data));
```

```
            zkClient.writeData(child01Path, "update datas");
            data=zkClient.readData(child01Path).toString();
            System.out.println(String.format("数据节点%s,数据为：%s",child01Path,data));
        }
        //递归删除节点
        zkClient.deleteRecursive(pathParant);
        exists = zkClient.exists(pathParant);
        System.out.println(String.format("节点%s,是否存在：%s",pathParant,exists));
    }
    //……省略其他两个方法
}
```

上面的代码中，使用 ZkClient 对 Zookeeper 进行操作。使用 createEphemeralSequential 方法创建临时有序节点，使用 createPersistent 方法创建永久节点，方法 exists 判断节点是否存在，readData 读取节点数据，writeData 写入节点数据，deleteRecursive 级联删除节点等。

（5）使用测试类进行测试

在 TestZKService 类中，添加如下测试代码，检验 ZkClient 的使用情况。

```
@RunWith(SpringRunner.class)
@SpringBootTest(classes = ZookeeperExampleApplication.class)
public class TestZKService {
    @Autowired
    ZookeeperDao zookeeperDao;
    @Test
    public void testZkClient() throws Exception {
        zookeeperDao.testZkCRUD();
    }
}
```

运行结果如下：

```
临时有序节点路径为：/ephemeralSequential-0000000084,数据为：epheSequentialDatas
临时有序节点路径为：/ephemeralSequential-0000000085,数据为：epheSequentialDatas
节点:/persis/child01,级联创建成功
临时节点:/ephemeral,创建成功
节点/persis,是否存在：false
数据节点/persis,数据为：rootDatas
child path is child02
数据节点:/persis/child02,数据:datas of child02
child path is child01
数据节点:/persis/child01,数据:datas of child01
数据节点/persis/child01,是否存在：true
数据节点/persis/child01,数据为：datas of child01
数据节点/persis/child01,数据为：update datas
节点/persis,是否存在：false
```

注入 ZookeeperDao 类的实例，然后调用 Dao 中的方法，从输出可见操作 Zookeeper 中的数据结果。

13.2.3 订阅子节点变化

ZkClient 的 subscribeChildChanges 方法用来订阅子节点变化，下面三个事件会触发订阅通知：

1）新增子节点。
2）减少子节点。
3）自身节点增删。

需要注意 subscribeChildChanges 不会监听节点内容的变化。

通过 ZookeeperDao 中的 testWathChildChange 方法来演示 Zookeeper 监听节点变化机制，具体如下：

```java
public void testWathChildChange(){
    try{
        String address = zookeeperModel.getAddress();
        ZkClient zkClient = new ZkClient(new ZkConnection(address),
                            zookeeperModel.getConnectionTimeout());
        String parentPath="/persist";
        boolean exists = zkClient.exists(parentPath);
        if(exists){
            zkClient.deleteRecursive(parentPath);
        }
        zkClient.subscribeChildChanges("/persist", new IZkChildListener() {
            @Override
            public void handleChildChange(String parentPath,
                            List<String> currentChilds) throws Exception {
                System.out.println(String.format("触发到监听事件:parentPath %s,
                    其所有子节点: %s",parentPath,currentChilds));
            }
        });
        zkClient.createPersistent(parentPath);
        Thread.sleep(1000);
        String child01Path=parentPath+"/child01";
        zkClient.createPersistent(child01Path, "child01 datas");
        System.out.println("----> create path: "+child01Path);
        Thread.sleep(1000);
        String child02path=parentPath+"/child02";
        zkClient.createPersistent(child02path, "child02 datas");
        System.out.println("----> create    path: "+child02path);
        Thread.sleep(1000);
        zkClient.writeData(child02path,"update child02 datas");
        System.out.println("----> update path:"+child02path);
        Thread.sleep(1000);
        zkClient.delete(child02path);
        System.out.println("----> delete path:"+child02path);
        Thread.sleep(1000);
        zkClient.deleteRecursive("/persist");
        System.out.println("----> delete path:/persist");
        Thread.sleep(4*1000);
        System.out.println("done");
    }catch (Exception e){
        e.printStackTrace();
    }
}
```

运行结果如下：

```
触发到监听事件:parentPath /persist, 其所有子节点: []
----> create path: /persist/child01
触发到监听事件:parentPath /persist, 其所有子节点: [child01]
----> create   path: /persist/child02
触发到监听事件:parentPath /persist, 其所有子节点: [child02, child01]
----> update path:/persist/child02
----> delete path:/persist/child02
触发到监听事件:parentPath /persist, 其所有子节点: [child01]
----> delete path:/persist
触发到监听事件:parentPath /persist, 其所有子节点: null
触发到监听事件:parentPath /persist, 其所有子节点: null
done
```

13.2.4 订阅节点的数据内容变化

ZkClient 的 subscribeDataChanges 方法用来订阅节点的数据内容变化。

通过 ZookeeperDao 中的 testDataChanges 方法来演示 Zookeeper 的监听节点数据机制，具体如下：

```java
public void testDataChanges(){
    try{
        String address = zookeeperModel.getAddress();
        ZkClient zkClient = new ZkClient(new ZkConnection(address),
                            zookeeperModel.getConnectionTimeout());
        String path="/persist";
        if(zkClient.exists(path)){
            zkClient.deleteRecursive(path);
        }
        zkClient.createPersistent(path, "datas");
        //对父节点添加监听子节点中数据的变化
        zkClient.subscribeDataChanges("/persist", new IZkDataListener() {
            @Override
            public void handleDataDeleted(String path) throws Exception {
                System.out.println("删除节点为:" + path);
            }

            @Override
            public void handleDataChange(String path, Object data) throws Exception {
                System.out.println(String.format("变更节点为:%s,
                                    变更内容为:%s" ,path,data));
            }
        });

        Thread.sleep(1000);
        zkClient.writeData(path, "update datas 01", -1);
        System.out.println("----> write datas:"+path);
        Thread.sleep(1000);

        String child02path=path+"/child02";
        zkClient.createPersistent(child02path, "child02 datas");
        System.out.println("----> create child path:"+child02path);
        Thread.sleep(1000);
```

```
                zkClient.writeData(child02path, "update child datas", -1);
                System.out.println("---->  update child datas:"+ child02path);
                Thread.sleep(1000);

                zkClient.writeData(path, "update datas 02", -1);
                System.out.println("---->  update datas:"+path);
                Thread.sleep(1000);

                zkClient.deleteRecursive(path);
                System.out.println("---->  delete    path:"+path);
                Thread.sleep(2*1000);
            }catch (Exception e){
                e.printStackTrace();
            }
        }
    }
```

运行结果如下：

```
---->  write datas:/persist
变更节点为:/persist, 变更内容为:update datas 01
---->  create child path:/persist/child02
---->  update child datas:/persist/child02
---->  update datas:/persist
变更节点为:/persist, 变更内容为:update datas 02
---->  delete    path:/persist
删除节点为:/persist
```

从上面的结果可以看出，当/persist 节点数据发生变化后，才会触发监听，其子节点数据的变化并不会被监听到。

13.3 服务注册与发现

根据上一节对 ZkClient 的学习，实现一个简单的服务注册与发现功能。实现思路如下：
- 服务提供方在启动后，在 Zookeeper 的指定父路径下注册服务，创建对应的 URL 临时节点，并将自己的服务名、IP 地址、端口、权重存放到临时节点的数据中。
- 服务调用方在启动服务后，到 Zookeeper 指定父路径下找到所有的子节点的数据并存放起来，订阅子节点变化，以便随时更新提供方服务列表。
- 当提供方出现宕机或者网络故障时，它对应的 URL 节点在 sessionTimeOut 后，就会被销毁，此时临时节点会删除，触发调用方的节点监听事件，所有调用方都会收到节点变化（watcher）的通知，调用方需要更新提供方服务列表并且继续监听。

13.3.1 服务注册

服务提供方在启动后，需要把自身的服务名、IP、端口、权重等信息以临时有序节点的形式存放到指定的父节点下面，具体实现如下：

```
public class Constants {
    public static final String parentZnodePath = "/javadevmap-Servers";
}
```

```java
/**
 * 服务提供方
 */
@Service
public class RemoteServer {
    private ZkClient zkClient = null;
    @Autowired
    ZookeeperModel zookeeperModel;

    @PostConstruct
    public void initMethod() {
        // 创建 zkclient
        zkClient = new ZkClient(zookeeperModel.getAddress(),
                        zookeeperModel.getConnectionTimeout());
    }
    /**
     * 在固定节点下面创建临时有序节点
     */
    public void registerServer(String serverName,String address, String port) throws Exception {
        // 先创建出父节点,用于被调用方监听
        if (!zkClient.exists(Constants.parentZnodePath)) {
            zkClient.create(Constants.parentZnodePath, null,
                    Ids.OPEN_ACL_UNSAFE, CreateMode.PERSISTENT);
        }
        Map<String,String> datas=new HashMap<>();
        datas.put("serverName",serverName);
        datas.put("host",address);
        datas.put("port",port);
        // 服务权重  0-100
        datas.put("weight",new Random().nextInt(100)+"");
        JSONObject jsonObject = JSONObject.fromObject(datas);
        String result = jsonObject.toString();
        //在指定路径下面创建临时节点
        String pathName = zkClient.create(
                Constants.parentZnodePath + "/" + serverName + "-",
                result, Ids.OPEN_ACL_UNSAFE,
                CreateMode.EPHEMERAL_SEQUENTIAL);
        System.out.println("[Server]>>>>>>" + serverName + " is register success! pathName = "+pathName);
    }
}
```

提供方服务在父节点/javadevmap-Servers 下面创建以服务名称开头的临时节点，在此临时节点中添加相应的服务相关信息。编写一个测试类 TestRegisterDemo，通过使用 startRemoteServer 方法模拟服务提供方程序，把此服务注册到 Zookeeper 上，启动后进行等待来模拟服务运行，具体内容如下：

```java
@Autowired
RemoteServer server;
@Test
public void startRemoteServer() throws Exception{
```

```java
            String serverName="orderServer";
            String address="192.168.1."+new Random().nextInt(255);   // 模拟 ip
            int port=1000+new Random().nextInt(599);                 //端口在 1000-1599 之间
            server.registerServer(serverName,address, port+"");      //注册服务
            System.out.println("[Server]>>>>>>" + serverName + " is Online ......");
            Thread.sleep(Long.MAX_VALUE);
    }
```

运行结果如下：

```
[Server]>>>>>>orderServer is register success! pathName = /javadevmap-Servers/orderServer-0000000003
[Server]>>>>>>orderServer is Online ......
```

13.3.2 服务发现

服务调用方在启动服务后，主动获取指定父节点下面所有的数据，并且订阅子节点变化事件，在此事件触发后，更新相应的服务提供方列表。这里定义 RemoteCient 类，来实现服务调用方，具体实现如下：

```java
/**
 * 服务调用方
 */
@Service
public class RemoteClient {
    private List<String> serList = null;
    private ZkClient zkClient = null;

    @Autowired
    ZookeeperModel zookeeperModel;
    @PostConstruct
    public void initMethod() {
        // 构建 zkclient
        zkClient = new ZkClient(zookeeperModel.getAddress(),
                        zookeeperModel.getConnectionTimeout());
    }

    public void subscribeChildChanges4Servers() throws Exception {
        zkClient.subscribeChildChanges(Constants.parentZnodePath, new IZkChildListener() {
            @Override
            public void handleChildChange(String parentPath, List<String> currentChilds)
                throws Exception {
                System.out.println(String.format("-->触发到监听事件:parentPath %s,
                                 其所有子节点: %s", parentPath, currentChilds));
                updateServerList(currentChilds);
                System.out.println("<<<<<<<<<<<<<<<<<<<<<监听结束");
            }
        });
    }


    /**
     * 主动获取 server list 数据
```

```java
    */
    public void getServerList() throws Exception {
        System.out.println(">>>>>>>>>>>>>>>>>>>>");
        // 先创建出父节点,用于被调用方监听
        if (!zkClient.exists(Constants.parentZnodePath)) {
            zkClient.create(Constants.parentZnodePath, null,
                    ZooDefs.Ids.OPEN_ACL_UNSAFE, CreateMode.PERSISTENT);
        }
        List<String> children = zkClient.getChildren(Constants.parentZnodePath);
        if (null == children || children.size() == 0) {
            return;
        }
        updateServerList(children);
        System.out.println("<<<<<<<<<<<<<<<<<<<<");
    }

    private void updateServerList(List<String> currentChilds) {
        ArrayList<String> serverList = new ArrayList<String>();
        for (String child : currentChilds) {
            String path = Constants.parentZnodePath + "/" + child;
            String data = zkClient.readData(path);
            serverList.add(new String(data));
        }
        serList = serverList;
        // 打印更新后的服务器列表信息
        for (String server : serverList) {
            System.out.println(String.format("服务数据:%s",server));
        }
    }
}
```

在 TestRegisterDemo 类中添加测试方法,模拟服务调用方获取并且监听服务提供方节点,具体如下:

```java
@Autowired
RemoteClient client;
@Test
public void startRemoteClient()throws Exception{
    client.getServerList();
    client.subscribeChildChanges4Servers();
    Thread.sleep(Long.MAX_VALUE);
}
```

如果上一节的服务没有停止的话,启动客户端,能够获取对应的服务列表信息,运行结果如下:

```
服务数据:
{"port":"1388","host":"192.168.1.100","serverName":"orderServer","weight":"30"}
```

每当通过测试类的 startRemoteServer 方法启动服务提供方程序时,会触发服务调用方子节点变化事件,服务调用方在监听事件中及时更新列表数据,当停止运行 startRemoteServer 方法关闭服务提供方程序时,服务调用方也会更新对应的列表数据。

第 14 章 FastDFS

业务开发过程中，或多或少都会接触到文件的存储和下载，这些文件在服务器中存储时，需要一个稳定、易扩容、高可用的环境。FastDFS 很好地支持了这些需求，本章使用 FastDFS 实现文件储存的相关操作。

14.1 FastDFS 基本介绍

FastDFS 是一个开源的高性能分布式文件系统，它的主要功能包括：文件存储、文件同步和文件访问，具备了大容量和负载均衡的能力。特别适合以文件为载体的在线服务。

14.1.1 FastDFS 概述

FastDFS 是用 C 语言实现的，目前提供了 C，Java，PHP 语言的支持。另外，FastDFS 可看作是基于文件的 key-value 存储系统，也可称为分布式文件存储服务。

FastDFS 有两个重要角色：跟踪器（Tracker）和存储节点（Storage）。

- Tracker：跟踪服务器，是 FastDFS 的协调者，起负载均衡的作用。记录集群中所有存储组（group ⊖）和存储服务器（Storage Server）的状态信息，是客户端和数据服务器交互的纽带。
- Storage：存储服务器，文件和文件属性（metadata）都保存到存储服务器上。为避免单个目录下的文件数过多，存储节点（Storage）在第一次启动时，会在每个数据存储目录里创建两级子目录，每级有 256 个，总共 65536 个文件夹，文件会以 hash 的方式被路由到对应的某个子目录下，然后将文件数据直接作为一个本地文件存储到该目录中。

FastDFS 特别适合作为中小文件（4KB <文件大小<500MB）载体的在线服务。其提供了针对文件的上传、下载、删除、设置文件属性等相关功能。

14.1.2 FastDFS 上传和下载过程

FastDFS 提供了基本文件访问接口，例如 upload、download、append、delete 等，以客户端的方式提供给开发者使用。下面介绍 FastDFS 上传和下载交互过程。

（1）FastDFS 上传过程：

1）存储服务器会定时向跟踪服务器（Tracker Server）上传自身状态信息。

2）FastDFS 客户端（client）提交上传请求到 Tracker。

3）跟踪服务器根据请求查询可用的存储服务器，并返回给客户端对应的存储服务器的地址和端口。

4）客户端直接根据返回的存储服务器相关信息，连接存储服务器进行上传操作。

⊖ group 组也可称为卷。同一组内服务器上的文件是完全相同的，同一组内的存储服务器是对等的，文件上传、下载、删除等操作可在任意一台存储服务器上进行。

5）存储服务器根据客户端传递过来的信息生成文件名（file_id）并将内容写入磁盘，写入成功后，返回给客户端文件路径相关信息。

6）客户端收到存储服务器返回的信息后，来存储对应的文件信息。

（2）FastDFS 下载过程：

1）客户端向跟踪服务器发起文件下载请求。

2）跟踪服务器根据客户端传递的参数分配可用的存储服务器，并返回存储服务器地址和端口。

3）客户端根据返回的文件名到存储服务器上查找文件。

4）存储服务器根据客户端传递过来的参数，返回给客户端对应文件内容。

14.2 Spring Boot 集成 FastDFS

创建一个 Spring Boot 工程，工程名使用 FastDFSExample，具体创建工程方法可参照第 7 章。下面使用 Spring Boot 工程整合 FastDFS 进行文件相关操作。

（1）添加依赖

为了整合 FastDFS，需要添加 fastdfs-client-java 依赖，后面的章节需要使用页面进行演示，所以添加 spring-boot-starter-thymeleaf 起步依赖。具体如下：

```
<dependency>
    <groupId>org.springframework.boot</groupId>
    <artifactId>spring-boot-starter-thymeleaf</artifactId>
</dependency>
<dependency>
    <groupId>org.csource</groupId>
    <artifactId>fastdfs-client-java</artifactId>
    <version>1.25</version>
</dependency>
```

（2）修改配置文件

在 resources 文件夹下面创建一个名为 fdfs_client.conf 的配置文件。其包含连接 Tracker 服务器超时时间、socket 连接超时时间、Tracker 服务器地址和端口、是否开启防盗链功能等，具体配置如下：

```
connect_timeout = 30
network_timeout = 30
charset = UTF-8
http.tracker_http_port = 8080
# token 防盗链功能；no 为关闭此功能
http.anti_steal_token = no
tracker_server = 47.95.113.117:22122
```

上面的配置文件中，http.anti_steal_token 表示是否开启防盗链功能，这里暂时不使用；如果有多台跟踪服务器，可以在配置文件中添加多个 tracker_server 键值对。

（3）上传工具类

FastDFS 的上传与下载等功能，主要通过 TrackerClient 工具类实现。创建一个类 FastDFS ClientUtils，初始化 FastDFS 相关工具类，具体如下：

```java
public class FastDFSClientUtils {
    private static Logger logger = LoggerFactory.getLogger(FastDFSClientUtils.class);
    /** 配置文件信息 */
    private static final String confFileName="fdfs_client.conf";
    private static TrackerClient trackerClient =null;
    static {
        try {
            String filePath = new ClassPathResource(confFileName).getFile().getAbsolutePath();
            ClientGlobal.init(filePath);
            trackerClient = new TrackerClient();
        } catch (Exception e) {
            e.printStackTrace();
            logger.error("FastDFS Client 初始化失败", e);
        }
    }

    private static TrackerServer getTrackerServer() throws IOException {
        TrackerServer trackerServer = trackerClient.getConnection();
        return trackerServer;
    }

    private static StorageClient getTrackerClient() throws IOException {
        TrackerServer trackerServer = getTrackerServer();
        StorageClient storageClient = new StorageClient(trackerServer, null);
        return storageClient;
    }

    public static String getTrackerUrl() throws IOException {
        return "http://" + getTrackerServer().getInetSocketAddress().getHostString()
                + ":" + ClientGlobal.getG_tracker_http_port() + "/";
    }
}
```

使用以上代码，类 FastDFSClientUtils 实现了初始化 FastDFS 工具类的能力。接下来通过文件上传和下载来演示 FastDFS 的使用。

14.2.1 文件上传

本节通过页面上传一个文件，来演示 FastDFS 如何保存文件。

（1）上传页面

新建一个上传页面 uploadfile.html，用来与用户进行交互，具体如下：

```html
<!DOCTYPE html>
<html xmlns:th="http://www.thymeleaf.org">
<body>
```

```html
<h2>Spring Boot FastDFS 文件上传</h2>
<form method="POST" action="/uploadAction" enctype="multipart/form-data">
    <input type="file" name="file" /><br/>
    <input type="submit" value="上传文件" />
</form>
</body>
</html>
```

新建一个 Controller 类 FastdfsController,通过 index 方法定位到 uploadfile.html 页面,具体如下:

```java
@Controller
public class FastdfsController {
    private static final Logger logger = LoggerFactory.getLogger(FastdfsController.class);
    @GetMapping("/uploadpage")
    public String index() {
        return "uploadfile";
    }
}
```

启动服务后,访问路径http://localhost:18095/uploadpage,就能直接跳转到上传页面,如图 14-1 所示。

图 14-1 上传页面

为了方便接收前端传递过来的参数,创建一个 model 类 FileInfoModel,具体如下:

```java
public class FileInfoModel {
    private String name;
    private byte[] content;
    /** 文件扩展名*/
    private String extName;

    public FileInfoModel() {
    }
    public FileInfoModel(String name, byte[] content,String extName) {
        this.name = name;
        this.content = content;
        this.extName=extName;
    }
    //..省略 get 与 set 方法
}
```

(2)上传功能实现

FastDFS 的上传功能主要通过 TrackerClient 实现,在上一节的工具类 FastDFSClientUtils 中添加 uploadFile 方法,实现文件的上传功能,具体如下:

```java
public static String[] uploadFile(FileInfoModel file, Map<String,String> extMap) {
    try {
        logger.info("文件名：" + file.getName() + "文件大小:" + file.getContent().length);
        NameValuePair[] metaArr  =null;
        // 添加额外数据
        if(null != extMap){
            metaArr = new NameValuePair[extMap.size()];
            int index=0;
            for(String key:extMap.keySet()){
                metaArr[index] = new NameValuePair(key, extMap.get(key));
            }
        }
        long startTime = System.currentTimeMillis();
        String[] uploadFileResults = null;
        StorageClient storageClient = null;

        storageClient = getTrackerClient();
        uploadFileResults = storageClient.upload_file(file.getContent(),
                                                     file.getExtName(), metaArr);
        logger.info("上传文件耗时:" + (System.currentTimeMillis() - startTime) + " ms");
        if (uploadFileResults == null && storageClient != null) {
            logger.error("上传文件失败，错误码:" + storageClient.getErrorCode());
        }
        String groupName = uploadFileResults[0];
        String remoteFileName = uploadFileResults[1];
        logger.info("上传文件成功： " + "group_name:" + groupName + ",
                    remoteFileName:" + " " + remoteFileName);
        return uploadFileResults;
    } catch (Exception e) {
        e.printStackTrace();
    }
    return null;
}
```

uploadFile 方法有两个参数，第一个参数 FileInfoModel，是自定义的 model 类，用于承接前端传递过来的数据，第二个参数 Map<String,String>用来存放扩展属性相关的数据，以 key-value 对的方式存放在存储服务器（storage）上的同名文件中。

接下来在 FastdfsController 类中实现一个接收前端上传请求的方法 fileUploadAction，具体如下：

```java
@PostMapping("/uploadAction")
public String fileUploadAction(@RequestParam("file") MultipartFile file,
                               RedirectAttributes redirectAttributes) {
    try {
        if (file.isEmpty()) {
            redirectAttributes.addFlashAttribute("message", "请选择要上传的文件");
            return "redirect:uploadStatus";
        }
        String[] arrs = saveFile(file);
        String groupName = arrs[0];
        String remoteFileName = arrs[1];
        //拼接路径
```

```java
            String path = FastDFSClientUtils.getTrackerUrl() + groupName + "/" + remoteFileName;
            redirectAttributes.addFlashAttribute("message",
                    "' 成功上传文件：'" + file.getOriginalFilename() + "'");
            redirectAttributes.addFlashAttribute("path", path);
            redirectAttributes.addFlashAttribute("groupName", groupName);
            redirectAttributes.addFlashAttribute("remoteFileName", remoteFileName);
            String fileName = file.getOriginalFilename();
            String ext = fileName.substring(fileName.lastIndexOf(".") + 1);
            if(ext.equalsIgnoreCase("jpg")){
                redirectAttributes.addFlashAttribute("ext", ext);
            }
    } catch (Exception e) {
        e.printStackTrace();
        logger.error("上传文件失败!! ");
    }
    return "redirect:/uploadStatus";
}
//上传文件方法
public String[] saveFile(MultipartFile multipartFile) throws IOException {
    try {
        String[] fileAbsolutePath;
        String fileName = multipartFile.getOriginalFilename();
        String ext = fileName.substring(fileName.lastIndexOf(".") + 1);
        byte[] file_buff = multipartFile.getBytes();// 获取文件流
        FileInfoModel file = new FileInfoModel(fileName, file_buff, ext);
        Map<String,String> data=new HashMap<>();
        data.put("test","javadevmap");
        // 上传文件
        fileAbsolutePath = FastDFSClientUtils.uploadFile(file,data);
        if (fileAbsolutePath == null) {
            logger.error("上传文件失败，请重新上传! ");
        }
        return fileAbsolutePath;
    } catch (Exception e) {
        logger.error("upload file Exception!", e);
    }
    return null;
}
```

在上面的代码中，通过用户传递过来的 MultipartFile 类实例读取文件流，然后通过 FastDFSClientUtils 工具类的 uploadFile 方法进行文件上传，返回给前端文件在 FastDFS 中的存放文件名。

为了观察 FastDFS 是否上传成功，新建页面 uploadStatus.html。文件上传后页面重定向到 uploadStatus.html，在重定向页面进行相关下载路径、组名、文件名等数据展示，具体如下：

```html
<!DOCTYPE html>
<html lang="en" xmlns:th="http://www.thymeleaf.org">
<body>
    <h1>Spring Boot FastDFS 上传成功页面</h1>
<div th:if="${message}">
    <h2 th:text="${message}"/>
</div>
```

```html
        <div th:if=""${path}"">
            下载路径: <h2 th:text="${path}"/>
            组名称: <h2 th:text="${groupName}"/>
            文件名: <h2 th:text="${remoteFileName}"/>
        </div>
        <div th:if="${ext}">
            <img th:src="${path}"/>
        </div>
    </body>
</html>
```

启动项目,访问 http://localhost:18095/uploadpage,上传一张图片,效果如图 14-2 所示。

图 14-2 上传成功页面

本节实现了 FastDFS 的文件上传功能。同时也返回给前端一个完整的 url 路径,可直接复制到浏览器中查看文件。

14.2.2 文件下载

如果不想通过 Web 直接下载文件,也可以通过接口的形式进行文件下载。上一节演示了如何进行文件上传,本节在上一节的基础上实现 FastDFS 文件的下载功能。

(1) 下载方法

FastDFS 的文件下载功能,通过工具类 StorageClient 的 download_file 方法实现。在上一节的 FastDFSClientUtils 类中实现一个 downloadFile 下载方法,具体如下:

```java
public static byte[] downloadFile(String groupName, String remoteFileName) {
    try {
        StorageClient storageClient = getTrackerClient();
        byte[] fileByte = storageClient.download_file(groupName, remoteFileName);
        return fileByte;
    } catch (Exception e) {
        e.printStackTrace();
```

```
        }
        return null;
```

上面方法中，StorageClient 通过组名和文件名就可以进行文件下载操作。

（2）Controller 下载请求方法

在上一节的 FastdfsController 类中添加 download 方法，用来接收前端提交的下载请求（组名和文件名），具体如下：

```
@RequestMapping(value = "/download")
public ResponseEntity<byte[]> download(HttpServletRequest request, @RequestParam("groupName") String groupName, @RequestParam("remoteFileName") String remoteFileName, Model model) throws Exception {
    //下载文件路径
    HttpHeaders headers = new HttpHeaders();
    //下载显示的文件名，解决中文名称乱码问题
    String filename = remoteFileName.substring(remoteFileName.lastIndexOf("/")+1);
    String downloadFileName = new String(filename.getBytes("UTF-8"), "iso-8859-1");
    //通知浏览器以 attachment（下载方式）打开图片
    headers.setContentDispositionFormData("attachment", downloadFileName);
    //application/octet-stream ：二进制流数据（最常见的文件下载）。
    headers.setContentType(MediaType.APPLICATION_OCTET_STREAM);
    byte[] file_buff = FastDFSClientUtils.downloadFile(groupName, remoteFileName);
    return new ResponseEntity<byte[]>(file_buff, headers, HttpStatus.CREATED);
}
```

（3）下载页面

在上一节的上传成功页面 uploadStatus.html 中，添加如下代码，使此页面能够提交文件下载请求，具体如下：

```
<!DOCTYPE html>
<html lang="en" xmlns:th="http://www.thymeleaf.org">
<body>

<h1>Spring Boot FastDFS 上传成功页面</h1>
    //... 上一节页面内容

<form method="POST" action="/download" >
    <input type="hidden" name="groupName" th:value="${groupName}" /><br/>
    <input type="hidden" name="remoteFileName" th:value="${remoteFileName}" /><br/>
    <input type="submit" value="下载文件" />
</form>
</body>
</html>
```

启动项目，访问 http://localhost:18095/uploadpage 链接，上传一个文件后，点击下载按钮，即可实现文件的下载。

第 15 章　ElasticSearch

随着业务数据的增多，在站内搜索信息可能是一个新的挑战。如果直接查询数据库，前提是要知道数据存放在哪个表中，有哪些字段，但是实际上大部分情况是只知道要搜索的内容，而不知道具体的位置。同时站内搜索对搜索耗时有较高的要求，如何优化查询也是要面临的一系列问题。ElasticSearch 的出现正是为了解决上面问题的。本章介绍如何使用 ElasticSearch 实现搜索相关的业务需求。

15.1　ElasticSearch 基本介绍

ElasticSearch[一]是一个基于 Lucene[二]的搜索服务器。它提供了一个分布式多用户能力的全文搜索引擎，基于 RESTful web 接口。ElasticSearch 是用 Java 开发的，使用 Apache 许可条款，开放源码，是当前流行的企业级搜索引擎。

15.1.1　ElasticSearch 概述

ElasticSearch 是一种文档型数据库，提供了存储服务、搜索服务、大数据准实时分析等能力。ElasticSearch 具有如下结构：
- 集群：是一个或多个节点的集合，用来保存应用的全部数据并提供基于全部节点的集成式索引和搜索功能。每个集群都需要有一个唯一的名称。
- 节点：是一个集群中的单台服务器，用来保存数据并参与整个集群的索引和搜索操作。每个节点都可以配置其名称。节点会加入指定名称的集群中。
- 索引（Index）：相当于数据库，用于定义文档类型的存储，在同一个索引中，同一个字段只能定义一个数据类型。
- 文档类型（Type）：相当于数据库中的表，用于描述文档中各个字段的定义，不同的文档类型，能够存储不同的字段，用于不同类型的查询请求。
- 文档（Document）：相当于关系表的数据行，存储数据的载体，包含一个或多个存有数据的字段：
  - 字段（Field）：文档的一个键值对。
  - 词（Term）：表示文本中的一个单词。
  - 标记（Token）：表示在字段中出现的词，由该词的文本、偏移量（开始和结束）以及类型组成。

15.1.2　分片与副本的关系

当系统中有大量的文档时，由于内存、硬盘等硬件资源的限制，如果存放在一起，很难及

[一] ElasticSearch 官网是 https://www.elastic.co/cn/。
[二] Lucene 是一款高性能的、可扩展的信息检索（IR）工具库。

时响应客户端的请求，显然一个节点是不够用的。ElasticSearch 采用了分片和副本的模式，即将数据分成较小的部分，称之为分片（shard）。每个分片可以放在不同的服务器上，因此，数据可在集群节点中传播，当查询的索引分布在多个分片上时，ElasticSearch 会把查询发送给每个相关的分片，将结果合并在一起。主分片（primary shard）默认是 5 个分片。

ElasticSearch 为了提高查询的吞吐量，使用副本机制。副本为一个分片的精确复制，每个分片可以有零到多个副本。ElasticSearch 可以有许多相同的分片，其中主分片具备更改索引等操作的能力，其余的为副本分片（replica shard）。在主分片丢失或主分片所在服务器无法访问时，将副本提升为新的主分片。每个分片的副本默认为 1 个。

15.1.3　ElasticSearch 主要特性

ElasticSearch 有以下主要特性。
- 安装方便：没有其他依赖，下载后安装非常简单。
- Json：输入/输出格式为 Json，不需要定义 Schema，快捷方便。
- RESTful：基本所有操作都可通过 HTTP 接口进行。
- 分布式：节点对外表现对等，加入节点自动均衡。
- 多租户：可根据用途不同创建对应的索引；可以同时操作多个索引。
- 准实时：从开始进行文档索引到可以被检索只有轻微延时。
- 支持插件机制（分词插件、同步插件、Hadoop 插件、可视化插件等）。

基于上面的特性，ElasticSearch 常用于全文搜索领域，构建业务的搜索功能模块，多是垂直领域的搜索，数据量级一般在千万级以上。

15.2　ElasticSearch 基本用法

本书重点不在 ElasticSearch 环境的安装，ElasticSearch 软件环境采用 Docker 容器部署。具体部署命令参照第 19 章。本章使用的 ElasticSearch 版本为 2.4.0。

ElasticSearch5.x 和 ElasticSearch2.x 的区别不是很大，但是由于 ElasticSearch5.x 集成了 Lucene 6.x，其中最重要的特性就是 Dimensional Point Fields，即多维浮点字段，ElasticSearch 里面相关的字段如 date、numeric、ip 和 Geospatial 都将大大提升性能。磁盘空间少一半；索引时间少一半；查询性能提升 25%；提供对 IPv6 的支持。更多新特性，大家可到官网上查阅。

ElasticSearch 安装完成后，可以在浏览器或者 Postman 中，输入 http://{Elastic SearchIP}:9200，查看当前 ElasticSearch 信息。如图 15-1 所示。

返回的信息包含当前集群的名称"elasticsearch"、当前 ElasticSearch 的版本 2.4.0 以及 Lucene 的版本 5.5.2 等。

图 15-1　ElasticSearch 信息

15.2.1 索引操作

在创建索引之前，需要了解 ElasticSearch 的 RESTful API 的调用风格，管理和使用 ElasticSearch 服务时，常用 HTTP 请求方式见表 15-1。

表 15-1　HTTP 请求方式

HTTP 请求方式	含　义
GET 请求	获取 ElasticSearch 服务器中的对象
POST 请求	更新 ElasticSearch 服务器中的对象
PUT 请求	在 ElasticSearch 服务器上创建对象
DELETE 请求	删除 ElasticSearch 服务器中的对象

下面手动创建一个索引。打开 Postman 工具，按照前面所讲 ElasticSearch 的 Restful API 操作规则，以 put 方式创建一个名为 productindex[一]的索引。如图 15-2 所示。

执行删除索引，以 delete 方式执行即可，如图 15-3 所示。

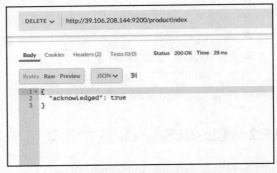

图 15-2　创建索引　　　　　　　　　　　图 15-3　删除索引

当然 Spring Data ElasticSearch[二]也提供了工具类 ElasticsearchTemplate 进行索引的相关操作。例如使用 createIndex()方法创建索引和使用 deleteIndex()方法删除索引。

例如创建索引：

```
@Autowired
private ElasticsearchTemplate esTemplate;
@Test
public void testIndex() {
    esTemplate.createIndex(Product.class);          //创建索引
}
```

15.2.2 索引映射 mappings

索引的 mappings 定义了文档的每个字段的数据类型：声明一个变量为 String 类型的字段，此字段只能存储 String 类型的数据。同语言的数据类型相比，mappings 还有一些其他的含

　㊀ ElasticSearch 中索引的名称不能有大写字母。

　㊁ Spring Data 与 ElasticSearch 进行整合，让操作变得简单。通过两者进行整合，用户可以像操作关系型数据库一样操作 ElasticSearch，CURD、排序、分页操作统统一步到位。

义，ElasticSearch 不仅可以根据 mappings 判断一个字段中是什么类型的值，还可以根据 mappings 来索引数据以及判断数据能否被搜索到。

接下来可以通过 Postman 工具查看刚才创建的 productindex 索引的 mappings 信息。如图 15-4 所示。

由于 ElasticSearch 中只创建了一个 productindex 索引，所以 mappings 的内容为空。接下来为 productindex 索引增加一个 product 类型，product 类型中含有 productName、price、brand、createTime 四个字段。mappings 定义字段用 properties 关键词，里面的 type 有以下几种类型，见表 15-2。

图 15-4　获取 mappings 信息

表 15-2　mappings 中的数据类型

类　型	含　义
string	文本字符类型
数值类型	Byte\short\integer\long\float\double
date	日期类型
Boolean	布尔类型
ip	IP 类型，以数字形式简化 IPv4 地址

例如，请求 url 为http://39.106.208.144:9200/productindex/product/_mapping?pretty[一]，以 post 方式请求，请求体为 Json，具体内容如下，效果如图 15-5 所示。

```
{
    "product": {
        "properties": {
            "productName": {
                "type": "string"
            },
            "price": {
                "type": "double"
            },
            "brand": {
                "type": "string"
            },
            "createDate": {
                "type": "date"
            }
        }
    }
}
```

mappings 支持再次添加字段操作，按照上面的格式，在请求体中添加要增加的字段以及类型即可，但是不支持修改已增加索引的字段类型。例如修改 price 类型由 double 类型变成

[一] 在任意的查询字符串中增加 pretty 参数，会让 Elasticsearch 美化输出（pretty-print）Json 响应，便于阅读。

String 类型，发送请求如图 15-6 所示。

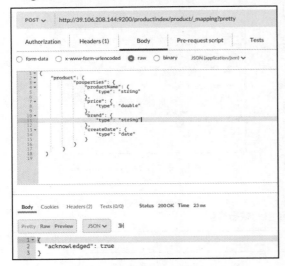

图 15-5　创建 product 类型

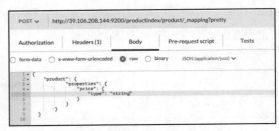

图 15-6　修改字段类型

返回结果如下：

```
{
    "error": {
      "root_cause": [
        {
          "type": "illegal_argument_exception",
          "reason": "mapper [price] of different type, current_type [double], merged_type [string]"
        }
      ],
      "type": "illegal_argument_exception",
      "reason": "mapper [price] of different type, current_type [double], merged_type [string]"
    },
    "status": 400
}
```

通过结果可以看到映射是不支持修改字段类型的。

15.2.3　ElasticSearch 之 Head 插件

在学习 ElasticSearch 的过程中，需要通过一些工具如 Head 插件查看 ElasticSearch 的运行状态以及数据。

ElasticSearch-Head 是一个图形化的集群操作和管理工具，可以对集群进行傻瓜式操作。可以通过插件形式把它集成到 ElasticSearch，也可以安装成一个独立应用。

Head 插件可参考第 19 章自行安装，这里就不再赘述。安装完成后打开浏览器，输入插件安装的 IP 地址，例如：http://{ip}:9200/_plugin/head/，界面如图 15-7 所示。

在地址栏输入 ElasticSearch 服务器的 IP 地址和端口，点击连接按钮（connect）就可以连接到 ElasticSearch 集群。连接后的视图如图 15-8 所示。在界面中，可以看到 ElasticSearch 集群的基本信息，例如节点情况、索引情况等。

主分片与副本的区别是粗细边框（主分片是粗边框），如图 15-9 所示。

第 15 章 ElasticSearch

图 15-7　Head 页面

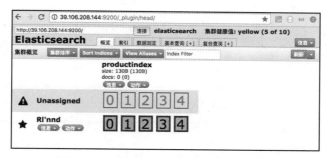

图 15-8　Head 页面查看 ElasticSearch 信息

每个索引下面有信息（info）和动作（action）两个按钮。信息可以查看索引的状态和 mapping 的定义。动作可对索引进行新建别名、刷新、Flush 刷新、优化、网关快照、测试分析器、关闭、删除操作。如图 15-10 所示。

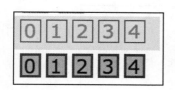

图 15-9　分片信息

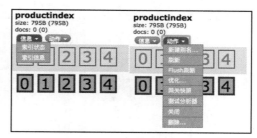

图 15-10　索引菜单

在 Head 的索引页签中可以新建索引、查看索引大小以及查看文档数量等，如图 15-11 所示。

图 15-11　索引操作

在 Head 的数据浏览（Browser）页签中可以针对选定的索引进行查询数据相关操作，如图 15-12 所示。

在 Head 的基本查询（Structured Query）页签中可以针对选定的索引进行基本查询操作，如图 15-13 所示。

Head 的复合查询（Any Request）页签类似一个 Restful 客户端，可以对 ElasticSearch 执行

321

相关操作，如图 15-14 所示。

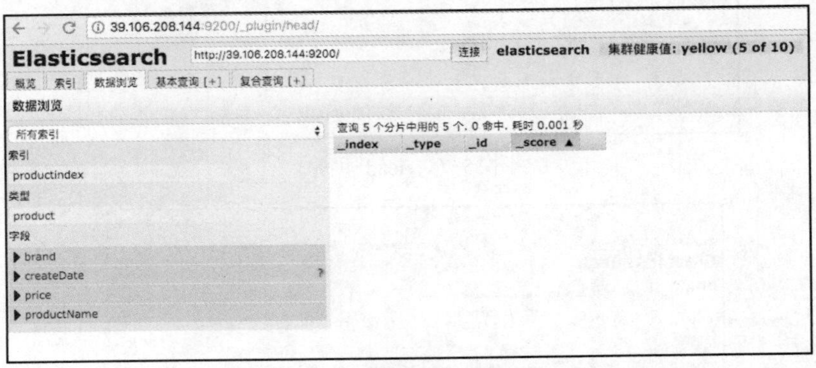

图 15-12　数据浏览

图 15-13　数据查询

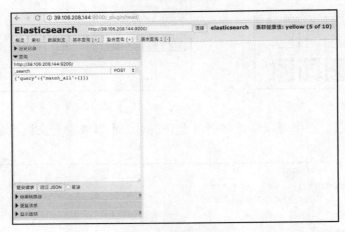

图 15-14　复合查询

Head 插件使用起来非常简单，使用此工具基本可以完成对 ElasticSearch 所支持的增删改查等相关操作。此工具的具体使用这里就不再演示，大家只要熟悉使用方法即可。

15.2.4　ElasticSearch 中文插件集成

ElasticSearch 内置的分词器○对中文的支持并不友好，它把中文拆分为单个字来进行全文

○ 分词器：接受一个字符串作为输入，将这个字符串拆分成独立的词或语汇单元（token）（可能会丢弃一些标点符号等字符），然后输出一个语汇单元流（token stream）。

检索，这就常常造成指定的文档没有被搜索到，中文搜索不能满足实际业务需求的情况。大部分情况归因于分词器和映射 mappings 的定义存在问题。

为了解决此问题，可以先对 ElasticSearch 分析过程进行调试。ElasticSearch 提供了 _analyze 和 _explain 这两个专用的 REST API。

- _explain 用来帮助分析文档的相关性评分。
- _analyze 可以帮助分析每一个字段(field)或者某个分析器（analyzer）/分词器（tokenizer）。

例如使用_analyze 分析文本"java 程序员"，使用 Postman 工具进行演示。分析效果如图 15-15 所示。

http://39.106.208.144:9200/_analyze?pretty&analyzer=standard&text="java 程序员"

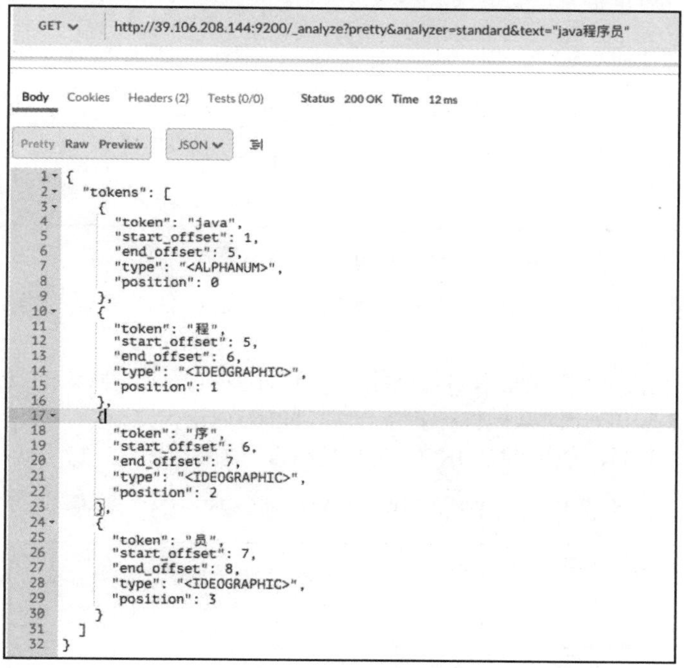

图 15-15　默认分词结果

ElasticSearch 中的默认分词器将"java 程序员"拆分成了四个字，而实际希望它能拆分成"java"、"程序员"两个词。

（1）IK-Analysis 分词器

这里介绍一款比较常用的中文分词器 IK-Analysis[○]，它是针对 ElasticSearch 的分词器扩充的中文分词插件。此插件的安装方法可以参考第 19 章自行安装，这里介绍其用法。使用安装好的 IK 分词器，再次针对"java 程序员"进行分析，IK 有两个分析器：ik_smart[○] 和 ik_max_word[○]，这里演示使用 ik_smart。分析结果如图 15-16 所示。

○ IK-Analysis 的网址是 https://github.com/medcl/elasticsearch-analysis-ik。

○ ik_smart 会做最粗粒度的拆分，例如会将"java 程序员"拆分为"java,程序员"。

○ ik_max_word 会将文本做最细粒度的拆分，例如会将"java 程序员"拆分为"java,程序员,程序,序,员"，会穷尽各种可能的组合。

http://39.106.208.144:9200/_analyze?pretty&analyzer=ik_smart&text="java 程序员"

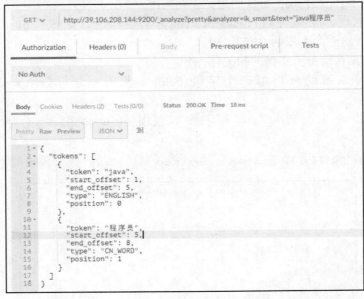

图 15-16　IK 分词器结果

通过输出的效果，可以看到使用 IK 分词器后，将短语拆分成需要的格式。

（2）自定义扩展词

很多行业都有一些特定的专业术语，网络上也有许多流行语，默认的 IK 分词器显然没法全面覆盖它们，需要在 IK 分词器插件中扩展配置自己的词语。

例如"丑橘"这个词，使用默认 IK 分析"丑橘"，效果如图 15-17 所示。

http://39.106.208.144:9200/_analyze?pretty&analyzer=ik_smart&text="丑橘"

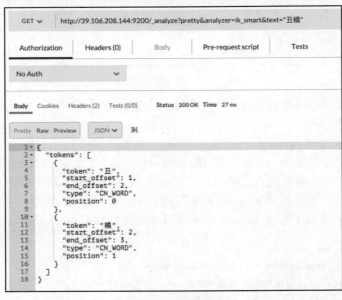

图 15-17　不支持的分词

显然 IK 的词典里面没有"丑橘"这个词，所以它被拆分成单个字。通过查看 IK 的配置文件{安装路径}/elasticsearch/config/analysis-ik/IKAnalyzer.cfg.xml，找到自定义词语的配置文件。IKAnalyzer.cfg.xml 具体内容如下：

```xml
<?xml version="1.0" encoding="UTF-8"?>
<!DOCTYPE properties SYSTEM "http://java.sun.com/dtd/properties.dtd">
<properties>
    <comment>IK Analyzer 扩展配置</comment>
    <!--用户可以在这里配置自己的扩展字典 -->
    <entry key="ext_dict">custom/mydict.dic;custom/single_word_low_freq.dic</entry>
    <!--用户可以在这里配置自己的扩展停止词字典-->
    <entry key="ext_stopwords">custom/ext_stopword.dic</entry>
    <!--用户可以在这里配置远程扩展字典 -->
    <!-- <entry key="remote_ext_dict">words_location</entry> -->
    <!--用户可以在这里配置远程扩展停止词字典-->
    <!-- <entry key="remote_ext_stopwords">words_location</entry> -->
</properties>
```

根据文件内的提示，需要在 custom/mydict.dic 文件里面增加自定义词汇，将"丑橘"添加进来。打开 mydict.dic 文件，添加"丑橘"词语，保存后，重启一下 ElasticSearch 服务。再次使用 IK 分析"丑橘"，效果如图 15-18 所示。

图 15-18　自定义后的分词效果

由输出结果可见 IK 分词器已经匹配到了"丑橘"这个词。

15.2.5　ElasticSearch 中文检索示例

本节简单演示如何使用 IK 分词器实现检索功能。希望把检索到的关键词用红色字体显示出来（<font color='red'>XXX</font>），来凸显关键词功能。

创建一个名为 javadevmap-news 的索引，设置它的分析器用 IK，使用 ik_smart 分词器，并创建名为 news 的类型，它只有一个 message 字段，并指明使用 ik_smart 分词器。这里需要以 put 方式请求操作。创建索引和类型的步骤如下。

请求路径为：

```
http://39.106.208.144:9200/javadevmap-news
```

提交内容为：

```
{
    "settings" : {
        "analysis" : {
            "analyzer" : {
                "ik" : {
                    "tokenizer" : "ik_smart"
                }
            }
        }
    },
    "mappings" : {
        "news" : {
            "dynamic" : true,
            "properties" : {
                "message" : {
                    "type" : "string",
                    "analyzer" : "ik_smart"
                }
            }
        }
    }
}
```

使用 Postman 工具进行提交，如图 15-19 所示。

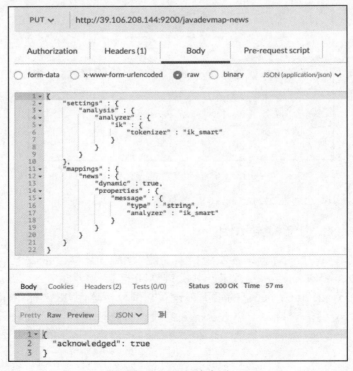

图 15-19　创建索引

执行完上面的操作后，使用 post 方式向索引添加一些数据，方便接下来的查询。

请求路径为：http://39.106.208.144:9200/javadevmap-news/news/1
提交内容为：{"message" : "航拍西班牙田野艳丽美景色彩柔美如缎带"}
请求路径为：http://39.106.208.144:9200/javadevmap-news/news/2
提交内容为：{"message" : "无人机航拍："天空之眼""}

使用 Postman 进行搜索演示，搜索"无人机"，以 post 方法提交 Json 数据请求。具体如下：

请求路径为：http://39.106.208.144:9200/javadevmap-news/news/_search?pretty

提交内容为：

```
{
    "query" : { "match" : { "message" : "无人机" }},
    "highlight" : {
        "pre_tags" : ["<font color='red'>"],
        "post_tags" : ["</font>"],
        "fields" : {
            "message" : {}
        }
    }
}
```

搜索结果如下：

```
{
    "took": 4,
    "timed_out": false,
    "_shards": {
        "total": 5,
        "successful": 5,
        "failed": 0
    },
    "hits": {
        "total": 1,
        "max_score": 0.13424811,
        "hits": [
            {
                "_index": "javadevmap-news",
                "_type": "news",
                "_id": "2",
                "_score": 0.13424811,
                "_source": {
                    "message": "无人机航拍："天空之眼""
                },
                "highlight": {
                    "message": [
                        "<font color='red'>无人机</font>航拍："天空之眼""
                    ]
                }
            }
        ]
    }
}
```

返回字段的含义见表 15-3。

表 15-3　返回字段含义

字段名	含义
took	ElasticSearch 查询花费的时间，单位为毫秒
time_out	标识 ElasticSearch 查询是否超时
_shards	描述 ElasticSearch 查询分片的信息，如共查询了多少个分片、成功的分片数量、失败的分片数量等
hits	ElasticSearch 搜索的结果，total 是全部的满足的文档数目，hits 是返回的实际数目
_score	文档的分数信息，跟排名相关度有关
highlight	高亮，针对检索结果中的关键词进行高亮显示

从上面的结果可见，可以通过 ElasticSearch 实现搜索能力，并且对关键词进行高亮显示。

15.3　SpringBoot 集成 ElasticSearch

本节以商品搜索功能为例讲解 ElasticSearch，通过 ElasticSearch 创建一个 Product 的索引，并进行增删改查相关操作。演示的程序并不是使用 ElasticSearch 的 Java 客户端直接访问 ElasticSearch 服务器，而是通过 Spring Data 的 ElasticSearch 组件与 ElasticSearch 服务交互。Spring Data ElasticSearch 已经提供了对于索引中类型（type）基本操作的支持。

创建一个 Spring Boot 工程，工程名为 ElasticSearchExample，使用 Spring Boot 工程整合 ElasticSearch 进行数据操作。

15.3.1　整合 ElasticSearch

（1）添加依赖

整合 ElasticSearch，需要添加起步依赖 spring-boot-starter-data-elasticsearch。具体如下：

```xml
<dependency>
    <groupId>org.springframework.boot</groupId>
    <artifactId>spring-boot-starter-data-elasticsearch</artifactId>
</dependency>
```

（2）修改配置文件

在配置文件 application.yml 中添加相关配置，这里需要填写计算机名称，以及集群节点的 IP 和端口，具体配置如下：

```yaml
server:
  port: 18093
spring:
  application:
    name: elasticsearch-example
  data:
    elasticsearch:
      cluster-name: elasticsearch
      cluster-nodes: 39.106.208.144:9300
      repositories:
        enabled: true
```

只要以上两步，即可完成 ElasticSearch 与 Spring Boot 的工程整合。

15.3.2 ElasticSearch 操作数据

可以通过实体类来进行类型的 mapping 映射。首先定义商品的数据结构，创建包 com.javadevmap.elasticexample.model，然后新建 Product 类，具体如下：

```java
@Document(indexName = "java-dev-map", type = "product")
public class Product {
    @Id
    private String id;
    private String productName;
    private Integer price;
    private String brand;
    private String productDesc;
    @Field(type = FieldType.String, index = FieldIndex.not_analyzed)
    private String productPic;

    public Product() {
    }

    public Product(String id, String productName, Integer price, String brand) {
        this.id = id;
        this.productName = productName;
        this.price = price;
        this.brand = brand;
    }
    //…省略 get 与 set 方法

    @Override
    public String toString() {
        return "Product{" +
                "id='" + id + '\'' +
                ", productName='" + productName + '\'' +
                ", price=" + price +
                ", brand='" + brand + '\'' +
                ", productDesc='" + productDesc + '\'' +
                ", productPic='" + productPic + '\'' +
                '}';
    }
}
```

在 Product 类上使用 Spring Data Elasticsearch 的注解@Document(indexName = "java-dev-map", type = "product")，此注解声明 Product 类为被索引的文档，属性 indexName 声明了索引的名称；属性 type 声明了索引中文档的类型。字段 id 上的注解@Id 即文档的主键，是唯一标识。Product 类中字段 productPic 的注解 @Field(type = FieldType.String, index = FieldIndex.not_analyzed)，声明了字段 productPic 的数据类型为 String，同时该字段的值不做分析，不做分析的字段在搜索时是会进行完全匹配的。字段 productName、price、brand、productDesc 没有添加@Field 注解，ElasticSearch 会根据其 Java 类型自动确定数据类型，同时对字段的值进行文本分析，因此可以支持全文检索。

ElasticSearch 常用于实体 Bean 的注解见表 15-4。

表 15-4　Bean 注解

注　解	含　义
@Document	文档对象（索引信息、文档类型）
@Id	文档主键，是唯一标识
@Field	每个文档的字段配置（类型、是否分词、是否存储、分词器）

在定义完 Product 实体 Bean 后，创建相应的仓库与 ElasticSearch 进行交互。Spring Data 为开发者简化了与数据源的操作交互方式。只需按照接口的方式声明所要执行的操作即可，具体的实现由 Spring Data 自动生成。Spring Data 提供了对常见的创建、查询、更新和删除操作以及分页和排序的支持，显著降低了开发人员操作 ElasticSearch 的门槛。

```java
public interface ProductRepository extends ElasticsearchRepository<Product, String> {
}
```

这里读者会发现不需要实现任何方法，当前 ProductRepository 已经具备了基本的增删改查能力。编写一个测试类进行验证。

```java
@RunWith(SpringRunner.class)
@SpringBootTest(classes = ElasticSearchExampleApplication.class)
public class TestProductRepository {
    @Autowired
    ProductRepository productRepository;

    @Test
    public void test(){
        Product product = new Product("1001", "JavaDevMap learn Elasticsearch", 67d, "计算机网络");
        Product saveBean = productRepository.save(product);
        System.out.println("save id is :"+saveBean.getId());
        Product findBean = productRepository.findOne(saveBean.getId());
        System.out.println("findBean is :"+findBean);
        findBean.setBrand("update brand");
        productRepository.save(findBean);
        Product updateBean = productRepository.findOne(findBean.getId());
        System.out.println("updateBean is "+updateBean);
        productRepository.delete(updateBean.getId());
        Product searchBean = productRepository.findOne(findBean.getId());
        System.out.println("delete search result is "+searchBean);
    }
}
```

运行结果如下：

```
save id is :1001
findBean is :Product{id='1001', productName='JavaDevMap learn Elasticsearch', price=67.0, brand='计算机网络', productDesc='null', productPic='null'}
updateBean is Product{id='1001', productName='JavaDevMap learn Elasticsearch', price=67.0, brand='update brand', productDesc='null', productPic='null'}
delete search result is null
```

当这些默认方法不能覆盖所有业务的数据操作需求时，就需要扩展方法。扩展使用方法时，就好像用英文直译要做的事情一样来定义方法名，例如想通过 brand 来查找 ElasticSearch

中某条数据，只要定义一个方法 findByBrand，这样就具备了通过 brand 查找的能力。下面在 ProductRepository 中定义几个方法来扩充 ElasticSearch 操作的能力。

```
public interface ProductRepository extends ElasticsearchRepository<Product, String> {
    Page<Product> findByProductName(String productName, Pageable pageable);
    List<Product>  findByBrand(String brand);
    List<Product>  findByPriceLessThan(double price);
    List<Product>  findByPriceGreaterThan(double price);
}
```

就如上面所写，只要把想做的事按照 JPA 的规则命名一个方法，即完成了 ElasticSearch 操作语句的建立，使用以上规则（参见 7.5.2 节），就可以自由组装接口方法来扩展 ElasticSearch 的操作能力。

15.4　SpringBoot 集成 Java Rest Client

ElasticSearch 的访问支持多种语言，在官网上可以看 ElasticSearch 使用标准的 RESTful API 和 JSON，构建和维护了很多其他语言的客户端，例如 Java, Python, .NET 和 PHP 等。

Java Rest Client 是相对于 Java API 更加轻量级的客户端，具有依赖少、自动负载均衡、使用方便、支持异步调用、权限认证、超时及失败重试等特点，而且 Java Rest Client 兼容所有的 ElasticSearch 版本并且完全遵守 Restful API 风格。下面简要演示其使用方法。

（1）添加依赖

在 pom 文件中添加如下依赖：

```xml
<dependency>
    <groupId>org.elasticsearch.client</groupId>
    <artifactId>elasticsearch-rest-client</artifactId>
    <version>6.2.4</version>
</dependency>
```

添加完上面的依赖，就可以在项目中使用 Java Rest Client 工具了。

（2）初始化客户端

Java Rest Client 主要的工具类为 RestClient，此工具类用于对 ElasticSearch 进行相关的数据操作。代码如下：

```java
private static RestClient restClient;
public static RestClient getRestClient(){
    return RestClient.builder(new HttpHost("39.106.208.144", 9200, "http"))
            .setRequestConfigCallback(
                    new RestClientBuilder.RequestConfigCallback() {
                        @Override
                        public RequestConfig.Builder customizeRequestConfig(
                                RequestConfig.Builder requestConfigBuilder) {
                            return requestConfigBuilder
                                    .setConnectTimeout(5000)
                                    .setSocketTimeout(60000);
                        }
                    }).setMaxRetryTimeoutMillis(60000).build();
}
```

```
@Before
public void getRest(){
    restClient=getRestClient();
}
```

注意由于一般线上业务 ElasticSearch 为集群模式，所以 RestClient 在创建的时候，可以传递多个 HttpHost，这样 RestClient 也会进行相应的负载均衡。

为防止业务数据的泄露，实际业务中的 ElasticSearch 一般会添加 Http 基本认证，那么就需要换一种连接 ElasticSearch 的方式，使用 CredentialsProvider 的方式连接 ElasticSearch，具体使用方法如下：

```
CredentialsProvider credentialsProvider = new BasicCredentialsProvider();
credentialsProvider.setCredentials(AuthScope.ANY,
                new UsernamePasswordCredentials("用户名", "密码"));
RestClient.builder(new HttpHost("39.106.208.144",9200,"http"))
    .setHttpClientConfigCallback(
        new RestClientBuilder.HttpClientConfigCallback() {
            @Override
            public HttpAsyncClientBuilder customizeHttpClient(
                HttpAsyncClientBuilder httpClientBuilder) {
                returnhttpClientBuilder.setDefaultCredentialsProvider(credentialsProvider);
            }
        }).setMaxRetryTimeoutMillis(60000).build();
```

（3）查询 ElasticSearch 信息

```
@Test
publicvoid testEsInfo()throwsException{
    String endpoint = "/";
    Map<String, String> params = Collections.singletonMap("pretty", "true");
    Response response = restClient.performRequest("GET", endpoint,params);
    System.out.println(EntityUtils.toString(response.getEntity()));
}
```

使用如上方法，会输出 ElasticSearch 的版本信息等，等同于访问 http://{ElasticSearchIP}:9200。

（4）创建索引

创建一个名为 java-dev-map-rest 的索引，设置它的分析器用 IK，使用 ik_smart 分词器；并创建名为 news 的类型，它只有一个 message 字段，此字段同样使用 ik_smart 分词器。这里需要以 put 方式提交请求。创建索引和类型的方法如下：

```
@Test
public void testCreateIndex() throws Exception{
    String method = "PUT";
    String endpoint = "/java-dev-map-rest";
    HttpEntity entity = new NStringEntity(
            "{\n" +
            "   \"settings\" : {\n" +
            "       \"analysis\" : {\n" +
            "           \"analyzer\" : {\n" +
            "               \"ik\" : {\n" +
            "                   \"tokenizer\" : \"ik_smart\"\n" +
```

```
                                        "                            }\n" +
                                        "                        }\n" +
                                        "                    }\n" +
                                        "                },\n" +
                                        "            \"mappings\" : {\n" +
                                        "                \"news\" : {\n" +
                                        "                    \"dynamic\" : true,\n" +
                                        "                    \"properties\" : {\n" +
                                        "                        \"message\" : {\n" +
                                        "                            \"type\" : \"string\",\n" +
                                        "                            \"analyzer\" : \"ik_smart\"\n" +
                                        "                        }\n" +
                                        "                    }\n" +
                                        "                }\n" +
                                        "            }\n" +
                                        "}", ContentType.APPLICATION_JSON);

    Response response = restClient.performRequest(method,endpoint,
                            Collections.<String, String>emptyMap(),entity);
    System.out.println(EntityUtils.toString(response.getEntity()));
}
```

（5）添加文档数据

```
@Test
public void testCreateDocument()throws Exception{
    Map<String, String> params = Collections.singletonMap("pretty", "true");
    String method = "PUT";
    String endpoint = "/java-dev-map-rest/news/1";
    HttpEntity entity = new NStringEntity(
            "{\"message\" : \"航拍西班牙田野艳丽美景色彩柔美如缎带\" }",
            ContentType.APPLICATION_JSON);
    Response response = restClient.performRequest(method,endpoint,params,entity);
    System.out.println(EntityUtils.toString(response.getEntity()));
    endpoint = "/java-dev-map-rest/news/2";
    entity = new NStringEntity("{\"message\" : \"无人机航拍：“天空之眼”\" }",
            ContentType.APPLICATION_JSON);
    response = restClient.performRequest(method,endpoint, params,entity);
    System.out.println(EntityUtils.toString(response.getEntity()));
}
```

（6）查询数据

希望查询关键词"无人机"，并以红色字体显示出关键字（`<font color='red'>`关键词`</font>`），来凸显关键词功能。

```
@Test
public void testGetDocsByParams() throws Exception {
    String method = "POST";
    String endpoint = "/java-dev-map-rest/news/_search?pretty";
    HttpEntity entity = new NStringEntity("{\n" +
            "    \"query\" : { \"match\" : { \"message\" : \"无人机\" }},\n" +
            "    \"highlight\" : {\n" +
            "        \"pre_tags\" : [\"<font color='red'>\"],\n" +
```

```
                "        \"post_tags\" : [\"</font>\"],\n" +
                "        \"fields\" : {\n" +
                "            \"message\" : {}\n" +
                "        }\n" +
                "    }\n" +
                "}", ContentType.APPLICATION_JSON);

        Response response = restClient.performRequest(method,endpoint,
                            Collections.<String, String>emptyMap(),entity);
        System.out.println(EntityUtils.toString(response.getEntity()));
    }
```

执行上面的程序，可见 Java Rest Client 使用非常方便。在实际业务中，可以先在 Postman 等工具中测试好 ElasticSearch 网络请求，然后复制到 NStringEntity 中，就可以完成相应的业务操作。

第 16 章 定 时 任 务

业务系统常常由于某些业务的需求，要在特定时刻或者特定时间间隔中做一些事情。例如研发一个电商平台，需要实时查看当前订单成交总额情况，这就需要用定时任务不停地查看系统内的订单并且统计总额；或者对于用户还未支付的订单，想在订单有效期内提醒用户，这就需要定时查看系统内的未支付订单，并且监控订单的有效时间，在失效前提醒用户支付；或者运营人员期望每天生成一个统计表，用来展示前一天平台内所有用户的购买情况，这也需要用一个定时任务每天按时启动统计。以上这些应用场景都是定时任务的用武之地。

16.1 Spring Boot 定时任务

Spring Boot 工程可以轻松实现单个服务的定时任务，只要在方法上添加@Scheduled 注解，此方法就会根据注解中指定的定时规则去执行。

16.1.1 单线程定时任务

根据之前章节的讲解，创建一个 Spring Boot 工程，命名为 ElasticJobExample，接下来几节将在此工程中分别演示@Scheduled 形式的定时任务和 ElasticJob 形式的分布式任务。

（1）单个任务

在工程中添加类 TimedTask，在此类中实现如下内容：

```java
@Component
public class TimedTask {
    @Scheduled(cron="0/10 * * * * ?")
    public void task1() {
        System.out.println(Thread.currentThread().getName() + " task1: " + new Date());
    }
}
```

运行服务结果如下：

```
pool-3-thread-1 task1: Wed May 23 14:41:40 CST 2018
pool-3-thread-1 task1: Wed May 23 14:41:50 CST 2018
pool-3-thread-1 task1: Wed May 23 14:42:00 CST 2018
…
```

由输出可见，task1 方法每隔 10s 打印一次线程和时间信息，@Scheduled 注解标明此方法是一个定时任务，cron 属性指定了定时任务执行的时间周期。cron 属性不仅能够指定时间间隔，还能指定具体时间。cron 的用法会在后面介绍。

（2）多个任务

在 TimedTask 类中，还可以定义一个由@Scheduled 注解标注的方法，这样在此类中就有两个定时任务，例如在类中添加如下方法，指定每 5s 执行一次任务。

```
@Scheduled(cron="0/5 * * * * ?")
public void task2() {
    System.out.println(Thread.currentThread().getName() + " task2: " + new Date());
}
```

运行服务结果如下：

```
pool-3-thread-1 task2: Wed May 23 14:47:55 CST 2018
pool-3-thread-1 task2: Wed May 23 14:48:00 CST 2018
pool-3-thread-1 task1: Wed May 23 14:48:00 CST 2018
pool-3-thread-1 task2: Wed May 23 14:48:05 CST 2018
pool-3-thread-1 task2: Wed May 23 14:48:10 CST 2018
pool-3-thread-1 task1: Wed May 23 14:48:10 CST 2018
```

由输出可见，task2 每隔 5s 执行一次任务，task1 每隔 10s 执行一次任务，但是你会发现这两个任务使用的是同一个线程，那么某一任务如果执行时间过长，会对另一任务造成什么影响呢？

（3）单线程多任务的相互影响

修改 task1 方法，在方法中添加一个 sleep 方法，模拟任务执行耗时，观察对两个定时任务的影响。

```
@Scheduled(cron="0/10 * * * * ?")
public void task1() {
    System.out.println(Thread.currentThread().getName() + " task1: " + new Date());
    try {
        TimeUnit.SECONDS.sleep(10);
    } catch (Exception e) {
        e.printStackTrace();
    }
}
```

运行服务结果如下：

```
pool-3-thread-1 task1: Wed May 23 15:01:20 CST 2018
pool-3-thread-1 task2: Wed May 23 15:01:30 CST 2018
pool-3-thread-1 task2: Wed May 23 15:01:35 CST 2018
pool-3-thread-1 task1: Wed May 23 15:01:40 CST 2018
pool-3-thread-1 task2: Wed May 23 15:01:50 CST 2018
pool-3-thread-1 task2: Wed May 23 15:01:55 CST 2018
```

由输出可见，此种情况，由于 task1 执行耗时过长，会导致 task2 需要等待 task1 执行完之后，才能开始计时和执行任务，可见两个定时任务在单线程的情况下是相互阻塞的。

16.1.2 多线程定时任务

使用多线程的方式执行任务，可以避免任务间的相互干扰。下面介绍使用注解和线程池两种方式实现多线程定时任务。

（1）注解实现多线程

在 TimedTask 类上添加注解@EnableAsync，在两个定时任务的方法上添加注解@Async。执行程序可以看到如下输出。

```
SimpleAsyncTaskExecutor-1 task2: Wed May 23 15:34:50 CST 2018
SimpleAsyncTaskExecutor-2 task1: Wed May 23 15:34:50 CST 2018
```

```
SimpleAsyncTaskExecutor-3 task2: Wed May 23 15:34:55 CST 2018
SimpleAsyncTaskExecutor-4 task1: Wed May 23 15:35:00 CST 2018
SimpleAsyncTaskExecutor-5 task2: Wed May 23 15:35:00 CST 2018
SimpleAsyncTaskExecutor-6 task2: Wed May 23 15:35:05 CST 2018
SimpleAsyncTaskExecutor-7 task2: Wed May 23 15:35:10 CST 2018
SimpleAsyncTaskExecutor-8 task1: Wed May 23 15:35:10 CST 2018
```

由输出可见，此种方式总是会新创建一个线程用于执行定时任务，并且在 task1 任务中，方法的耗时完全不影响任务的时间间隔。

（2）线程池实现多线程

去掉在 TimedTask 类中添加的多线程注解，在工程中添加一个配置类 ScheduledConfig，此类用来创建线程池，具体内容如下：

```
@Configuration
public class ScheduledConfig implements SchedulingConfigurer {
    @Override
    public void configureTasks(ScheduledTaskRegistrar scheduledTaskRegistrar) {
        scheduledTaskRegistrar.setScheduler(setTaskExecutors());
    }

    @Bean
    public Executor setTaskExecutors(){
        return Executors.newScheduledThreadPool(3);
    }
}
```

运行服务结果如下：

```
pool-1-thread-1 task2: Wed May 23 16:04:05 CST 2018
pool-1-thread-2 task2: Wed May 23 16:04:10 CST 2018
pool-1-thread-2 task1: Wed May 23 16:04:10 CST 2018
pool-1-thread-1 task2: Wed May 23 16:04:15 CST 2018
pool-1-thread-2 task2: Wed May 23 16:04:20 CST 2018
pool-1-thread-2 task2: Wed May 23 16:04:25 CST 2018
pool-1-thread-3 task1: Wed May 23 16:04:30 CST 2018
pool-1-thread-1 task2: Wed May 23 16:04:30 CST 2018
pool-1-thread-1 task2: Wed May 23 16:04:35 CST 2018
```

由输出可见，task2 的执行没有受到 task1 耗时的影响，task1 本身的计时受到了自己运行时间的影响。

16.1.3 用定时任务实时统计

上面演示了定时任务的使用方法，但是要想真正理解一种技术还需要在实战中进行应用。下面模拟一种情况，让定时任务实时统计电商平台当天的订单流水。

（1）创建一个数据库表

创建一个简单的数据库订单表 order_job，用于记录一个电商平台的订单信息，此表中每个字段的含义如下：

- id：订单 ID。
- price：订单的价格。
- userid：此订单的购买者。

- status：订单支付状态，0 表示未支付，在本节中为了使逻辑简单明了所以没有使用此字段，后面会使用此字段。
- createtime：订单创建时间。
- statis：订单是否已经计入统计汇总，本节没有使用此字段。

创建完以上数据库表后，在工程中添加 Mybatis、起步依赖及工程配置，以实现操作数据库的能力。

（2）添加自定义 mapper

由于要对订单流水进行统计操作，所以需要自定义 mapper 用来操作数据库，在 resources/mybatis/manual 目录下，添加文件 OrderManualMapper.xml，此文件的内容如下：

```xml
<?xml version="1.0" encoding="UTF-8" ?>
<!DOCTYPE mapper PUBLIC "-//mybatis.org//DTD Mapper 3.0//EN" "http://mybatis.org/dtd/mybatis-3-mapper.dtd" >
<mapper
    namespace="com.javadevmap.elasticjobexample.model.mapper.OrderManualMapper">
    <resultMap id="StatisResultMap"
        type="com.javadevmap.elasticjobexample.model.OrderStatis">
        <result column="total" property="priceTotal" jdbcType="DOUBLE" />
        <result column="cou" property="count" jdbcType="INTEGER" />
    </resultMap>

    <sql id="Example_Where_Clause">
        …
    </sql>

    <select id="getOrderStatis"
    parameterType="com.javadevmap.elasticjobexample.model.OrderJobExample"
        resultMap="StatisResultMap">
        select COALESCE(SUM(price),0) as total,COUNT(1) as cou from order_job
        <if test="_parameter != null">
            <include refid="Example_Where_Clause" />
        </if>
    </select>
</mapper>
```

在上面的 SQL 语句中，select 方法主要查询当天订单的总金额和订单总数。

添加 resultMap 对应的数据类型，此类型用于承接数据库查询到的数据。

```java
public class OrderStatis {
    private Double priceTotal;
    private Long count;
    //...省略 get 与 set 方法
    @Override
    public String toString() {
        return "pricetotal = " + priceTotal + " count = " + count;
    }
}
```

添加 OrderManualMapper 接口，用于映射 select 方法。

```
public interface OrderManualMapper {
    public OrderStatis getOrderStatis(OrderJobExample example);
}
```

(3）添加数据操作类

添加接口类 OrderDao[一]和实现类 OrderDaoImpl，用于操作自定义 mapper 及设置逻辑查询条件。

```
@Repository
public class OrderDaoImpl implements OrderDao{
    @Autowired
    private OrderManualMapper manualMapper;

    @Override
    public OrderStatis getStatis() {
        OrderJobExample example = new OrderJobExample();
        OrderJobExample.Criteria criteria = example.createCriteria();
        DateFormat format = new SimpleDateFormat("yyyy-MM-dd");
        Date date = null;
        try {
            date = format.parse(format.format(new Date()));
        } catch (Exception e) {
            e.printStackTrace();
        }
        criteria.andCreatetimeGreaterThanOrEqualTo(date);
        OrderStatis statis = manualMapper.getOrderStatis(example);
        return statis;
    }
}
```

在 getStatis 方法中，获取当天时间作为查询条件，然后使用自定义 mapper 查询当天平台内的订单总体情况并且返回。

（4）添加定时任务

添加一个定时任务，此任务每 10s 执行一次，查询并且显示平台内的订单总体信息。

```
@Component
public class TimedTask {
    @Autowired
    private OrderDao dao;

    @Scheduled(cron="0/10 * * * * ?")
    public void getStatis() {
        System.out.println(dao.getStatis());
    }
}
```

[一] 本章展示的代码中省略了 OrderDao 接口类的代码。

执行程序后，向数据库中添加一个订单，可以看到如下输出：

> pricetotal = 0.0 count = 0
> …
> pricetotal = 25.6 count = 1

使用此定时任务，可以实时查看平台内的订单情况，类似于双十一实时查看订单交易额。

16.2 Cron 配置

Cron 表达式由 6 到 7 个时间元素组成，每个字符表示一个时间含义，从左到右（用空格隔开）依次表示为：

秒 分 小时 月份中的日期 月份 星期 年份

（1）各字段允许的值及允许的特殊字符见表 16-1。

表 16-1 Cron 元素

字 段	允 许 值	允许的特殊字符
秒（Seconds）	0~59 的整数	, - * /　四个字符
分（Minutes）	0~59 的整数	, - * /　四个字符
小时（Hours）	0~23 的整数	, - * /　四个字符
日期（DayofMonth）	1~31 的整数（与月份有关）	, - * ? / L W C　八个字符
月份（Month）	1~12 的整数或者 JAN-DEC	, - * /　四个字符
星期（DayofWeek）	1~7 的整数或者 SUN-SAT（1=SUN）	, - * ? / L C #　八个字符
年(可选)（Year）	1970~2099	, - * /　四个字符

（2）各特殊字符的含义如下：

1）*：表示匹配该域的任意值，假如在 Minutes 域使用*，则表示每分钟都会触发事件。

2）?：只能用在 DayofMonth 和 DayofWeek 两个域。如果在其中一个域设置了值，那么另一个域需要使用"?"符号，因为它们会互相影响。

3）-：表示范围，例如在 Minutes 域使用 5-20，表示从 5 分到 20 分每分钟触发一次。

4）/：表示起始时间开始触发，然后每隔固定时间触发一次，例如在 Minutes 域使用 5/20，则表示第 5 分钟触发一次，每隔 20 分钟触发一次，即 25，45 分别触发一次。

5），：表示列出枚举值。例如：在 Minutes 域使用 5,20，则表示在 5 和 20 分钟触发一次。

6）L：表示最后，只能出现在 DayofWeek 和 DayofMonth 域，如果在 DayofWeek 域使用 6L，意味着在最后一个星期五触发。

7）W：表示有效工作日（周一到周五），只能出现在 DayofMonth 域，系统将在离指定日期最近的有效工作日触发事件。例如：在 DayofMonth 使用 5W，如果 5 日是星期六，则将在最近的工作日：星期五，即 4 日触发。如果 5 日是星期天，则在 6 日（星期一）触发；如果 5 日在星期一到星期五中的某天，就在 5 日触发。请注意 W 的最近寻找不会跨过月份。

8）LW：这两个字符可以连用，表示在某个月最后一个工作日。

9）#：用于确定每个月第几个星期几，只能出现在 DayofWeek 域。例如 4#2 表示某月的第二个星期三。

（3）表达式举例见表 16-2

表 16-2　Cron 表达式

表　达　式	含　　义
0 0 2 1 * ? *	表示在每月的 1 日凌晨 2 点
0 15 9 ? * MON-FRI	表示周一到周五每天上午 9:15
0 0 9,14,16 * * ?	表示每天上午 9 点，下午 2 点，4 点
0 0/30 9-17 * * ?	表示朝九晚五时间内每半小时
0 0 12 * * ?	表示每天中午 12 点
0 15 9 * * ? 2018	表示 2018 年的每天上午 9:15
0 * 14 * * ?	表示在每天下午 2 点到下午 2:59 期间的每 1 分钟
0 0-5 14 * * ?	表示在每天下午 2 点到下午 2:05 期间的每 1 分钟
0 10,45 14 ? 3 WED	表示每年三月的每个星期三的下午 2:10 和 2:45
0 15 9 15 * ?	表示每月 15 日上午 9:15
0 15 9 L * ?	表示每月最后一日的上午 9:15
0 15 9 ? * 6L	表示每月的最后一个星期五上午 9:15
0 15 9 ? * 6L 2018-2025	表示 2018 年至 2025 年的每月的最后一个星期五上午 9:15
0 15 9 ? * 6#3	表示每月的第三个星期五上午 9:15

如果对 cron 表达式不够熟悉，可以在网上搜索在线的 cron 表达式生成器。

16.3　ElasticJob 介绍

在上面定义的数据库中，记录了订单的基本信息，其中包含了用户是否已经支付的状态，订单允许用户有 48 小时的支付时间⊖，如果超过 48 小时未支付则订单失效。为了让用户有一个良好的购物体验或者为了促使订单能够成交，还是希望用户能够在规定时间内支付订单，这就需要在订单失效前提示用户支付订单。为了达到此目的，一般会使用定时任务轮流查询未支付订单，然后提示用户支付。

使用前面介绍的定时任务方法，确实能够实现此目的，但是存在以下问题：

- 当只启动一个定时任务程序时，虽然最终能够达到通知用户的目的，但是时效性是个问题，单一程序遍历列表并且实现通知的时间较长。
- 当为了提高执行速度，启动多个定时任务程序时，集群中的服务无法知道其他服务是否已经执行了通知逻辑，即多个定时任务之间如何协作是个问题。

针对以上两个问题，ElasticJob 使用了非常简单的方法就实现了多个定时任务间的协作，从而提高了执行速度。它的原理非常简单，对要执行的任务设定分片总数，然后多个任务实例通过 ElasticJob 获取各自的分片信息，根据自己的分片信息执行对应的数据部分，从而实现了横向扩容的能力。

⊖ 因为订单是占用商品库存的，如果用户长时间不支付会使商品数量被未生效订单占用但是又无法卖出，这是不符合商家利益的，所以订单都会有一个超时时间。

ElasticJob 是一个开源的分布式任务框架,它使用 Zookeeper 作为各分片任务的信息管理中心,各任务节点的分片信息可以通过 Zookeeper 进行查看。

ElasticJob 的作用是根据分片总数为每个定时任务的实例分配准确的片段值。ElasticJob 不负责定时任务获取到分片值之后的业务逻辑处理。举例来说,现在设定 ElasticJob 的总分片值是 2,同时启动两个定时任务服务 A 和 B,ElasticJob 会保证给 A 和 B 分配对应的 1 和 2 这两个值,至于 A 拿到了 1 之后应该怎么执行逻辑,这是编程者要自己定义的。例如在 A 服务中拿到了 1 这个分片,可以选择数据库中 userid 对总数 2 取模后等于 1 的数据进行处理。

ElasticJob 支持几种任务模式,下面主要介绍 Simple 类型作业和 Dataflow 类型作业。这两种作业模式都需要在工程中引入如下依赖。

```xml
<dependency>
    <groupId>com.dangdang</groupId>
    <artifactId>elastic-job-lite-core</artifactId>
    <version>2.1.5</version>
</dependency>
<dependency>
    <groupId>com.dangdang</groupId>
    <artifactId>elastic-job-lite-spring</artifactId>
    <version>2.1.5</version>
</dependency>
```

引入依赖后,在 yml 配置文件中添加注册中心 Zookeeper 的信息:

```
regCenter:
    serverList: 39.106.10.196:2181
    namespace: jdmelasticjob
```

添加注册信息配置类:

```java
@Configuration
@ConditionalOnExpression("'${regCenter.serverList}'.length() > 0")
public class RegistryCenterConfig {
    @Bean(initMethod = "init")
    public ZookeeperRegistryCenter regCenter(@Value("${regCenter.serverList}") final String serverList, @Value("${regCenter.namespace}") final String namespace) {
        return new ZookeeperRegistryCenter(new ZookeeperConfiguration(serverList, namespace));
    }
}
```

添加任务记录配置:

```java
@Configuration
public class JobEventConfig {
    @Resource
    private DataSource dataSource;

    @Bean
    public JobEventConfiguration jobEventConfiguration() {
        return new JobEventRdbConfiguration(dataSource);
    }
}
```

完成如上几步,此工程就实现了 ElasticJob 的基本信息配置,下面就可以根据不同的任务类型编写不同的代码了。

16.4 简单任务

使用简单任务只需要配置此任务的总体分片规则和时间调度规则，然后实现任务的配置类，最后在真正执行任务逻辑的类中继承 SimpleJob 接口类，并且实现其 execute 方法，这样就完成了简单任务的全部工作。

（1）时间及分片设置

在 yml 文件中添加如下信息：

```yml
simpleJob:
    cron: 0 0 0/1 * * ?
    shardingTotalCount: 3
    shardingItemParameters: 0=A,1=B,2=C
```

使用此配置，设定任务总体分片数为 3，每个小时执行一次定时任务。

（2）添加配置类

```java
@Configuration
public class SimpleJobConfig {
    @Resource
    private ZookeeperRegistryCenter regCenter;

    @Resource
    private JobEventConfiguration jobEventConfiguration;

    @Bean
    public SimpleJob simpleJob() {
        return new SpringSimpleJob();
    }

    @Bean(initMethod = "init")
    public JobScheduler simpleJobScheduler(final SimpleJob simpleJob,
                    @Value("${simpleJob.cron}") final String cron,
                    @Value("${simpleJob.shardingTotalCount}") final int shardingTotalCount,
        @Value("${simpleJob.shardingItemParameters}") final String shardingItemParameters) {
        return new SpringJobScheduler(simpleJob, regCenter,
                        getLiteJobConfiguration(simpleJob.getClass(),
                        cron, shardingTotalCount, shardingItemParameters),
                        jobEventConfiguration);
    }

    private LiteJobConfiguration getLiteJobConfiguration(final Class<? extends SimpleJob> jobClass, final String cron, final int shardingTotalCount,final String shardingItemParameters) {
            return LiteJobConfiguration.newBuilder(
                    new SimpleJobConfiguration(JobCoreConfiguration.newBuilder(
                    jobClass.getName(), cron,
                    shardingTotalCount).shardingItemParameters(shardingItemParameters).build(),
                    jobClass.getCanonicalName())).overwrite(true).build();
    }
}
```

此类根据之前设置的配置项和新创建的定时任务对象启动调度执行。

(3) 定义定时任务执行逻辑

```java
public class SpringSimpleJob implements SimpleJob{
    @Autowired
    private OrderDao dao;

    @Override
    public void execute(ShardingContext shardingContext) {
        int total = shardingContext.getShardingTotalCount();
        int cur = shardingContext.getShardingItem();
        System.out.println(String.format("SpringSimpleJob------Thread ID: %s, 任务总片数: %s, 当前分片项: %s", Thread.currentThread().getId(), total, cur));
        System.out.println("SpringSimpleJob: " + Thread.currentThread().getId() + " cur = " + cur + " list is "+ dao.getTimeoutUserId(total, cur));
    }
}
```

此类是真正定时任务执行业务逻辑的地方，在 execute 方法中打印了总分片数和当前分片数，并且使用数据操作类 OrderDao 获取即将支付超时的用户 id。

(4) 基于当前分片数的数据处理

在 OrderDaoImpl 类中添加如下方法，此方法接收两个参数：总分片数和当前分片数，然后根据这两个参数进行数据查询，获取用户 id。

```java
public List<Long> getTimeoutUserId(int total,int cur){
    Map<Object, Object> map = new HashMap<Object, Object>();
    map.put("total",total);
    map.put("cur", cur);
    List<Long> list = manualMapper.getUnpaidUser(map);
    return list;
}
```

(5) 自定义 mapper

```xml
<select id="getUnpaidUser" parameterType="Map" resultType="java.lang.Long">
    select DISTINCT userid from order_job where status=0 and
    mod(userid,#{total,jdbcType=INTEGER})=#{cur,jdbcType=INTEGER} and
    TIMESTAMPDIFF(Hour,createtime,NOW())>=47
</select>
```

自定义 Mapper 类映射的接口方法为

```java
public List<Long> getUnpaidUser(Map<Object, Object> map);
```

此语句的目的是查询 userid 对总分片数取模等于此任务的分片值的即将到期未支付订单的 userid。

(6) 运行情况演示

运行一个定时任务实例，当达到时间规则时，可以看到如下输出：

```
SpringSimpleJob------Thread ID: 51, 任务总片数: 3, 当前分片项: 2
SpringSimpleJob------Thread ID: 50, 任务总片数: 3, 当前分片项: 1
SpringSimpleJob------Thread ID: 49, 任务总片数: 3, 当前分片项: 0
SpringSimpleJob: 50 cur = 1 list is [7, 10, 1, 4, 37]
```

```
SpringSimpleJob: 49 cur = 0 list is [6, 3, 33, 12, 9, 15, 18]
SpringSimpleJob: 51 cur = 2 list is [2, 5, 8, 23, 11, 98, 17]
```

由上面的输出可见,返回的 userid 对 3 取模等于它的分片数,即各个分片根据研发者定义的逻辑,处理自己分片相应的数据内容。

如果再启动一个任务实例,那么新启动的实例就会分担全部分片中某些分片的任务,这样就实现了横向扩展的目的。

16.5 流式任务

流式任务与简单任务不同的地方是此任务的逻辑实现类要继承 DataflowJob 接口类,此接口类中包含两个方法: fetchData 和 processData。fetchData 方法负责数据抓取,processData 方法负责对抓取到的数据进行处理。当流式处理数据时,只有当 fetchData 方法返回 null 或空 List 时作业才停止,否则作业会一直执行下去。

流式任务比较适用于在某一时刻,利用流式任务不停执行的特性,进行不间断的大量数据处理的工作。下面编写一个例子,让流式任务在每天 2 点统计前一天每个用户的购买情况。

(1) 时间及分片设置

在 yml 文件中添加如下内容。

```
dataflowJob:
    cron: 0 0 2 * * ?
    shardingTotalCount: 3
    shardingItemParameters: 0=A,1=B,2=C
```

使用此配置,设定任务总体分片数为 3,每天 2 点执行任务。

(2) 添加配置类

```
@Configuration
public class DataflowJobConfig {
    @Resource
    private ZookeeperRegistryCenter regCenter;

    @Resource
    private JobEventConfiguration jobEventConfiguration;

    @Bean
    public DataflowJob dataflowJob() {
        return new SpringDataflowJob();
    }

    @Bean(initMethod = "init")
    public JobScheduler dataflowJobScheduler(final DataflowJob dataflowJob,
    @Value("${dataflowJob.cron}") final String cron,
    @Value("${dataflowJob.shardingTotalCount}") final int shardingTotalCount,
    @Value("${dataflowJob.shardingItemParameters}") final String shardingItemParameters) {
        return new SpringJobScheduler(dataflowJob, regCenter,
                        getLiteJobConfiguration(dataflowJob.getClass(), cron,
                        shardingTotalCount, shardingItemParameters), jobEventConfiguration);
    }
```

```java
            private LiteJobConfiguration getLiteJobConfiguration(final Class<? extends DataflowJob> jobClass, final String cron, final int shardingTotalCount, final String shardingItemParameters) {
                return LiteJobConfiguration.newBuilder(
                        new DataflowJobConfiguration(JobCoreConfiguration.newBuilder(
                                jobClass.getName(), cron,
                                shardingTotalCount).shardingItemParameters(shardingItemParameters).build(),
                                jobClass.getCanonicalName(), true)).overwrite(true).build();
            }
        }
```

此类根据之前设置的配置项和新创建的定时任务对象启动调度执行。

（3）定义定时任务执行逻辑

```java
        public class SpringDataflowJob implements DataflowJob {
            @Autowired
            private OrderDao dao;

            @Override
            public List fetchData(ShardingContext shardingContext) {
                int total = shardingContext.getShardingTotalCount();
                int cur = shardingContext.getShardingItem();
                System.out.println(String.format("SpringDataflowJob fetchData ------Thread ID: %s, 任务总片数: %s, 当前分片项: %s", Thread.currentThread().getId(),total, cur));
                return dao.getStatisList(total, cur);
            }

            @Transactional
            @Override
            public void processData(ShardingContext shardingContext, List data) {
                int total = shardingContext.getShardingTotalCount();
                int cur = shardingContext.getShardingItem();
                System.out.println(String.format("SpringDataflowJob processData ------Thread ID: %s, 任务总片数: %s, 当前分片项: %s", Thread.currentThread().getId(), total, cur));

                //todo sth;

                dao.completeStatis(data);
            }
        }
```

SpringDataflowJob 类定义了定时任务执行的逻辑，其中 fetchData 方法用于分批次从数据库中查询前一天的订单列表；processData 方法根据查询到的数据进行统计，代码中的注释部分"todo sth;"表示可以把统计数据存入另外一个统计表中，这里囿于篇幅没有截取代码，可以到随书附带的工程中查看。processData 方法最后调用 dao.completeStatis(data)方法，用于对已经处理完的数据设置标记位，避免再次被 fetchData 方法查询出已经处理过的数据。

（4）数据查询及处理

在 OrderDaoImpl 中添加如下方法，getStatisList 方法负责查询用户的未统计订单，completeStatis 方法用于对已经统计完的数据设置标记位。

```java
        @Autowired
        private OrderJobMapper mapper;①
```

---

① 此 mapper 是使用 Mybatis 自动生成的 mapper。

```java
public List<OrderJob> getStatisList(int total,int cur){
    Map<Object, Object> map = new HashMap<Object, Object>();
    map.put("total",total);
    map.put("cur", cur);
    List<Long> userlist = manualMapper.getStatisUser(map);

    if(!userlist.isEmpty()) {
        Map<Object, Object> listmap = new HashMap<Object, Object>();
        listmap.put("list",userlist);
        List<OrderJob> list = manualMapper.getStatisOrder(listmap);
        return list;
    }
    return null;
}
public void completeStatis(List<OrderJob> list) {
    for (OrderJob orderJob : list) {
        orderJob.setStatis(true);
        mapper.updateByPrimaryKey(orderJob);
    }
}
```

（5）自定义的 mapper

在查询未统计订单时，使用了自定义的 SQL 查询方法，代码如下。

```xml
<select id="getStatisUser" parameterType="Map" resultType="java.lang.Long">
    select DISTINCT userid from order_job where statis=0 and
    mod(userid,#{total,jdbcType=INTEGER})=#{cur,jdbcType=INTEGER} and
    to_days(createtime)=to_days(DATE_SUB(CURDATE(),INTERVAL 1 DAY)) limit
    10
</select>

<select id="getStatisOrder" parameterType="Map" resultMap="BaseResultMap">①
    select id, price, userid, status, createtime, statis from order_job
    where statis=0 and
    to_days(createtime)=to_days(DATE_SUB(CURDATE(),INTERVAL 1 DAY)) and
    userid in
    <foreach item="item" index="index" collection="list" open="("
        separator="," close=")">
        #{item}
    </foreach>
</select>
```

自定义 Mapper 类映射的接口方法为：

```java
public List<Long> getStatisUser(Map<Object, Object> map);
public List<OrderJob> getStatisOrder(Map<Object, Object> map);
```

至此，使用 DataflowJob 形式的定时任务编写完成，它会在每天 2 点统计前一天的业务数据，可以让你在第二天上班时看到前一天详细的业务情况，给业务的发展以数据的支持。

---

① 此处用于演示，所以没有对订单的支付状态进行筛选，实际业务中的查询逻辑要比这个复杂；BaseResultMap 的定义和 Mybatis 自动生成的相同，记得将 Mybatis 自动生成的 BaseResultMap 的定义引入进此文件。

第 17 章　RabbitMQ

对于某些时限要求不高的业务，或者为了降低后端服务压力的情况，可能会用到消息队列。消息队列就像一个仓储或者转运中心，某些服务需要处理一些事务，但是又不急于得到返回，或者能够处理此事务的服务由于各种原因不能立刻返回，这种情况下就会把这个事务放入消息队列，然后由能够处理此事务的服务从消息队列中获取需要执行的事务再执行。所以消息队列在此种情形下达到了解耦、暂存、削峰的目的。

RabbitMQ 是实现了 AMQP[⊖]协议的消息中间件的一种，易用性、扩展性、高可用性较好，同时支持多种客户端，如 Python、Java、PHP、C 等。本章介绍 RabbitMQ 的主要用法。

17.1　队列传递字符串

新建两个 Spring Boot 服务，一个服务用于向消息队列中添加消息，另一个服务负责从消息队列中获取消息并且进行处理。发送消息的服务命名为 RabbitMQSender，处理消息的服务命名为 RabbitMQReceiver。服务间传递的消息是一个 String 类型的字符串。

17.1.1　消息队列基本配置

引入消息队列非常简单，只要添加消息队列的工程依赖，并且配置消息队列的连接信息即可。在两个服务中配置同样的基础信息，具体如下。

（1）添加消息队列依赖

在两个工程的 pom 文件中，添加如下依赖：

```
<dependency>
    <groupId>org.springframework.boot</groupId>
    <artifactId>spring-boot-starter-amqp</artifactId>
</dependency>
```

（2）添加消息队列配置

在 yml 文件中，添加如下 RabbitMQ 的连接配置：

```
spring:
  rabbitmq:
    host: 39.106.10.196
    port: 5673
    username: guest
    password: guest
```

如果此服务在一个 Spring Cloud 集群中，并且集群中已经通过 RabbitMQ 进行了信息的搜集，例如把链路监控信息通过 RabbitMQ 传到 Zipkin 中，那么上面的配置和 Zipkin 的消息队列的配置可以共用，不用重复添加。

⊖ 即 Advanced Message Queuing Protocol，高级消息队列协议，是应用层协议的一个开放标准，为面向消息的中间件设计。

17.1.2 发送方配置及使用

在发送方,需要配置消息队列中的具体接收消息的队列,然后向此队列发送 String 类型的消息。

(1) 队列配置

在工程中添加配置类 RabbitmqConfig,然后添加如下内容:

```
@Configuration
public class RabbitmqConfig {
    //message queue ************************
    @Bean
    public Queue StringQueue() {
        return new Queue("StringQueue");
    }
}
```

这样,就定义了 RabbitMQ 的队列名称为 StringQueue。

(2) 添加发送逻辑

新建一个专门用于向 RabbitMQ 发送消息的类 RabbitmqSender,在此类中添加向 RabbitMQ 发送消息的方法,具体如下:

```
@Repository
public class RabbitmqSender {
    @Autowired
    private AmqpTemplate rabbitTemplete;

    public String sendString() {
        rabbitTemplete.convertAndSend("StringQueue","string message send");
        return "string send ok!";
    }
}
```

新建一个 Controller 类,此类中提供一个接口,当用 HTTP 请求调用此接口时,此接口会使用如上方法向 RabbitMQ 中发送一条消息。

```
@RestController
@RequestMapping(value="/rabbitsender")
public class RabbitMQController {
    @Autowired
    private RabbitmqSender sender;

    @RequestMapping(value="/sendMessage",method=RequestMethod.GET)
    public Result<String> sendMessage() {
        String ret = sender.sendString();
        Result<String> result = new Result<>(ResultCode.OK, ret);
        return result;
    }
}
```

启动发送方服务,调用如上接口,可以在 RabbitMQ 的 StringQueue 队列中看到相应变化,如图 17-1 所示。

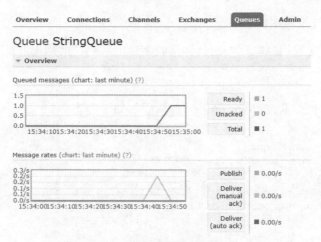

图 17-1　消息队列接收消息

17.1.3　接收方配置及使用

在接收方，需要做的是把接收逻辑与 RabbitMQ 中的队列绑定，当队列中有待处理消息时，接收方从中获取消息并且进行处理。

（1）队列配置

在工程中添加配置类 RabbitmqConfig，然后添加如下内容：

```
@Configuration
public class RabbitmqConfig {
    //message queue ************************
    @Bean
    public Queue StringQueue() {
        return new Queue("StringQueue");
    }
}
```

（2）添加队列监听，并且获取消息进行处理

添加一个队列监听类 MessageReceiver，此类通过@RabbitListener○注解监听指定的队列，并且通过@RabbitHandler 注解标注的方法获取队列信息进行处理。

```
@Component
@RabbitListener(queues = "StringQueue")
public class MessageReceiver {
    @RabbitHandler
    public void process(String message) {
        System.out.println("messageReceiver: " + message);
    }
}
```

上面的方法中，获取消息队列中的消息，并且输出至控制台。启动此接收服务，可以看到 RabbitMQ 中的变化，如图 17-2 所示。

○ 在代码中可以点击此注解查看其注解内容，其使用范围不只是作用于类之上。

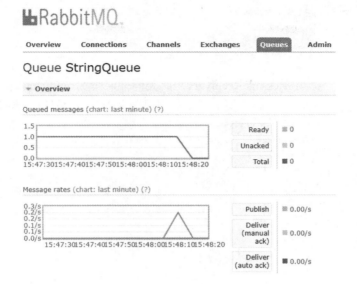

图 17-2　消息队列消息消耗

并且在控制台输出如下内容：

messageReceiver: string message send

17.1.4　多对多实现

消息队列不仅可以应用于一对一的场景，还可以实现多对多的通信。可以有多个消息发送方向同一队列发送消息，可以有多个消息接收方从同一队列中获取消息，这样就可以实现消息队列的多对多。这里简单演示此种情况。

（1）发送方改造

添加一个 RabbitmqSender2 类，此类添加一个发送方法，此方法需要传入一个 Index 参数用于消息计数，在前面的 RabbitmqSender 类中也同样添加此方法。

```
@Repository
public class RabbitmqSender2 {
    @Autowired
    private AmqpTemplate rabbitTemplete;

    public String sendString(int index) {
        rabbitTemplete.convertAndSend("StringQueue","string message send " + index);
        return "string send ok!";
    }
}
```

修改 Controller 类，注入 RabbitmqSender2 的类实例，添加一个批量发送的方法。

```
@RequestMapping(value="/sendMultiMessage",method=RequestMethod.GET)
public Result<String> sendMultiMessage() {
    for(int i=0;i<10;i++) {
        sender.sendString(i);
        sender2.sendString(i);
    }
```

```
            Result<String> result = new Result<>(ResultCode.OK, "OK");
            return result;
    }
```

（2）接收方改造

在接收方添加另一个类 MessageReceiver2，此类同样监听 StringQueue 队列。

```
@Component
@RabbitListener(queues = "StringQueue")
public class MessageReceiver2 {
    @RabbitHandler
    public void process(String message) {
        System.out.println("messageReceiver2: " + message);
    }
}
```

（3）效果演示

首先启动发送方，然后调用发送方的多发方法，可以在 RabbitMQ 中看到相应变化，如图 17-3 所示。

图 17-3　消息队列中收到的消息

启动接收方程序，可以在控制台看到如下信息输出：

```
messageReceiver: string message send 0
messageReceiver: string message send 0
messageReceiver: string message send 1
messageReceiver2: string message send 1
messageReceiver: string message send 2
messageReceiver2: string message send 2
messageReceiver: string message send 3
messageReceiver2: string message send 3
…
```

17.2 队列传递对象

使用 RabbitMQ 不仅可以在程序间实现消息的传递，还可以传递对象实例，只不过双方间传递的对象需要具备相同的 package 包名、对象类名，并且此类需实现序列化接口。下面来演示此种情况。

17.2.1 发送方配置及使用

首先配置 RabbitMQ 中的队列，作为发送方与接收方数据传递的通道，在 Rabbitmq Config 类中添加如下内容。

```
//object queue ***************************
@Bean
public Queue ObjectQueue() {
    return new Queue("ObjectQueue");
}
```

在 com.javadevmap.model 路径下添加 UserModel 类，此类是对象传递的数据结构，通过实现 Serializable 接口实现序列化。

```
public class UserModel implements Serializable{
    private static final long serialVersionUID = 1L;
    public String name;
    public int age;
    public String address;

    public UserModel(String name,int age,String address) {
        this.name = name;
        this.age = age;
        this.address = address;
    }
    @Override
    public String toString() {
        return "name = " + name + " age = " + age + " address = " + address;
    }
}
```

在发送方的 RabbitmqSender 类中添加如下方法，用于向队列中发送对象数据。

```
public String sendObject() {
    UserModel user = new UserModel("javadev", 20, "java");
    rabbitTemplete.convertAndSend("ObjectQueue",user);
    return "object send ok!";
}
```

在 Controller 类中添加接口，用于接收 HTTP 请求并调用消息队列的发送方法向 RabbitMQ 中发送对象消息。

```
@RequestMapping(value="/sendObject",method=RequestMethod.GET)
public Result<String> sendObject() {
    String ret = sender.sendObject();
    Result<String> result = new Result<>(ResultCode.OK, ret);
    return result;
```

353

```
}
```

17.2.2 接收方配置及使用

在接收方的 RabbitmqConfig 类中添加需要监控的队列。

```
//object queue ****************************
@Bean
public Queue ObjectQueue() {
    return new Queue("ObjectQueue");
}
```

复制发送方的 UserModel 类实现，并且放入相同的 package 路径下，如果 package 不同会出现无法解析的错误。然后添加队列监听及消息处理类 ObjectReceiver，其具体实现如下。

```
@Component
@RabbitListener(queues = "ObjectQueue")
public class ObjectReceiver {
    @RabbitHandler
    public void process(UserModel user) {
        System.out.println("ObjectReceiver: " + user.toString());
    }
}
```

这里监听 ObjectQueue 队列，如果队列中存在数据则获取数据并且打印。启动发送方及接收方服务，然后调用发送方的接口，可以在接收方看到如下输出，这样就完成了对象数据的传递。

```
ObjectReceiver: name = javadev age = 20 address = java
```

17.3 队列传递 Json 数据

如上一节所示，使用对象传递的方式，在不同服务间传递数据，需要在相同的路径下创建相同名字的类。这种做法无疑会增加服务间的耦合，并且会对编程工作造成一些不必要的麻烦。所以可以使用另一种方案，在发送方把对象数据转为 Json 格式，通过 String 类型进行传递，在接收方接收 String 类型的数据，再把 Json 数据转化为对象，这样既可以实现对象数据的传递，也可以降低服务间的耦合。

17.3.1 发送方配置及使用

在发送方的 RabbitmqConfig 类中添加队列信息。

```
//object json queue
@Bean
public Queue ObjectJsonQueue() {
    return new Queue("ObjectJsonQueue");
}
```

添加发送方的方法，使其通过 ObjectJsonQueue 队列传递对象的 Json 数据。

```
ObjectMapper mapper = new ObjectMapper();
public String sendObjectJson() {
```

```
        try {
                UserModel user = new UserModel("javadev", 20, "java");
                String msg = mapper.writeValueAsString(user);
                rabbitTemplete.convertAndSend("ObjectJsonQueue",msg);
                return "object json send ok!";
        } catch (Exception e) {
                e.printStackTrace();
                return "object json send fail!";
        }
}
```

在 Controller 类中添加接口方法，使其可以调用发送方法发送 Json 数据。

```
@RequestMapping(value="/sendObjectJson",method=RequestMethod.GET)
public Result<String> sendObjectJson() {
        String ret = sender.sendObjectJson();
        Result<String> result = new Result<>(ResultCode.OK, ret);
        return result;
}
```

17.3.2 接收方配置及使用

在接收方的 RabbitmqConfig 类中添加需要监听的队列。

```
//object json queue
@Bean
public Queue ObjectJsonQueue() {
        return new Queue("ObjectJsonQueue");
}
```

在接收方添加队列监听和处理逻辑。

```
@Component
@RabbitListener(queues = "ObjectJsonQueue")
public class ObjectJsonReceiver {
        ObjectMapper mapper = new ObjectMapper();

        @RabbitHandler
        public void process(String message) {
                System.out.println("ObjectJsonReceiver: " + message);
                try {
                        UserModel user = mapper.readValue(message, UserModel.class);
                        System.out.println(user);
                } catch (Exception e) {
                        e.printStackTrace();
                }
        }
}
```

在上面的方法中，获取队列中的 String 类型的消息，并且打印；然后把其转化为对象数据，再进行输出。可在控制台看到如下信息。

```
ObjectJsonReceiver: {"name":"javadev","age":20,"address":"java"}
name = javadev age = 20 address = java
```

17.4 Topic 模式

之前使用 RabbitMQ 的方式，都是在发送方直接向绑定的队列发送消息，如果有一种情况，希望发送方的消息可以同时向不同的队列发送，那么逐个队列的调用确实是一种解决方案，但是不够优雅。RabbitMQ 使用了交换机的方式，可以在发送方通过交换机发送消息，达到以上的目的，甚至可以配置交换机的发送规则，有选择地发送至不同的队列。

17.4.1 Topic 模式讲解

首先在发送方的 RabbitmqConfig 类中添加如下配置。

```
//topic *****************************
@Bean
public Queue TopicMessageQueue() {
    return new Queue("Topic.MessageQueue");
}

@Bean
public Queue TopicAllQueue() {
    return new Queue("Topic.AllQueue");
}

@Bean
public TopicExchange TopicExchange() {
    return new TopicExchange("TopicExchange");
}

@Bean
public Binding bindingExchangeMessage(Queue TopicMessageQueue,TopicExchange TopicExchange) {
    return BindingBuilder.bind(TopicMessageQueue).
                    to(TopicExchange).with("Topic.MessageQueue");
}

@Bean
public Binding bindingExchangeAll(Queue TopicAllQueue,TopicExchange TopicExchange) {
    return BindingBuilder.bind(TopicAllQueue).to(TopicExchange).with("Topic.#");
}
```

上面的代码中，创建了两个队列 Topic.MessageQueue 和 Topic.AllQueue，创建了一个交换机 TopicExchange；然后通过 bind 方法把队列和交换机进行绑定，此绑定的意义如图 17-4 所示。其中 P 表示生产者，C 表示消费者。

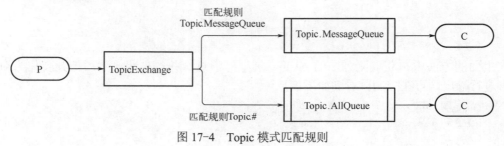

图 17-4　Topic 模式匹配规则

在发送方，向 TopicExchange 发送消息，同时要指定队列选择信息，匹配规则就会根据队列选择信息发送到不同的队列。例如指定的队列选择信息是 Topic.MessageQueue，那么上图中的两个队列都能够正确匹配，即两个队列都能够收到信息。如果指定的队列选择信息是 Topic.other，那么上图中只有 Topic.AllQueue 能够匹配此规则，即只有 Topic.AllQueue 能够收到消息。

在编写匹配规则时，需要用到*和#通配符，*可以匹配零个或一个词，#可以匹配零个或多个词。

通过 RabbitMQ 的可视化页面，查看 Exchange 的队列绑定情况，如图 17-5 所示。

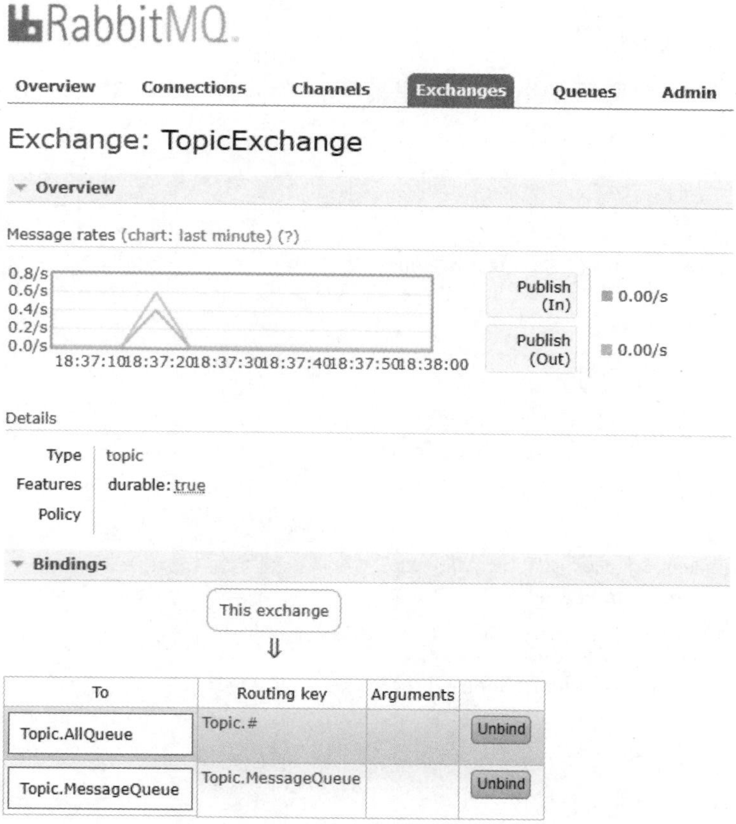

图 17-5　消息队列绑定

17.4.2　发送方配置及使用

在发送方添加队列、交换机及绑定配置后，在 RabbitmqSender 类中添加如下方法。

```
public String sendTopic() {
    rabbitTemplete.convertAndSend("TopicExchange",
                    "Topic.MessageQueue", "topic message");
    rabbitTemplete.convertAndSend("TopicExchange", "Topic.other", "topic all");
    return "topic send ok!";
}
```

以上代码的目的，是向交换机 TopicExchange 发送消息，convertAndSend 方法的第二个参数指定了队列选择信息，此队列选择信息会由交换机绑定队列时使用的匹配规则进行匹配，找到合适的队列发送消息。

在 Controller 类中，添加如下接口，用以调用发送方方法。

```
@RequestMapping(value="/sendTopic",method=RequestMethod.GET)
public Result<String> sendTopic() {
    String ret = sender.sendTopic();
    Result<String> result = new Result<>(ResultCode.OK, ret);
    return result;
}
```

17.4.3 接收方配置及使用

在接收方首先配置监听的队列，这里监听两个队列 Topic.MessageQueue 和 Topic.AllQueue，所以要在 RabbitmqConfig 类中添加如下代码。

```
//topic *****************************
@Bean
public Queue TopicMessageQueue() {
    return new Queue("Topic.MessageQueue");
}

@Bean
public Queue TopicAllQueue() {
    return new Queue("Topic.AllQueue");
}
```

添加两个队列监听处理类，分别实现如下。

```
@Component
@RabbitListener(queues = "Topic.AllQueue")
public class TopicAllReceiver {
    @RabbitHandler
    public void process(String message) {
        System.out.println("topicAllReceiver: " + message);
    }
}

@Component
@RabbitListener(queues = "Topic.MessageQueue")
public class TopicMessageReceiver {
    @RabbitHandler
    public void process(String message) {
        System.out.println("topicMessageQueue: " + message);
    }
}
```

启动发送方和接收方服务，然后调用发送接口，可以在接收方的控制台看到如下输出。

```
topicAllReceiver: topic message
topicMessageQueue: topic message
topicAllReceiver: topic all
```

对于 topic message 消息，由于指定的队列选择信息是 Topic.MessageQueue，所以两个队列规则都能够匹配此信息，监听的两个队列都能够打印 topic message 消息；对于 topic all 消息，由于指定的队列选择信息是 Topic.other，所以 Topic.MessageQueue 规则无法匹配此信息，只有 Topic.#规则可以匹配，结果只有 Topic.AllQueue 队列收到了消息。

17.5　Fanout 模式

Fanout 模式与 Topic 模式不同，不需要匹配队列规则，只要绑定到 Fanout 交换机上的队列都能够收到发送到交换机上的消息。

17.5.1　发送方配置及使用

在发送方，只需要创建好队列，将队列绑定到 FanoutExchange 上，然后向 Fanout Exchange 发送消息，所有绑定的队列都可以收到消息。

（1）配置队列及绑定

在 RabbitmqConfig 中，添加如下代码。

```
//fanout *****************************
@Bean
public Queue AQueue() {
    return new Queue("AQueue");
}

@Bean
public Queue BQueue() {
    return new Queue("BQueue");
}

@Bean
public FanoutExchange FanoutExchange() {
    return new FanoutExchange("FanoutExchange");
}

@Bean
Binding bindingExchangeA(Queue AQueue,FanoutExchange FanoutExchange) {
    return BindingBuilder.bind(AQueue).to(FanoutExchange);
}

@Bean
Binding bindingExchangeB(Queue BQueue,FanoutExchange FanoutExchange) {
    return BindingBuilder.bind(BQueue).to(FanoutExchange);
}
```

（2）添加发送方法

完成队列和交换机的配置后，需要在 RabbitmqSender 类中添加如下发送方法，此方法向 FanoutExchange 交换机发送消息。

```
public String sendFanout() {
    rabbitTemplete.convertAndSend("FanoutExchange", "all", "fanout send");
    return "fanout send ok！";
}
```

为了方便发送的调用，在 Controller 类中添加发送调用接口，具体如下：

```
@RequestMapping(value="/sendFanout",method=RequestMethod.GET)
public Result<String> sendFanout() {
    String ret = sender.sendFanout();
    Result<String> result = new Result<>(ResultCode.OK, ret);
    return result;
}
```

17.5.2 接收方配置及使用

接收方只需要监听队列，不关注队列的绑定关系，这是和发送方不同的地方。在接收方首先配置 FanoutExchange 绑定的两个队列 AQueue 和 BQueue，在 Rabbitmq Config 文件中添加如下代码。

```
//fanout *****************************
@Bean
public Queue AQueue() {
    return new Queue("AQueue");
}

@Bean
public Queue BQueue() {
    return new Queue("BQueue");
}
```

然后添加两个队列的监听处理类，具体如下：

```
@Component
@RabbitListener(queues = "AQueue")
public class AQueueReceiver {
    @RabbitHandler
    public void process(String message) {
        System.out.println("AQueueReceiver: " + message);
    }
}

@Component
@RabbitListener(queues = "BQueue")
public class BQueueReceiver {
    @RabbitHandler
    public void process(String message) {
        System.out.println("BQueueReceiver: " + message);
    }
}
```

启动发送方和接收方服务，然后调用发送方的 Fanout 模式接口，可以在接收方看到如下输出。

```
AQueueReceiver: fanout send
BQueueReceiver: fanout send
```

由控制台输出可见，Fanout 模式会向所有绑定到此交换机的队列发送消息。

第 18 章 ELK

在一个使用 Spring Cloud 进行微服务化的系统集群内，包含不同能力的服务，同一服务还包含多个程序实例，这些实例可能运行在不同的服务器中。如果集群中某一服务出现了问题需要查看日志，可能要到不同的服务器中逐个查看日志文件，这明显是个很麻烦的事情。使用 ELK 就能够很好地解决此问题。

ELK 是 ElasticSearch、Logstash、Kibana 三个软件的聚合，使用这三个软件最后达到日志搜集、存储、分析等目的。其中 ElasticSearch 提供存储及搜索引擎的能力；Logstash[⊖]是日志搜集、分析、过滤、输出的工具；Kibana 为日志分析提供友好的 Web 界面，可以对日志进行可视化的搜索、汇总和分析。

日志搜集的全流程大概如图 18-1 所示。大体的思路就是先把日志保存到某种介质，然后使用工具把分散的日志搜集起来，对搜集到的信息进行加工、过滤然后放入某个存储介质，最后通过一个可视化的页面进行分析展示。在实际业务中可以对图中的某些模块进行修改，例如使用 Filebeat 或者引入消息队列，但是日志搜集、存储、展示的思路不变。

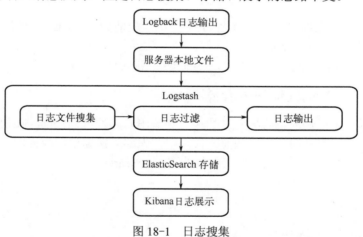

图 18-1　日志搜集

由于本书之前已经演示过 ElasticSearch 的具体用法，所以不再赘述，本章主要介绍 Logstash 和 Kibana 的用法。

18.1　Logstash 使用

Logstash 主要做的事情就是获取数据、数据加工、输出数据，所以 Logstash 主要有 input、filter、output 三个角色，这三个角色对应着 Logstash 配置文件的三段配置。Logstash 的

⊖ 在日志搜集方面，可以在各个服务器上使用 Filebeat 作为专门的日志搜集工具，Filebeat 可以把搜集到的数据发送给消息队列用于 Logstash 读取，或者直接由 Filebeat 发送给 Logstash。

安装可以参考第 19 章，本节介绍 Logstash 的配置和效果展示。

18.1.1　Logstash 概要介绍

可以对 Logstash 进行最简单的配置，例如让 Logstash 从控制台读取信息，并且输出到控制台，配置如下。

```
input { stdin { } } output { stdout { } }
```

当使用此配置执行的 Logstash 向控制台输入一串信息后，Logstash 会进行搜集和输出。例如向控制台输入：javadevmap，可以得到如下输出。

```
{
    "@version" => "1",
    "host" => "c819060a5310",
    "@timestamp" => 2018-05-28T09:38:43.139Z,
    "message" => "javadevmap"
}
```

以上就完成了 Logstash 最简单的搜集和输出能力。对于 Logstash 来讲，input 和 output 是必要的，filter 是非必要的，但是 filter 的数据处理能力能够完成很多事情，例如对非规范数据进行处理或者对某些特殊字段进行处理。Logstash 三个模块主要能力如下：

- input：从数据源获取数据，数据源包含 File 文件、syslog 系统日志、redis、beats（Filebeats）。
- filter：负责处理数据与转换，包含 grok 正则匹配能力、mutate 事件转换、drop 事件丢弃、clone 事件复制、geoip 处理 IP 等。
- output：输出到目标介质，包含 elasticsearch、file、statsd 等。

18.1.2　文件搜集及 ElasticSearch 存储

Logstash 可以使用插件的形式配置它的输入和输出，并且可以在同一模块中配置多个插件实现多输入源或多输出的目的。由于之前的代码中，大部分日志都已经打印至文件，所以这里主要介绍文件形式的日志搜集，输出的目的地为 ElasticSearch。

对 Logstash 进行如下配置。

```
input {
    file {
        path => "/logs/*/*"
        start_position => "beginning"
    }
}

output {
    elasticsearch {
        hosts => ["172.17.238.238:9200"]
        index => "logstash-%{+YYYY.MM.dd}"
    }
}
```

在此配置中，设置 input 的输入方式是文件读取，使用 file 插件，在插件中使用 path 参数设置文件读取的目录，这里使用正则的方式匹配符合规则的文件地址；start_position 设置了文件开始读取的位置，beginning 表示从文件头读取。在 output 中，设置了 ElasticSearch 的地址和 index 的命名方式。

登录 ElasticSearch，可以查看日志搜集后的信息，这里截取其中一条。

```
{
    "_index": "logstash-2018.05.29",
    "_type": "logs",
    "_id": "AWOpxzCn2OudMVXehqwQ",
    "_version": 1,
    "_score": 1,
    "_source": {
        "@version": "1",
        "host": "serverA",
        "path": "/logs/SpringCloudEureka/2018-05-29.0.log",
        "@timestamp": "2018-05-29T02:43:13.523Z",
        "message": "2018-05-29 10:42:09.577+0800 INFO o.s.b.c.e.t.TomcatEmbeddedServletContainer [SpringCloudEureka, , , ], [ main ] : Tomcat initialized with port(s): 18001 (http)"
    }
}
```

18.1.3　使用 Json 格式日志

从上一节的输出可以看到，在程序中使用 Logback 打印的数据，仅仅是 Logstash 数据内容中的 message 字段的内容，这个 message 内容是按照日志在 Logback 中设置的打印格式进行输出的，如果使用 Kibana 进行内容检索或者进行日志分析，这种格式就很难进行分析和制图了，所以需要对 log 内容进行处理。

处理这种内容可以使用 grok 插件进行正则匹配，grok 会根据设置的规则在 message 信息中截取符合规则的字段，这样就能实现数据内容的处理。还有一种方法是在打印日志时，直接输出 Json 格式，这样就不需要正则匹配的处理，也节省了 Logstash 的计算过程。

（1）Logback 设置 Json 格式打印

在工程的 pom 文件中添加如下依赖。

```
<dependency>
    <groupId>net.logstash.logback</groupId>
    <artifactId>logstash-logback-encoder</artifactId>
    <version>5.0</version>
</dependency>
```

在 logback-spring.xml 的日志设置文件中，修改输出到文件的编码方式，改为使用 LogstashEncoder 方式。

```
<appender name="FILE" class="ch.qos.logback.core.rolling.RollingFileAppender">
    //其他内容不变...
    <encoder charset="UTF-8" class="net.logstash.logback.encoder.LogstashEncoder">
        <customFields>{"servicename":"${APP_Name}"}</customFields>
    </encoder>
</appender>
```

（2）Logstash 接收 Json 数据

修改 Logstash 的配置，在 input 中使用 codec 参数设置接收的内容为 Json 格式。

```
input {
    file {
        path => "/logs/*/*"
        codec => "json"
        start_position => "beginning"
    }
}
```

（3）输出展示

启动 Logstash 搜集新打印的日志后，可以在 ElasticSearch 中看到如下数据内容，Logback 打印的 Json 数据可以在 Logstash 中识别。

```
{
    "_index": "logstash-2018.05.29",
    "_type": "logs",
    "_id": "AWOp0q-12OudMVXehrHm",
    "_version": 1,
    "_score": 1,
    "_source": {
        "path": "/logs/SpringCloudEureka/2018-05-29.0.log",
        "@timestamp": "2018-05-29T02:54:44.206Z",
        "level": "INFO",
        "thread_name": "main",
        "level_value": 20000,
        "@version": "1",
        "host": "serverA",
        "servicename": "SpringCloudEureka",
        "logger_name": "org.apache.catalina.core.StandardEngine",
        "message": "Starting Servlet Engine: Apache Tomcat/8.5.29"
    }
}
```

18.1.4　使用 filter 处理数据

在实际业务中，可能期望根据用户的年龄和地域进行统计分析，这种情况下使用 filter 处理用户当前 IP 然后解析出具体的地址是一个不错的选择。例如在代码中，使用 log.info()方法打印一个 Json 格式的日志{"userIp":"47.95.113.117","age":30}，日志的总体打印内容如下：

```
{
    "@timestamp": "2018-05-29T15:57:54.812+08:00",
    "@version": "1",
    "message": "{\"userIp\":\"47.95.113.117\",\"age\":30}",
    "logger_name": "com.javadevmap.serviceprovider.controllers.UserController",
    "thread_name": "hystrix-provider-usercontroller-2",
    "level": "INFO",
    "level_value": 20000,
    "X-Span-Export": "true",
    "X-B3-SpanId": "3c7551c6dc8a5d5b",
    "X-B3-TraceId": "3c7551c6dc8a5d5b",
    "servicename": "SpringCloudServiceProvider"
}
```

在 Logstash 配置文件中添加如下配置，就可以实现对 log.info()方法中打印数据的 Json 解析，并且可以根据 IP 识别地址。

```
//省略 input…
filter{
json {
source => "message"
}
geoip {
source => "userIp"
}
}
//省略 output…
```

通过 Logstash 识别后，除了正确地解析了 message 信息，还在生成的数据中添加了 geoip 数据段，在此数据段下包含一系列地理位置的内容。

18.2 Kibana 使用

Kibana 是一个日志聚合展示的可视化 Web 工具，可以在 Kibana 中查看单条日志的情况，可以根据特定条件聚合某类日志信息，从而实现饼图等可视化视图的分析展示。Kibana 的安装请参看第 19 章，本节介绍 Kibana 的几种基本使用方法。

（1）基本配置

使用 Docker 安装 Kibana 需要注意 ElasticSearch 的地址，如果使用了错误的地址会出现如图 18-2 所示的错误提示信息。

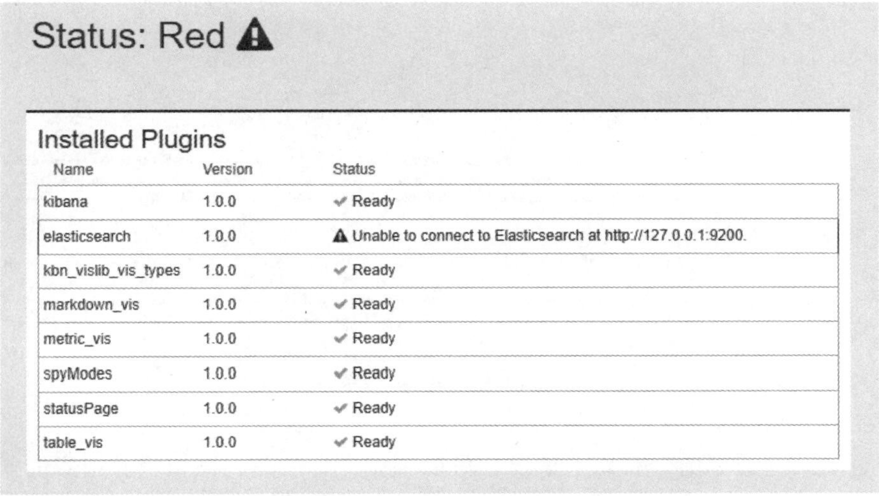

图 18-2　Kibana 配置错误提示

正确配置 ElasticSearch 地址，并且进入 Kibana 之后，可以在"Settings"选项中选择要检索的 index，本节使用正则匹配"logstash-*"，如图 18-3 所示。

（2）简单搜索

在 Kibana 中，进入"Discover"页签，可以在输入框输入简单的查询条件，并且在右上角配置正确的时间维度，Kibana 会根据查询条件进行搜索，并且在搜索到的内容中对搜索关键词高亮显示，如图 18-4 所示。

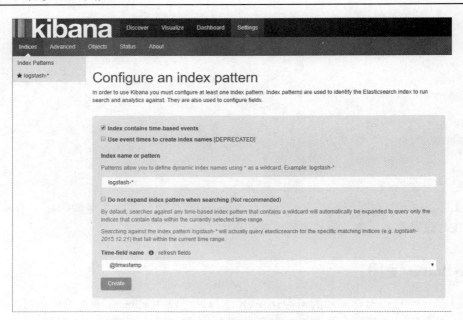

图 18-3　Kibana 起始页面

图 18-4　Kibana 搜索

（3）制作简单的饼图

如果想查看某种数据的分布比例情况，可以制作一个饼图。例如查看用户的年龄分布情况，可以选择"Visualize->Pie chart->From a new search"，然后配置检索的数据项和展示的个数，点击执行按钮后就可以看到统计情况，如图 18-5 所示。

（4）双层饼图

可以在饼图的基础上，选择子分析项，从而形成双层的饼图。在图 18-5 的左下方，点击"add sub-buckets"，再次输入一个分析条件，可以看到效果如图 18-6 所示。

可以选择保存此图片，在页面的右上方有一个保存按钮，点击此按钮，输入信息并保存即可。

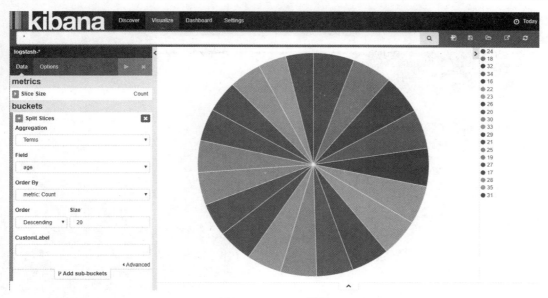

图 18-5　Kibana 饼图

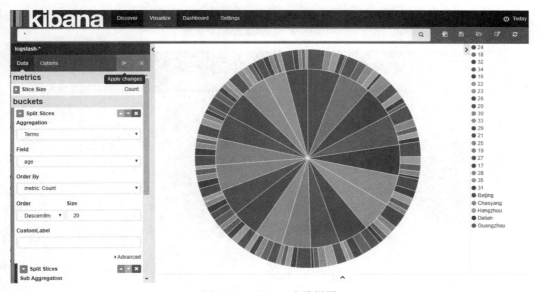

图 18-6　Kibana 多维饼图

（5）地图分布

Kibana 可以根据地理位置绘制一个用户的位置分布图。选择"Visualize->Tile map->From a new search"，然后在配置项中选择 Geohash 和 geoip.location，可见如图 18-7 所示的用户分布情况，此地图可以放大、缩小。

（6）聚合展示

Kibana 可以把之前保存的分析图聚合展示出来，选择"Dashboard"，然后根据页面的提示添加之前保存的分析图，可以看到如图 18-8 所示聚合展示页面。

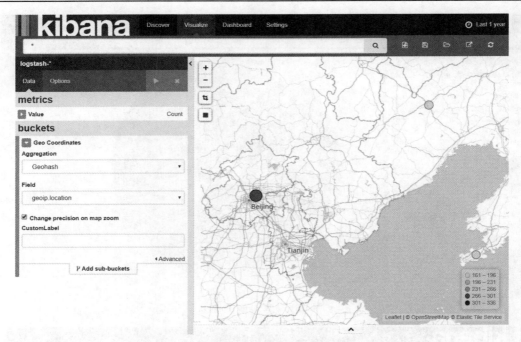

图 18-7　Kibana 地图

图 18-8　Kibana 聚合展示

第四篇 部 署 篇

对于一名研发人员来讲，前面的章节基本已经覆盖了日常工作的大部分内容，但是软件研发这个职业就是需要不断探索的，所以本篇将主要介绍 Docker 和 Jenkins 的使用以及服务集群的管理，让研发者能够从服务运行的角度了解自己的程序。

Docker 是目前非常流行的镜像技术，可以通过 Docker 运行任何一个组件，包括自己编写的程序。Docker 的特性会保证平台间的兼容性以及快速部署的能力。上一篇中使用的系统组件，有些安装和配置较为烦琐，但是如果仅是研发人员在功能环境中进行简单的使用，就没有必要麻烦运维人员了，使用 Docker 就可以非常简单地安装系统组件。在第 19 章，将介绍 Docker 的基本用法、命令以及组件的安装。

在研发或者生产环境上线过程中，尤其是服务联调阶段，把程序部署到相应环境是一个非常频繁的工作，而对于一个服务集群来讲，如果全部手动部署，工作量将会非常大的。使用 Jenkins 可以方便快捷地完成自动化编译和部署，在第 20 章，将会介绍 Jenkins 的基本使用。

如果想让集群内的程序具备 Docker 的特性，那么就要把程序生成为一个镜像，这就会面临镜像存储和镜像运行后容器管理的问题。在第 20 章将会介绍镜像仓库 Harbor 以及容器管理工具 Rancher 的使用。

希望通过本篇的学习，读者能够从集群运行的角度考虑大集群的管理和自动化的使用。

第 19 章　Docker

当把程序部署到服务器时，常常面临环境的问题，例如系统版本和程序不匹配、使用的 JDK 版本需要重新安装、系统的环境变量需要重新设置、新程序与服务器中已存在程序之间的冲突等问题。这些问题给服务集群的部署带来了一定的麻烦，每次上线可能由于几个环境问题要调试好久。解决以上问题就是 Docker 应用的场景。

Docker 是怎么解决这些问题的？其实看一张 Docker 的图片可能更容易理解。如图 19-1 所示，图片中的鲸可以想象为服务器，图片中的集装箱可以想象为一个一个的程序，Docker 的核心理念就是把程序的所有准备工作都放到集装箱中，当需要某个程序在某台服务器运行时，只要把这个集装箱放到服务器上就可以正常运行了。

图 19-1　Docker 图标

基于以上的出发点，Docker 包含的主要特性如下：
- 更高效的虚拟化，Docker 对程序进行隔离的虚拟化技术占用资源极少，虚拟化部分不会占用过多的系统开销。
- 更快的部署，只要服务器具备 Docker 容器运行的基础条件，就可以把 Docker 容器部署到此服务器中，而不用考虑其他环境因素。
- 简单的镜像生成，通过 Dockerfile 可以方便地生成 Docker 镜像，这个镜像会把程序以及程序运行的基础环境统一打包。
- 方便移植，Docker 的兼容性可以保证在任何平台上运行的 Docker 容器能够快速地在其他平台的 Docker 环境下使用。
- 可以使用 Docker 相关的管理工具，对程序的历史镜像版本、容器的启动及当前状态进行监控和管理。

19.1　Docker 基础环境搭建

要想在某台服务器中运行 Docker 容器，就必须为容器准备基础的 Docker 环境。本节使用 yum 命令安装和卸载 Docker 环境，并且为提高镜像的下载速度配置加速器。

19.1.1　Docker 环境安装

本节介绍 CentOS 7.3 版本 64 位系统中 Docker 环境的安装。
- 安装命令

```
$ yum install docker
```

- 查看服务状态

```
$ service docker status
```

- 启动服务

    $ service docker start

- 查看服务版本

    $ docker version

- 查看 docker 信息

    $ docker info

19.1.2 Docker 环境卸载

如果想卸载 Docker 环境，只要顺序执行如下命令即可完成。

- 查看安装的 Docker 组件

    $ yum list installed | grep docker

- 逐个删除 Docker 组件

    $ yum -y remove [name]

- 删除 Docker 遗留文件

    $ rm -rf /var/lib/docker

19.1.3 镜像加速

在使用 Docker 下载镜像并且运行容器之前，还有一项配置是必不可少的，即为 Docker 的镜像下载配置加速器。这里使用 DaoCloud 的加速器配置方法，在服务器执行如下命令：

    $ curl -sSL https://get.daocloud.io/daotools/set_mirror.sh | sh -s http://5b55f8e6.m.daocloud.io

执行命令后，会得到如下输出。

    $ docker version >= 1.12
    {"registry-mirrors": ["http://5b55f8e6.m.daocloud.io"],}
    Success.
    You need to restart docker to take effect: sudo systemctl restart docker

上面生成的地址中，末尾多了一个逗号，所以需要修改/etc/docker/目录下的 daemon.json 文件，改为

    {"registry-mirrors": ["http://5b55f8e6.m.daocloud.io"]}

然后执行如下命令，重启服务。

    $ systemctl daemon-reload
    $ systemctl restart docker

19.2 Docker 常用命令

在服务器中准备好了 Docker 环境，就可以通过 Docker 的命令获取镜像，并且运行容器了。Docker 命令主要是针对镜像和容器的操作，例如拉取镜像、创建镜像、运行容器。如果第一次接触镜像和容器可能会比较难以理解，简单来讲可以把镜像理解为程序的安装包，镜

像运行后就成为了容器，所以可以把容器理解为程序运行的进程。下面介绍 Docker 的常用命令。

19.2.1 针对镜像的命令

（1）搜索镜像

```
$ docker search [OPTIONS] NAME
```

使用如上命令，可以针对某个镜像的名字进行搜索。OPTIONS 是可选项，指定搜索的条件，例如 docker search –s 10 redis 的意思是搜索收藏数不小于 10 的 Redis 的镜像。

（2）拉取镜像

```
$ docker pull [OPTIONS] NAME
```

使用此命令可以拉取镜像，其中 NAME 中可以指定具体的 tag 标签，即可拉取相应 tag 的镜像，如果不指定此标签则拉取最新的镜像。

（3）显示当前镜像列表

```
$ docker images [OPTIONS]
```

此命令用于显示镜像列表，如果设置[OPTIONS]可以指定条件或者显示方式。

（4）删除镜像

```
$ docker rmi [OPTIONS] image
```

例如删除镜像 Redis，可以使用 docker rmi docker.io/redis 命令。

（5）给镜像打标签

```
$ docker tag NAME NAME/version
```

此命令在自定义镜像的版本管理时非常有用。

（6）登录镜像仓库

```
$ docker login[OPTIONS] [SERVER]
```

此命令可以通过 Server 指定具体的仓库地址，在后面演示集群部署时登录自定义仓库会用到。

（7）登出镜像仓库

```
$ docker logout[OPTIONS] [SERVER]
```

（8）推送镜像至仓库

```
$ docker push NAME
```

（9）将镜像保存成归档文件

```
$ docker save –o FILE IMAGE
```

例如使用 docker save –o redis.tar docker.io/redis 把镜像保存为文件。

（10）从归档文件中创建镜像

```
$ docker import FILE IMAGE
```

例如使用 docker import redis.tar javadevmap/redis 导入镜像后，可以查看镜像列表中多出一个镜像。

（11）使用 Dockerfile 创建镜像

```
$ docker build [OPTIONS] IMAGE PATH
```

19.2.2 针对容器的命令

（1）运行容器

```
$ docker run [OPTIONS] IMAGE
```

以运行 redis 容器为例，可以使用如下命令。

```
$ docker run --name jdm_redis -p 6379:6379 -d docker.io/redis
```

此命令运行了一个 Redis 容器，并且指定容器名字为 jdm_redis，服务器与容器的端口映射为 6379:6379，并且指定容器在后台运行。运行容器时可以指定的参数较多，这些参数直接指定容器的属性，所以比较重要，见表 19-1。

表 19-1　Docker 运行参数

参　数	含　义
-d	后台运行容器，并返回容器 ID
--name	指定容器名称
-p	指定容器端口映射
--net	指定容器网络类型，共有 bridge、host、none、container 四种类型
-h	指定容器的主机名
-v	指定容器的挂载卷，可以让容器使用宿主机的目录
-e	指定容器的环境变量
-m	指定容器的内存上限，格式是数字加单位，单位可以为 b,k,m,g

（2）查看运行中的容器

```
$ docker ps [OPTIONS]
```

不包含参数则查看当前运行容器，如果指定[OPTIONS]为-a，则查看所有容器（包括已经关闭的）。

（3）停止容器

```
$ docker stop CONTAINER
```

（4）启动被停止的容器

```
$ docker start CONTAINER
```

（5）重启容器

```
$ docker restart CONTAINER
```

（6）强制杀掉容器

```
$ docker kill CONTAINER
```

（7）删除容器

```
$ docker rm [OPTIONS] CONTAINER
```

[OPTIONS]使用-v 参数可以删除容器挂载的数据卷。

（8）创建容器但不启动

```
$ docker create[OPTIONS] IMAGE
```

（9）查看容器日志

```
$ docker logs[OPTIONS] CONTAINER
```

（10）进入运行的容器执行命令

```
$ docker exec [OPTIONS] CONTAINER
```

例如进入刚刚运行的 redis 容器，设置 redis 的密码，可以使用如下命令。

```
$ docker exec –it CONTAINERID /bin/bash
$ redis-cli
$ config set requirepass mypass
$ exit
$ exit
```

（11）获取容器/镜像的元数据

```
$ docker inspect IMAGE/CONTAINER
```

（12）查看容器运行的进程信息

```
$ docker top CONTAINER
```

（13）由容器创建镜像

```
$ docker commit [OPTIONS] CONTAINER IMAGE
```

（14）主机和容器间的数据复制

```
$ docker cp [OPTIONS] SRC DEST
```

例如向 redis 容器中复制数据，可以使用如下命令。

```
$ docker cp /logs CONTAINERID:/logs
```

19.2.3　使用 Dockerfile 创建镜像

把自己编写的 Java 服务运行在 Docker 上的前提是生成一个服务的镜像，可以使用 Dockerfile 文件配置镜像内容，然后就可以通过 docker build 命令生成镜像了。下面以之前编写的工程 SpringCloudServiceProvider 为例，生成此服务的镜像，并且了解 Dockerfile 文件的属性含义。

（1）生成服务 jar 包

首先编译工程，可以得到 SpringCloudServiceProvider-0.0.1-SNAPSHOT.jar 文件，这个文件之前介绍过，可以通过 java –jar 的方式直接启动，镜像文件也需要此文件才能启动服务。

（2）编写 Dockerfile 文件

Dockerfile 文件其实可以理解为一个配置文件，它描述了一个镜像包含的内容和需要启动的方法等，本例先创建 Provider 服务的 Dockerfile 文件，后面会详细讲解文件中字段的含义。

```
FROM java:8
```

```
ADD SpringCloudServiceProvider-0.0.1-SNAPSHOT.jar /data/run/
EXPOSE 18010
WORKDIR /data/run/
ENTRYPOINT ["java","-jar","SpringCloudServiceProvider-0.0.1-SNAPSHOT.jar","--spring.profiles.active=providerA"]
```

上面的代码中,用到了部分 Dockerfile 的属性。FROM 指定了基础的镜像;ADD 把 jar 文件放入镜像的/data/run 目录;EXPOSE 指定镜像对外暴露 18010 端口;WORKDIR 指定镜像内的工作目录为/data/run;ENTRYPOINT 配置了镜像的启动运行命令。

(3) 生成镜像

把上面的 jar 包和 Dockerfile 文件放入同一目录下,然后运行如下命令,注意命令中结尾有"."符号表示当前路径。

```
$ docker build -t springcloud/provider:0.0.1 .
```

运行命令后,可以看到显示如下内容。

```
Sending build context to Docker daemon 57.34 MB
Step 1 : FROM java:8
 ---> d23bdf5b1b1b
Step 2 : ADD SpringCloudServiceProvider-0.0.1-SNAPSHOT.jar /data/run/
 ---> e42cfd8b2d40
Removing intermediate container c60ab61ba06f
Step 3 : EXPOSE 18010
 ---> Running in cda6abe5f6a1
 ---> fbead3671920
Removing intermediate container cda6abe5f6a1
Step 4 : WORKDIR /data/run/
 ---> Running in dc9bdb55164a
 ---> 66f3e557c0b8
Removing intermediate container dc9bdb55164a
Step 5 : ENTRYPOINT java -jar SpringCloudServiceProvider-0.0.1-SNAPSHOT.jar --spring.profiles.active=providerA
 ---> Running in be49fd1b8fe4
 ---> e6ad17c121bc
Removing intermediate container be49fd1b8fe4
Successfully built e6ad17c121bc
```

(4) 运行镜像并查看服务情况

现在镜像已经生成,可以通过 docker images 命令查看镜像,会得到如下输出。

```
REPOSITORY          TAG    IMAGE ID       CREATED         SIZE
springcloud/provider 0.0.1  e6ad17c121bc   2 minutes ago   700.5 MB
```

这就是刚刚生成的镜像,可以使用 docker run 命令启动镜像。

```
$ docker run -d --name provider --net=host -v /logs:/logs e6ad17c121bc
```

上面的命令指定容器在后台运行,和主机共享网络,把镜像内的/logs 目录进行挂载,容器名为 provider。运行命令后,可以通过 docker ps 命令查看容器情况,如图 19-2 所示。

CONTAINER ID	IMAGE	COMMAND	CREATED	STATUS	PORTS	NAMES
ed076372f1e1	e6ad17c121bc	"java -jar SpringClou"	7 seconds ago	Up 6 seconds		provider

图 19-2 容器运行情况

（5）容器运行情况

通过 Postman 请求此容器服务，可以看到如图 19-3 所示的请求结果。

图 19-3　请求容器中的服务

查看 Eureka 服务注册信息，可以看到新的容器服务已经注册进来，如图 19-4 所示。

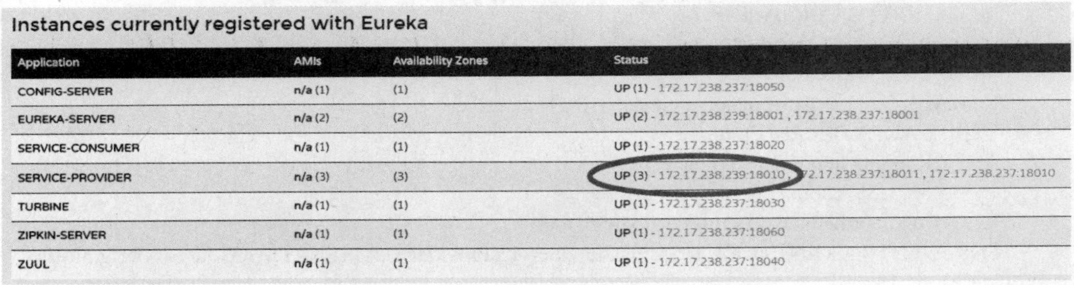

图 19-4　容器服务注册至 Eureka

（6）常用指令

上面使用 Dockerfile 的方式创建了镜像，在文件中使用了部分 Dockerfile 的指令设置镜像的内容和执行方法。下面简单介绍 Dockerfile 的常用指令，见表 19-2。

表 19-2　Dockerfile 指令

指　　令	格　　式	作用与用法
FROM	FROM <image>; FROM <image>:<tag>	FROM 指令必须是 Dockerfile 的第一条指令，使用此指令指定基础镜像，FROM 指令之后的其他指令都依赖于此镜像
ADD	ADD <src><dest>	用于复制指定的文件或 URL 至容器的目标地址
COPY	COPY <src><dest>	本地文件向容器目标地址复制
EXPOSE	EXPOSE <port> [<port>...]	声明容器运行时提供的端口，可声名多个端口；运行容器时可以使用"–p port:port"的方式映射主机和容器端口
WORKDIR	WORKDIR /path	为后续的 RUN、CMD、ENTRYPOINT 指令指定容器内的工作目录

（续）

指 令	格 式	作用与用法
ENTRYPOINT	ENTRYPOINT ["executable", "param1", "param2"]; ENTRYPOINT command param1 param2	设置容器启动时的执行命令，可以设置多个，但只有最后一个生效
RUN	RUN <command>; RUN ["executable", "param1", "param2"]	RUN <command>在 shell 终端运行，即/bin/sh –c；后者使用 exec 执行，例如 RUN ["/bin/bash", "-c", "echo hello"]
CMD	CMD ["executable","param1","param2"]; CMD ["param1","param2"]; CMD command param1 param2	启动容器时执行的命令，每个 Dockerfile 只能有一条 CMD 命令
ENV	ENV <key><value>	设置环境变量
VOLUME	VOLUME ["/data"]	指定挂载点
MAINTAINER	MAINTAINER<name>	指定维护者信息
LABEL	LABEL<key>=<value>	为镜像指定元数据

19.3 Docker 搭建功能组件

前面章节使用了大量的功能组件，对于研发人员来讲，搭建这些组件可能是一件非常麻烦和费时的事情，因为各种依赖和环境经常要配置好久，可能还会出错。大部分情况下研发人员想搭建一套功能环境都会求助于运维人员，现在有了 Docker 就可以让一切简单很多，通过 Docker 运行一个容器就可以启动一个功能组件。下面介绍使用 Docker 简单搭建组件的方法，虽然使用如下方法无法搭建真正高可用的功能组件，但是研发人员在功能环境使用已经足够了。

（1）搭建 MySQL

```
$ docker pull docker.io/mysql:5.6.35
$ docker run --name jdmmysql -p 3306:3306 -e MYSQL_ROOT_PASSWORD=mypass -v /etc/localtime:/etc/localtime -d docker.io/mysql:5.6.35
```

使用如上命令，可以完成 MySQL 的搭建，容器时间使用的是系统时间，使用 SQLyon 连接数据库时的密码是 mypass。

（2）搭建 MongoDB

```
$ docker pull docker.io/mongo
$ docker run --name jdmmongo -p 27017:27017 -d docker.io/mongo --auth
$ docker exec -it jdmmongo /bin/bash
$ mongo
$ use admin
$ db.createUser({user:"root",pwd:"mypass",roles:[{role:'root',db:'admin'}]})
$ exit
$ exit
```

配置完成后，可以使用 Robomongo 或 NoSQL Manager for MongoDB 工具查看 MongoDB，账号 root，密码 mypass。

（3）搭建 Redis

```
$ docker pull docker.io/redis
$ docker run --name jdmredis -p 6379:6379 -d docker.io/redis
$ docker exec -it jdmredis /bin/bash
```

```
$ redis-cli
$ config set requirepass mypass
$ exit
$ exit
```

搭建成功后，可以使用 Redis Desktop Manager 工具查看 Redis，Auth 为 mypass。

（4）搭建 Zookeeper

```
$ docker pull docker.io/zookeeper
$ docker run --name jdmzookeeper -p 2181:2181 -p 2888:2888 -p 3888:3888 -d docker.io/zookeeper
```

搭建成功后，可以使用 Zooinspector 工具查看 Zookeeper，并且修改其中的数据内容。

（5）搭建 RabbitMQ

```
$ docker pull docker.io/rabbitmq:3-management
$ docker run --name jdmrabbitmq -p 5672:5672 -p 15672:15672 -d rabbitmq:3-management
```

使用如上命令搭建成功后，可以访问 http://{ip}:15672 地址，输入账号和密码（均为 guest），登录 RabbitMQ 查看。

（6）搭建 ElasticSearch

搭建 ElasticSearch 需要注意环境的内存空间，新版本的 ElasticSearch 需要内存较大，如果自己试验时可以搭建老版本的 ElasticSearch，相对使用内存较小。

```
$ docker pull docker.io/elasticsearch:2.4.0
$ docker run --name jdmelasticsearch -p 9200:9200 -p 9300:9300 -d docker.io/elasticsearch:2.4.0
$ docker exec -it jdmelasticsearch /bin/bash
$ cd bin/
$ ./plugin install mobz/elasticsearch-head○
$ ./plugin install https://github.com/medcl/elasticsearch-analysis-ik/releases/download/v1.10.0/elasticsearch-analysis-ik-1.10.0.zip○
$ cp -r /etc/elasticsearch/analysis-ik /usr/share/elasticsearch/config/
$ cd /usr/share/elasticsearch/config/
$ chown -R elasticsearch.elasticsearch *
$ exit
$ docker restart jdmelasticsearch
```

使用如上命令，可以搭建 2.4.0 版本的 ElasticSearch，包括 ik 分词器。访问 http://{ip}:9200/_plugin/head/地址，可以看到 Web 管理页面。

（7）搭建 Kibana

```
$ docker pull docker.io/kibana:4.6.6
$ docker run --name jdmkibana -e ELASTICSEARCH_URL=http://{ESip}:9200 -p 5601:5601 -d docker.io/kibana:4.6.6
```

注意上面的命令中 ELASTICSEARCH_URL 属性要输入 ElasticSearch 的地址。使用如上方法搭建完毕，可以访问 http://{ip}:5601 地址登录 Kibana。

（8）搭建 Logstash

```
$ docker pull logstash
$ docker run -it --rm logstash -e 'input { stdin { } } output { stdout { } }'
```

○ 注意 ElasticSearch 在 5.X 版本的 head 插件已经不支持此方法安装。

○ 安装 ElasticSearch 2.4.0 版本对应的 ik 分词器版本为 1.10.0。

运行如上命令，然后在命令行输入测试内容 test logstash，可以得到如下输出。

```
{
    "@version" => "1",
    "host" => "696528decccb",
    "@timestamp" => 2018-05-14T06:31:56.482Z,
    "message" => "test logstash"
}
```

退出运行的容器，然后在/etc/logstash/config-dir 目录下创建 logstash.conf 文件，填写如下内容：

```
input {
    file {
        path => "/logs/*/*"
        codec => "json"
        start_position => "beginning"
    }
}

filter{
    json {
        source => "message"
    }
    geoip {
        source => "ClientIp"
    }
}

output {
    elasticsearch {
        hosts => ["{ESip}:9200"]
        index => "logstash-%{+YYYY.MM.dd}"
    }
}
```

以上内容是 logstash 的采集、过滤及输出设置。日志采集的位置是/logs/*/*；日志的过滤功能可以根据 ClientIp 匹配地址；最后输出到 ElasticSearch 中，根据日期定义 ElasticSearch 中的 index。完成设置后运行容器：

```
$ docker run --name jdmlogstash -v /logs:/logs -p 5500:5500 -d -v "$PWD":/config-dir logstash -f /config-dir/logstash.conf
```

（9）搭建 Jenkins

```
$ docker pull jenkins
$ mkdir /home/jdm/jenkins
$ chown -R 1000:1000 /home/jdm/jenkins
$ docker run -itd -p 8089:8080 -p 50000:50000 --name jdmjenkins --privileged=true -v /home/jdm/jenkins:/var/jenkins_home Jenkins
```

安装完成后，登录 http://{ip}:8089 地址，会进入 Jenkins 首次使用页面，需要输入密码，在服务器中，运行如下命令获取密码：

```
$ cat /home/jdm/jenkins/secrets/initialAdminPassword
```

输入密码后即可使用 Jenkins。

第 20 章 项目构建

本章介绍使用 Jenkins[①]进行软件的持续集成,然后自动构建程序的镜像,推送至 Harbor 私有仓库,最后用 Rancher 管理容器的运行。

20.1 Jenkins 基本介绍

在讲解 Jenkins 之前,需要了解什么是持续集成。持续集成是一种软件开发实践,其倡导团队开发成员协同工作,每次集成都是通过自动化的构建来验证,包括自动编译、发布和测试,从而尽快地发现集成错误,让团队能够更快进行业务开发。

持续集成中的任何一个环节都是自动完成的,无需太多的人工干预,有利于减少重复过程,节省了时间、费用和工作量。

Jenkins 原名 Hudson,是一个开源的持续集成工具。Jenkins 能实时监控集成中存在的错误,提供详细的日志文件和提醒功能,还能用图表形象地展示项目构建的趋势和稳定性。

Jenkins 的优点是:

- 易安装:仅需一个 war 包,从官网下载该文件后,直接运行。
- 易配置:Jenkins 提供友好的 GUI 配置界面。
- 代码版本管理支持:Jenkins 能从代码仓库(Git/Svn)中拉取代码。
- WebHook:可以关联提交业务代码的事件,触发 Jenkins 的自动构建项目。
- 集成 E-Mail/RSS/IM:当完成集成时,可通过这些工具实时通报集成结果。
- JUnit/TestNG 测试报告:以图表等形式提供详细的测试报表功能。
- 支持分布式构建:Jenkins 可以把集成构建等工作分配到多台计算机中完成。
- 集成记录信息:Jenkins 会保存每次集成构建产生的 jar 文件,以及每次集成构建的记录信息。
- 支持第三方插件:支持扩展插件,可以定制适合团队使用的工具。

20.2 Jenkins 基本设置

本节简要介绍 Jenkins 的安装方法,包含设置 Git 和 Maven,然后完成 Jenkins 环境的基本构建。

20.2.1 Jenkins 的安装

(1)安装 JDK

```
$ yum search jdk
```

[①] Jenkins 的官网是 https://jenkins.io/。

```
$ yum install java-1.8.0-openjdk-devel.x86_64 –y
$ java –version
```

JDK 安装完成后，注意 JDK 的安装地址/usr/lib/jvm，这个地址在配置 Jenkins 时会用到。

（2）安装 Git

```
$ yum install git –y
$ git version
```

（3）安装 Maven

```
$ cd /usr/local
$ wget http://mirror.bit.edu.cn/apache/maven/maven-3/3.5.3/binaries/apache-maven-3.5.3-bin.tar.gz ⊖
$ tar –zxvf apache-maven-3.5.3-bin.tar.gz
$ rm –rf apache-maven-3.5.3-bin.tar.gz
$ mv apache-maven-3.5.3 maven
```

配置环境变量，进入/etc/profile 文件，添加如下配置：

```
export M2_HOME=/usr/local/maven
export PATH=$PATH:$M2_HOME/bin
```

然后调用如下命令使配置生效，并且查看 Maven 版本：

```
$ source /etc/profile
$ mvn –version
```

（4）安装 Jenkins

```
$ wget –O /etc/yum.repos.d/jenkins.repo https://pkg.jenkins.io/redhat-stable/jenkins.repo
$ rpm ––import https://pkg.jenkins.io/redhat-stable/jenkins.io.key
$ yum install Jenkins
```

安装完成后，可以修改 Jenkins 设置，例如进入/etc/sysconfig/jenkins 文件，修改 JENKINS_PORT="8088"，这样就把 Jenkins 对外提供的端口修改为 8088。修改完配置后，使用如下命令启动 Jenkins：

```
$ service jenkins start
```

至此，Jenkins 的相关安装工作完成。

20.2.2 Jenkins 初次使用配置

使用上一节介绍的方法安装成功后，访问 Jenkins 的地址 http://{ip}:8088，可以看到如图 20-1 所示页面，此页面是初次使用 Jenkins 的提示页面，需要输入密码。

在图片中已经详细提示了 Jenkins 密码的路径，输入如下命令获取密码：

```
$ cat /var/lib/jenkins/secrets/initialAdminPassword
```

在页面中输入获取的密码，可以进入如图 20-2 所示页面。

选择"安装推荐的插件"，会进入一个安装页面，如图 20-3 所示，等待所有插件安装完成。

插件安装完成后，进入用户设置页面，如图 20-4 所示，进行简单的用户账号和密码设置。

⊖ 如果此地址不可下载，可以登录http://maven.apache.org/download.cgi选择其他下载地址。

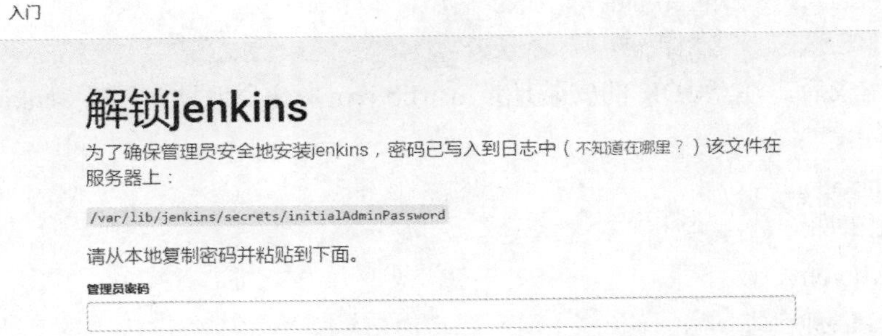

入门

图 20-1　解锁 Jenkins

图 20-2　选择插件安装

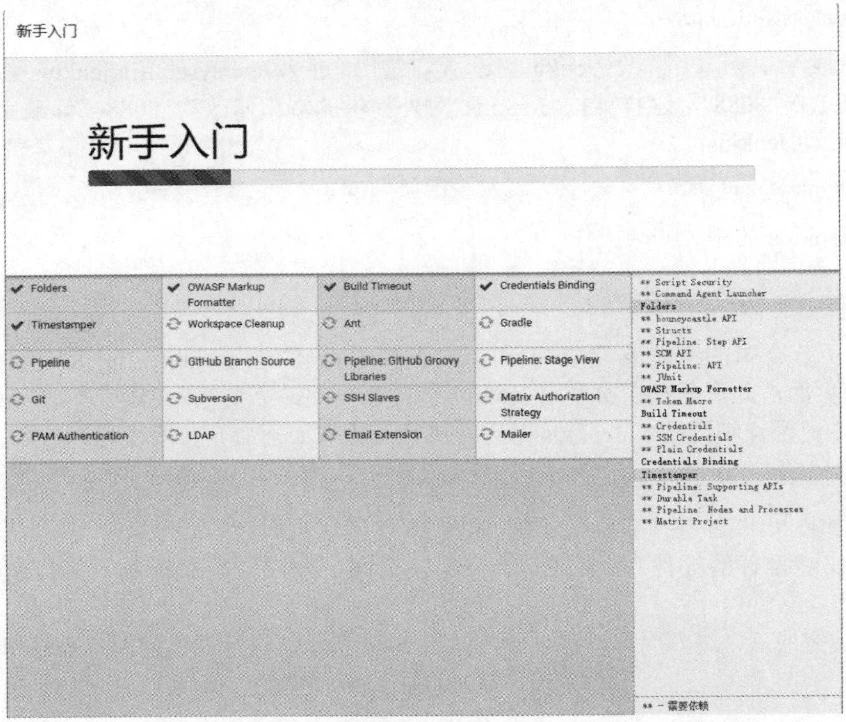

图 20-3　安装插件

第 20 章 项目构建

新手入门

创建第一个管理员用户

用户名：javadevmap

密码：●●●●●●●●

确认密码：●●●●●●●●

全名：javadevmap

电子邮件地址：your@email.com

图 20-4 创建用户

进行了上述操作，即完成了 Jenkins 基本的设置工作。

20.2.3 Jenkins 环境变量配置

在 Jenkins 中构建项目的基本流程为：Jenkins 从 Git/Svn 代码库获取最新代码，通过 Maven 编译、打包。为了完成此任务，需要 Jenkins 具备 Git 等相关组件的使用能力，并且通过 Maven 生成 Jar 包，这些配置就是对 Jenkins 环境变量的设置。

前面已经完成了 JDK、Git、Maven 的安装，下面要在 Jenkins 中配置这些组件的安装路径，以使 Jenkins 能够找到这些组件。

（1）配置 JDK 地址

进入"系统管理->全局工具配置"，找到 JDK 模块，进行配置，如图 20-5 所示。其中 JAVA_HOME 使用的地址是之前 JDK 的安装地址。

图 20-5 配置 JDK

（2）配置 Git 环境变量

同样在"系统管理->全局工具配置"中找到 Git 模块，进行设置，如图 20-6 所示。

（3）配置 Maven 环境变量

在"系统管理->全局工具配置"中，进行 Maven 的配置，File path 中填写的是 Maven 的配置文件地址，如图 20-7 所示；MAVEN_HOME 中填写的是 Maven 的地址，如图 20-8 所示。

进行如上配置后，即完成了构建工程相关插件的基础配置工作。

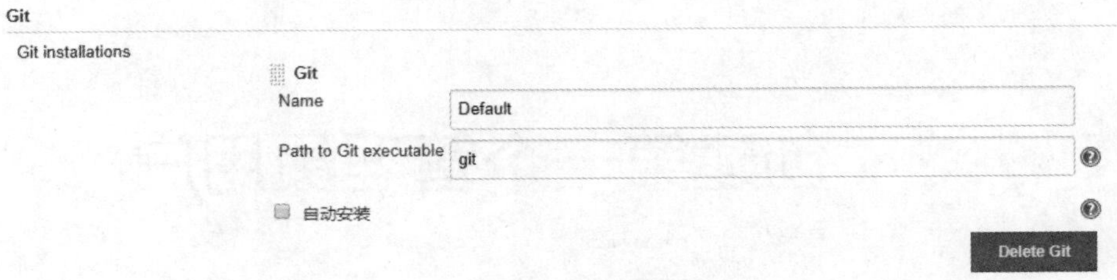

图 20-6　配置 Git

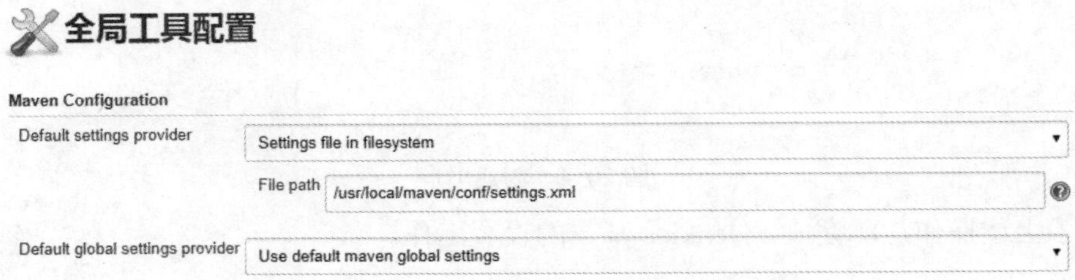

图 20-7　配置 Maven

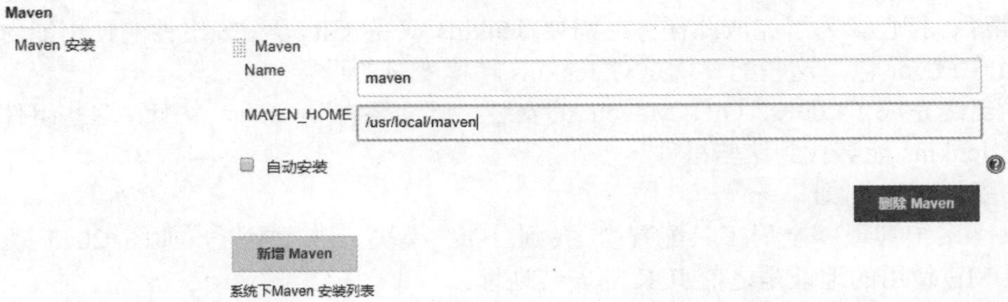

图 20-8　配置 Maven 地址

20.2.4　Jenkins 日志级别设置

Jenkins 的默认日志级别为 info，默认日志存放在/var/log/jenkins/jenkins.log，由于 Jenkins 默认会进行 DNS 查询，因此会出现很多的如下日志信息：

```
type: TYPE_IGNORE index 0, class: CLASS_UNKNOWN index
```

而且很快会使本地的日志文件变得很大。此时可以取消其日志打印。办法是：在"系统管理->系统日志->日志级别"中将名称为 javax.jmdns 的级别修改为 off，这样就关闭了 DNS 服务的日志开关，如图 20-9 所示。

20.2.5　安装常用插件

Jenkins 初次启动配置时，用户可以按照提示安装一些常用的推荐插件。但是针对 Maven 项目来说，还是不够的，需要安装其他几个相关的插件。

（1）安装 Maven 插件

在 Jenkins 中构建 Maven 项目，需安装 Maven 插件，因为刚安装好的 Jenkins 在新建任务的时候，并没有创建 Maven 工程的选项，如图 20-10 所示。

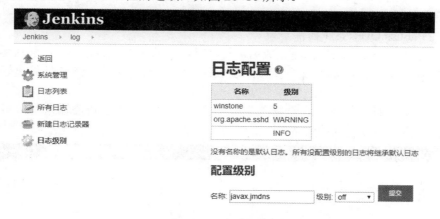

图 20-9　日志配置

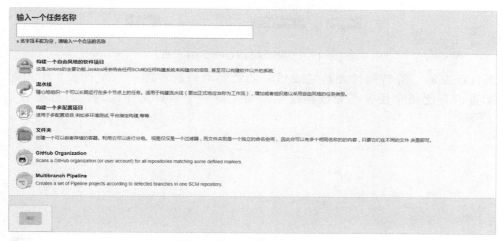

图 20-10　默认项目创建页面

由于 Jenkins 默认没有创建 Maven 项目任务的选项，所以需要在"系统设置->插件管理->可选插件"界面搜索安装。在右上角的输入框填写"Maven Integration"并搜索，可以看到搜索结果，如图 20-11 所示。

图 20-11　搜索 Maven 插件

这里勾选 Maven Integration 选项，然后点击直接安装即可。安装完成后，就能在新建任务时出现"构建一个 Maven 项目"的选项。

（2）安装 Publish Over SSH 插件

一般部署操作，需要在项目构建完成后，将项目生成的文件，通过 SSH 复制到目标服务器，然后执行命令或自定义 shell 脚本直接启动服务。而 Publish Over SSH 插件正好可以胜任此需求。

在"系统设置->插件管理->可选插件"界面搜索安装，在右上角的输入框填写"Publish Over SSH"并搜索，可以看到搜索结果，如图 20-12 所示。

图 20-12　安装 Publish Over SSH 插件

点击直接安装，等待插件安装完成后，需要到"系统管理->系统设置"中添加 SSH Server 配置，方便每个任务在完成构建后，将生成的可执行文件推送至目标服务器。如图 20-13 所示。

图 20-13　添加 SSH Server

- Name：SSH 标识的名字，自定义即可。
- HostName：需要连接 SSH 的主机名或 IP 地址。
- Username：SSH 连接登录目标 IP 服务器所使用的用户名。
- Remote Directory：用 SSH 连接后的远程根目录，这个目录是必须存在的，原因是 Jenkins 不会自动创建此目录。Jenkins 会将文件远程复制到该目录。（注意：SSH 连接的用户需要有权限才可以创建、删除、移动文件及文件夹）。

完成配置后，可以点击 Test Configuration 测试配置是否成功，如果出现 Success 表示成功。

20.3　构建 Maven 项目

接下来通过 Jenkins 构建第 9 章中的 Eureka 服务，实现 Eureka 服务的 Maven 打包以及发布运行操作。

20.3.1　Maven 构建设置

（1）创建 Maven 项目。

新建一个通过 Maven 构建的项目 SpringCloudEureka，然后点击确定。如图 20-14 所示。

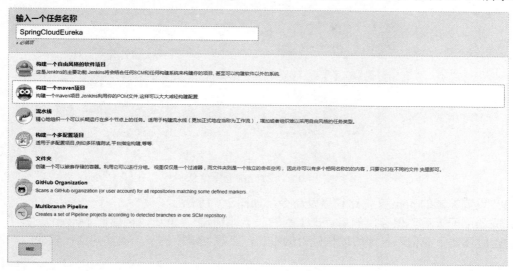

图 20-14　创建 Maven 项目

（2）填写项目名称与描述。

在接下来的页面中，填写项目名称为 SpringCloudEureka，描述为 Eureka Project。如图 20-15 所示。

图 20-15　输入工程信息

（3）设置构建的保存期限，如图 20-16 所示。

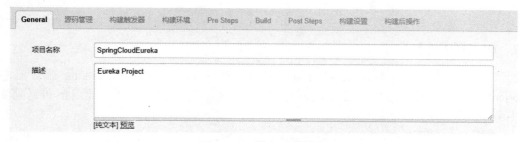

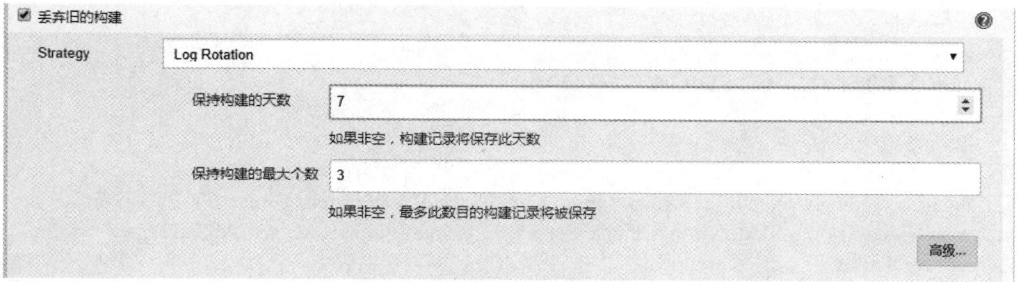

图 20-16　设置构建保存期限

(4)设置代码地址。

在这里填写 Git 的工程地址,并且在 Credentials 中配置 Git 的账号密码。如图 20-17 所示。

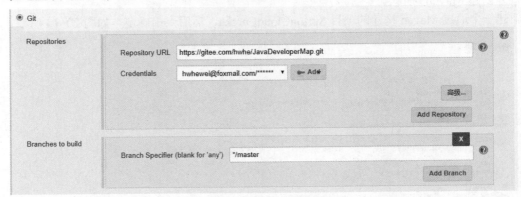

图 20-17 设置 Git 仓库信息

(5)设置编译的 pom 文件路径以及命令,如图 20-18 所示。

完成如上配置后,保存,然后点击工程的"立即构建",就可以构建此 Maven 工程了。第一次构建时需要下载依赖包,所以构建时间较长,下载完成后的下一次构建会比较快。

图 20-18 设置 pom 文件路径及命令

20.3.2 服务的执行

上一节已经完成了 Maven 工程的构建,构建后的 jar 包存放在 Jenkins 服务器中的 /var/lib/jenkins/workspace 路径下相应工程的目录中。下面要做的事情就是把此 jar 推送到要运行此服务的服务器并且启动运行。

(1)配置运行脚本

在服务要运行的服务器中添加运行脚本,此脚本的目的就是启动此服务。进入 /data/javadevmap/sc_shell 目录,新建文件 eureka_start.sh,在文件中填写如下内容:

```
#!/bin/sh
echo "eureka stop"
curl -X POST 172.17.238.239:18001/shutdown⊖
echo "eureka stop ok!"
sleep 20s
echo "eureka start!"
nohup  command>/data/javadevmap/Spring_Cloud/SpringCloudEureka/eureka.file 2>&1 java -jar –Xmx 128m /data/javadevmap/Spring_Cloud/SpringCloudEureka/SpringCloudEureka-0.0.1-SNAPSHOT.jar —spring. profiles. active=eurekaA &
```

⊖ 请注意服务的 /shutdown 端点关闭服务的耗时以及是否正确关闭了服务。

```
echo "eureka end!"
```
然后设置此文件的运行权限。
```
$ chmod u+x eureka_start.sh
```

（2）配置 Jenkins 构建后操作

进入 Jenkins 中的 SpringCloudEureka 工程的配置页，在"构建后操作"选项中选择"Send build artifacts over SSH"，然后进行配置，如图 20-19 所示。

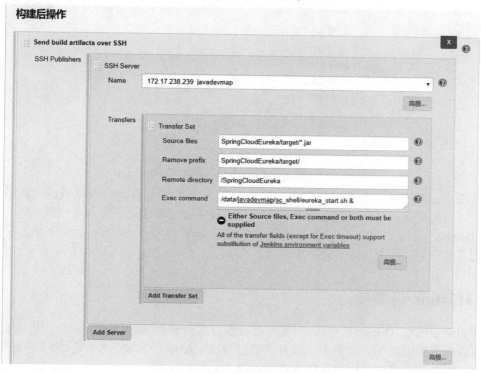

图 20-19　配置推送到的服务器及执行脚本

- **SSH Server-Name**：选择系统设置中已添加的 SSH 服务器。
- **Source files**：需要上传到应用服务器的文件（注意：相对于 Jenkins 工作空间的路径）。
- **Remove prefix**：去掉目录前缀（只能指定 Source files 中的目录）。
- **Remote directory**：目标服务器的文件夹。
- **Exec command**：远程服务器要执行的命令。在远程 SSH 传输执行后，才会执行这里配置的脚本命令，此脚本即上面所展示的运行 Eureka 服务的脚本。

（3）自动构建及服务运行

现在 Jenkins 自动构建已经配置完毕，进入 Jenkins 中的 SpringCloudEureka 任务中，点击"立即构建"，如图 20-20 所示，可见任务正在构建。

可以在控制台输出区查看 Jenkins 的构建过程，在此任务中，最后输出的内容为

```
eureka stop
{"message":"Shutting down, bye..."}eureka stop ok!
eureka start!
eureka end!
```

```
SSH: EXEC: completed after 20,223 ms
SSH: Disconnecting configuration [172.17.238.239_javadevmap] ...
SSH: Transferred 1 file(s)
Finished: SUCCESS
```

图 20-20　服务自动构建

待自动构建及服务运行完毕,可以登录目标服务器查看服务运行情况或通过浏览器查看 Eureka 的可视化页面。这样,整个服务的自动化部署流程就完成了。在服务启动的过程中,只要点击 Jenkins 的构建按钮即可,不需要其他操作即完成了一个服务的编译及运行,可见 Jenkins 对工作效率的提升是很明显的。

20.4　Harbor 镜像管理

工程如果要通过 Docker 运行,前提是把可执行文件通过 Dockerfile 的形式生成镜像。那么镜像生成后放到哪里呢？如果多台服务器要部署某一镜像,手动拖来拖去明显是低效的,这种情况下就需要一个私有的镜像仓库来存储已经生成好的镜像。

Harbor 是开源的企业级 Docker Registry 管理项目,它以 Docker 开源的 registry 为基础,提供了权限管理（RBAC）、LDAP、日志审核、管理界面、自我注册、镜像复制和中文支持等功能。

20.4.1　Harbor 安装

（1）安装 Docker-compose

首先确认服务器已经安装好了 Docker,然后执行如下命令安装 Docker-compose。

```
$ curl -L https://github.com/docker/compose/releases/download/1.21.2/docker-compose-`uname -s`-`uname -m` -o /usr/local/bin/docker-compose
$ chmod +x /usr/local/bin/docker-compose
$ docker-compose version
```

（2）下载 Harbor 安装文件并运行

```
$ mkdir /usr/local/harbor
$ cd /usr/local/harbor
$ wget https://storage.googleapis.com/harbor-releases/release-1.5.0/harbor-offline-installer-v1.5.0-rc2.tgz
```

```
$ tar -xvf harbor-offline-installer-v1.5.0-rc2.tgz
```

解压缩后,可以查看 harbor.cfg 文件,这是 Harbor 的配置文件。可以修改 hostname 属性为自己的 IP 地址,文件中包含 Harbor 的默认密码 Harbor12345。

执行./install.sh 脚本,Harbor 会自动下载相应组件并且安装运行。成功后执行 docker ps 命令,可以看到如图 20-21 所示输出。

图 20-21　Harbor 容器

访问http://{ip},输入账号 admin,密码 Harbor12345,可以进入 Harbor。

20.4.2　生成镜像并保存

现在 Harbor 仓库已经搭建完毕,下面做的事情就是通过 Jenkins 生成镜像,然后把镜像推送至 Harbor 仓库。

(1) 镜像推送前准备

在将要生成镜像的服务器中,先关掉 Harbor 仓库的安全认证。新建/etc/default/docker 文件,文件中填写如下内容:

```
--insecure-registry={your harbor ip}
OPTIONS='--selinux-enabled --log-driver=journald --signature-verification=false --insecure-registry= {your harbor ip}'
```

然后在/usr/lib/systemd/system/docker.service 文件中添加如下内容:

```
EnvironmentFile=-/etc/default/docker
```

使用如下命令重启 Docker:

```
$ systemctl daemon-reload
$ systemctl restart docker
```

(2) 设置 Jenkins

以之前编写的 SpringCloudServiceProvider 工程为例,在 Jenkins 中新建一个任务,这个任务的作用就是编译 Provider 工程,并且把此工程的 jar 文件推送至镜像生成服务器,然后运行此服务器中的一个脚本,生成 Provider 工程的镜像并且推送至 Harbor 仓库。此工程的编译部分与之前介绍 Jenkins 生成 jar 文件相同,只要配置好工程名和路径即可,推送和运行脚本部分有一点需要注意,就是 jar 文件的文件夹要和脚本文件夹相同,如图 20-22 所示。

(3) 镜像生成服务器配置

在 Jenkins 推送到的目录中,要准备好生成镜像的 Dockerfile 文件和可执行脚本 provider_image.sh。Dockerfile 文件的作用正如第 19 章所介绍,是镜像生成的配置;而 provider_image.sh 文件的作用是调用生成镜像的命令并推送镜像至 Harbor。

Dockerfile 文件配置如下:

```
FROM java:8
ADD SpringCloudServiceProvider-0.0.1-SNAPSHOT.jar /data/run/
EXPOSE 18010
WORKDIR /data/run/
ENTRYPOINT ["java","-jar","SpringCloudServiceProvider-0.0.1-SNAPSHOT.jar","--spring.profiles.
```

active=providerA"]

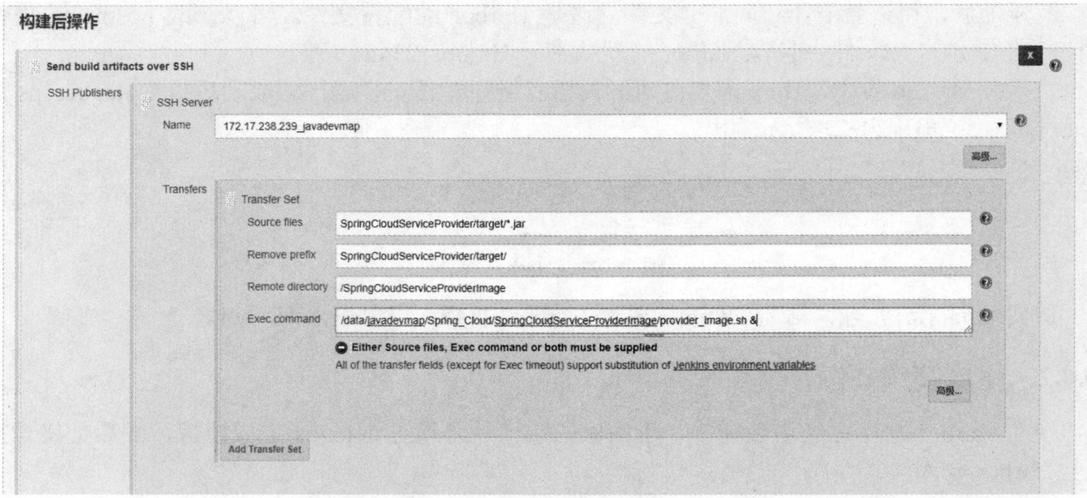

图 20-22　设置 Jenkins 推送目的地及脚本

provider_start.sh 文件配置如下：

```
#!/bin/sh
cd /data/javadevmap/Spring_Cloud/SpringCloudServiceProviderImage
Date_time=`date "+%F-%H-%M"`
docker build -t 172.17.238.239/springcloud/provider:0.0.1.${Date_time} .
docker login 172.17.238.239 --username admin --password Harbor12345
docker push 172.17.238.239/springcloud/provider:0.0.1.${Date_time}
```

添加完文件后，记得执行 chmod u+x provider_image.sh 命令设置文件权限。

（4）Harbor 仓库添加项目

在 Harbor 中，要先添加一个项目，用于承接推送到 Harbor 的镜像，此项目名称要和镜像工程名相同，这里创建项目名称为 springcloud。由于 Harbor 的可视化页面操作起来较为简单，这里就不再演示。

（5）执行效果

在 Jenkins 中执行"立即构建"，然后观察构建进度，可以发现项目编译完成后执行了镜像生成，并且把镜像推送至 Harbor 仓库，在 Harbor 的镜像仓库中可以看到镜像文件列表，如图 20-23、图 20-24 所示。

图 20-23　Harbor 仓库镜像

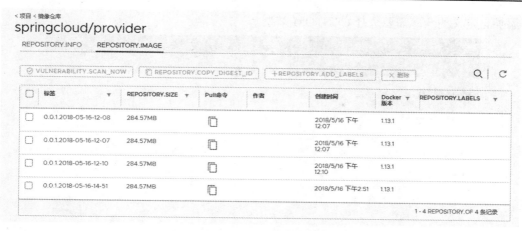

图 20-24　Harbor 仓库镜像文件

20.5　Rancher 容器管理

前面已经把自己的工程生成为镜像，并且保存到镜像仓库中，下一步就需要运行此镜像生成 Docker 容器，这样容器中运行的服务就可以对外提供业务能力，此服务通过 Eureka 注册中心成为整个 Spring Cloud 集群的一部分。

Rancher 是一个开源的企业级容器管理平台，下面使用 Rancher 管理容器，通过 Rancher 启动容器后，观察 Eureka 的服务注册情况，并且对 Spring Cloud 集群中的 Zuul 进行访问，观察容器中服务的负载情况。

20.5.1　Rancher 的安装及主机添加

（1）Rancher 的安装

Rancher 使用容器化安装，所以安装 Rancher 的服务器需要保证已经安装好了 Docker 环境。在服务器中执行如下命令即可完成安装。

```
$ docker run -d --restart=always -p 8080:8080 rancher/server
```

安装完成后，用浏览器访问http://{ip}:8080地址即可看到 Rancher 的可视化页面。

（2）添加主机

进入 Rancher 可视化页面后，选择"基础架构->主机->添加主机"，可以看到如图 20-25 所示页面。

可以修改主机注册地址为其他地址⊖，点击保存后，可以看到一个添加主机的引导页，如图 20-26 所示，此引导页描述得非常清楚，按照此引导页的说明添加主机即可。

按照提示，在需要加入 Rancher 集群的主机中执行图中的注册命令，注册完后，再点击 Rancher 的"基础架构->主机"，可见如图 20-27 所示主机实例。

（3）添加私有镜像仓库

选择"基础架构->镜像库"，如图 20-28 所示，按照页面提示添加刚创建的 Harbor 仓库。

⊖ 本书中选择了其他地址，把主机注册地址修改为一个内网地址，如果此处忘记修改，可以在"系统管理->系统设置"中修改主机注册地址。

记得添加自定义的仓库需要选择 CUSTOM 类型。

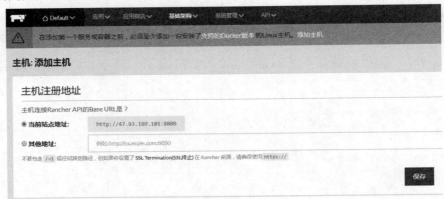

图 20-25 添加主机

图 20-26 添加主机引导页

20.5.2　Rancher 启动单一容器

下面使用 Rancher 启动一个容器，这个容器使用的镜像就是之前用 Provider 服务创建的镜像。

启动容器配置。在单台主机的列表下方，有一个"添加容器"选项，选择该选项后可见如图 20-29 所示页面。

在页面中可以配置容器名称、镜像地址以及拉取镜像规则。设置完这些内容后，选择网络标签，在网络选项中选择主机模式，如图 20-30 所示。

图 20-27　主机实例

图 20-28　添加镜像库

图 20-29　启动单一容器

图 20-30　设置网络模式

然后选择卷标签，添加卷，让容器内的卷可以映射到主机，如图 20-31 所示。
完成如上配置后，点击"创建"，可以看到容器启动了，如图 20-32 所示。

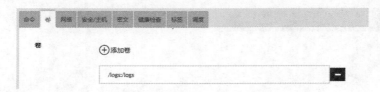

图 20-31　添加卷

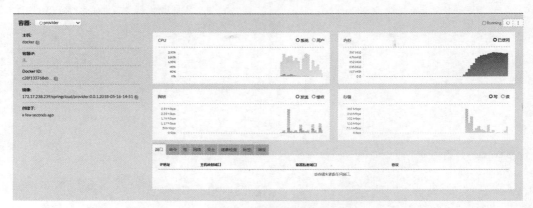

图 20-32　容器监控信息

进入此台主机的/logs 目录，可以看到容器中的服务日志已经挂载出来，执行 tail –f 命令可以和本机服务一样查看日志打印。使用 Postman 单独访问此服务，也可以正确返回数据。

这样单个容器的运行已经完全可以使用此方法进行操作，但是 Rancher 的目标肯定不只是运行单节点服务，下一节介绍快速部署大量服务实例。

20.5.3　Rancher 启动批量容器

在服务集群中，每个服务都会启动多个实例，当用 Docker 运行服务容器，用 Rancher 管理容器时，也希望 Rancher 能够提供简单、快速的服务启动及监控能力。下面就演示一种服务容器批量启动的方法。

（1）添加多台主机

使用上一节介绍的方法，在另外一台服务器中运行 Rancher 注册命令，把新服务器加入 Rancher 主机集群中。如图 20-33 所示。

（2）给主机添加标签

点击主机右上角的按钮，在弹出框中选择"编辑"，在弹出页中添加标签，例如给上面两台主机添加 springcloud=provider 标签。如图 20-34 所示。

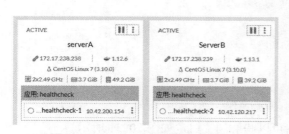

图 20-33　Rancher 管理的两台主机

图 20-34　设置主机标签

（3）在 Rancher 中创建应用和服务

在 Rancher 中选择"应用->添加应用"，这里输入应用名 springcloud。然后进入刚创建的应用中，选择"添加服务"。这里和启动单台容器不同的地方有两处，一是在数量选项中选择"总是在每台主机上运行一个此容器的实例"，如图 20-35 所示。

图 20-35　设置容器运行数量

另一个不同的地方是添加调度规则，进入"调度->添加调度规则"，这里配置所有具备 springcloud=provider 标签的主机启动此容器，如图 20-36 所示。

图 20-36　设置调度规则

配置完上述信息后，还要配置网络和卷，方法和启动单个容器相同，然后点击"创建"启动容器，可见两台主机会同时启动 Provider 服务容器。如图 20-37 所示。

图 20-37　服务启动

在此页面中，可以点击单个容器后面的多选框，选择"查看日志"，即可看到此服务的日志。在 Eureka 中可以看到新启动的两个 Provider 服务，如图 20-38 所示。

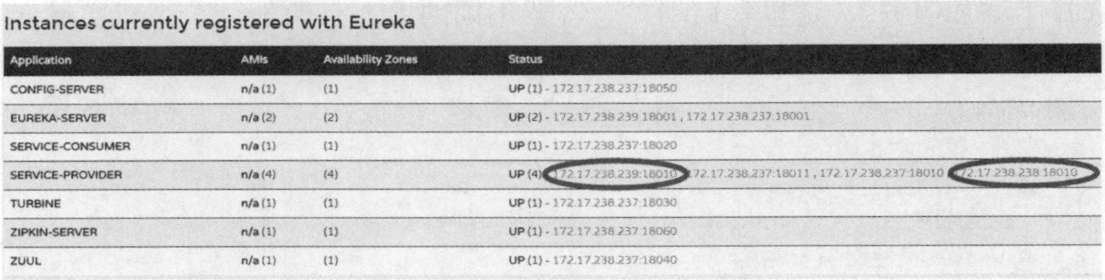

图 20-38　Eureka 服务注册列表

通过 Zuul 访问此服务，查看日志，可以看到几个 Provider 服务实例均匀承担请求压力。

20.5.4　服务更新

前面介绍了 Rancher 的批量部署，其实它还具备一个较为方便好用的服务升级能力。Rancher 的服务升级不只是简单的更新服务容器，还带了一个简便好用的回滚功能，当服务新版本不可用，需要回退的时候，可以一键实现回退。

（1）升级

进入新建的 springcloud 应用中，可以看到应用所包含的服务，单个服务的最右侧多选框，有一个升级按钮。如图 20-39 所示。

图 20-39　服务操作菜单

点击此按钮，在新的弹出页中，只要输入新的镜像地址，并且配置简单的升级规则，就可以实现服务升级，如图 20-40 所示。

图 20-40　服务升级规则

（2）回滚

服务升级后，查看此服务运行的容器，可见新老容器都存在，只是老容器是 Stopped 状态，如图 20-41 所示。这就方便了服务的回滚。

第 20 章 项目构建

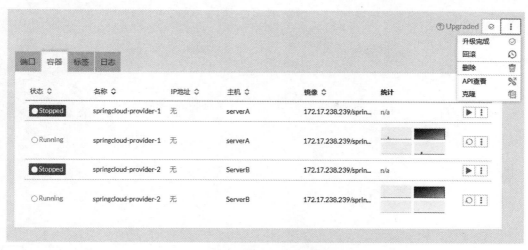

图 20-41 服务升级后选项

在此页面的右上角可以选择升级完成或者回滚，如果选择升级完成，Rancher 会删除旧的容器；如果选择回滚，Rancher 会关掉新容器，启动旧容器。

第五篇 工 具 篇

　　前面四篇已经把 Java 服务端研发所用到的语言、框架、组件和部署等内容都讲述完了，之所以在最后编写此篇内容，是由于其很难归入以上四类，并且这些工具的使用对一名研发人员来讲非常重要。

　　本篇仅有一章内容，介绍了日常使用的工具，这些工具的选择也是作者从研发技术角度进行考虑的。编写本篇之初选择的内容很多，包含 bug 管理工具、进度管理工具等，但是最后作者确定内容时，觉得这类工具应该属于项目管理的范畴，所以没有选录进来。当然工具还不止这些，还包含二维码生成工具、音视频转换工具等等，但是作者认为这类工具更偏向于某一技术，而不够通用，所以也没有选录进来，最后选择的内容就是目前读者看到的这些。

　　本篇介绍的工具有：Swagger 用于生成接口文档并且验证接口和调试；Jmeter 用于模拟真实的请求和压力；ab 用于验证服务本身的性能；VisualVM 用于查看服务的运行情况；JD-GUI 用于反编译程序，查看程序包的代码情况。

　　希望通过本篇的学习，读者能够在工作中及时检验服务的状态，提高自己的编码质量。

第 21 章 常 用 工 具

本章将介绍 Swagger、JMeter、ab、VisualVM、JD-GUI 工具的使用，这些工具会帮助你提高工作效率和质量。

21.1 Swagger

在之前编写的工程中，Controller 类中实现的方法都是为前台服务提供的能力接口，前台研发需要知道后端提供了什么样的接口，包含接口路径、采取什么调用方式、传递什么参数等信息，这样前台程序才能正确调用后端接口。

传统的做法是后台研发人员先设计接口，然后形成接口文档，继而进行后台业务的研发，但是这种方式存在流程上的和人为的几个问题。在后台研发过程中，发现某一接口需要改动，那么除了修改代码以外还需要修改接口文档并且通知前端人员；接口文档由于人为的原因编写错误或者滞后，或者接口不可用而无法及时验证；最主要的是对于研发人员来讲，相对于编写文档，大家都更喜欢写代码，频繁的修改文档会大大降低效率，而 Swagger 可以减少文档工作对研发效率的影响[○]。

Swagger 是一个接口文档的自动生成及测试软件，它的优点和缺点一样明显，或者说 Swagger 的缺点正是它优点的所在。Swagger 是一个强侵入的接口文档生成工具，需要在代码中编写接口的说明性注解，而这可能正符合某些研发人员不想专门写文档的心态。在代码中写一些注解就能形成文档，很多人还是能够接受的。Swagger 可以根据这些注解自动生成一个 Web 页面，页面中包含研发者编写的接口说明，并且可以直接调用接口进行测试。下面在 Spring Boot 工程中演示 Swagger 的主要用法。

21.1.1 Swagger 基本配置

使用之前编写的 SpringBootMybatis 工程，在工程中进行如下改造，引入 Swagger。

（1）添加依赖

在 pom 文件中，添加如下依赖：

```xml
<dependency>
    <groupId>io.springfox</groupId>
    <artifactId>springfox-swagger2</artifactId>
    <version>2.6.1</version>
</dependency>
<dependency>
    <groupId>io.springfox</groupId>
    <artifactId>springfox-swagger-ui</artifactId>
```

○ 虽然可能最后还是需要一份接口文档，但是可以在联调结束后统一提供，Swagger 可以使联调过程更加顺畅。

```
    <version>2.6.1</version>
</dependency>
```

（2）添加 Swagger 配置

新建一个类 SwaggerConfig，在此类中添加如下代码：

```
@Configuration
public class SwaggerConfig {
    @Bean
    public Docket createRestApi() {
        return new Docket(DocumentationType.SWAGGER_2)
                .apiInfo(apiInfo()).select()
                .apis(RequestHandlerSelectors.basePackage("com.javadevmap.mybatis.controllers"))
                .paths(PathSelectors.any())
                .build();
    }

    private ApiInfo apiInfo() {
        return new ApiInfoBuilder()
                .title("Spring Boot Mybatis 接口文档")
                .description("Spring Boot Mybatis Swagger2 UserService Interface")
                .termsOfServiceUrl("127.0.0.1:18089")
                .version("0.0.1")
                .build();
    }
}
```

（3）开启 Swagger

在启动类中添加注解@EnableSwagger2。经过这几步，就完成了 Swagger 的基本配置，下面就可以对每一个 Controller 实现接口文档的编写。

21.1.2 使用 Swagger 编写接口文档

在此工程的 UserController 类中，对类和方法添加注解，就可以形成一个带测试能力的接口文档。

```
@Api(value="用户服务")
@RestController
@RequestMapping("/user")
public class UserController {
    @ApiOperation(value="获取用户信息", notes="根据 id 获取用户信息")
    @ApiImplicitParam(name = "Id", value = "用户 Id",
                    required = true, dataType = "integer",paramType = "path")
    @RequestMapping(value="/{Id}",method=RequestMethod.GET)
    public Result<DomainUser> getUser(@PathVariable("Id") int id) {
        …
    }

    @ApiOperation(value="添加用户", notes="获取用户信息并保存")
    @ApiImplicitParams({
        @ApiImplicitParam(name = "user", value = "用户信息",
                    required = true, dataType = "DomainUser",paramType = "body")
    })
```

```
@RequestMapping(value="/add",method=RequestMethod.POST)
public Result<String> addUser(@RequestBody @Valid DomainUser user) {
    …
}

@ApiIgnore()
@RequestMapping(value="/ignore",method=RequestMethod.GET)
public Result<String> ignore() {
    …
}
}
```

在上面的代码中，新添加了一些注解，有些注解添加在类上，有些注解添加在方法上，有些注解能够一眼看出它是在说明方法参数。Swagger 就是通过这种注解的形式对接口进行说明继而生成文档的。Swagger 常用的注解及作用见表 21-1。

表 21-1　Swagger 注解

注解	作用
@Api	作用于类，说明接口类
@ApiOperation	作用于方法，提供接口说明
@ApiIgnore	作用于类、方法、参数，目的是忽略被标注内容
@ApiImplicitParam	作用于方法，表示单个请求参数
@ApiImplicitParams	作用于方法，表示多个请求参数
@ApiParam	作用于方法、参数、字段，描述参数
@ApiModel	作用于类，说明对象实体类
@ApiModelProperty	作用于方法、字段，描述对象的方法、字段
@ApiResponses	作用于类、方法，响应描述集
@ApiResponse	作用于方法，响应描述

21.1.3　Swagger 测试演示

完成对 Controller 类的配置后，启动此服务，然后通过浏览器访问此服务的地址，在地址后加上/swagger-ui.html 路径，例如 http://localhost:18089/swagger-ui.html，就可以看到 Swagger 的可视化界面，如图 21-1 所示。

图 21-1　Swagger 可视化页面

（1）GET 方法测试

点击页面中的/user/{Id}方法，可以看到接口说明，并且包含了返回类型的模板、参数的输入框等。可以在输入框中填写数据，然后点击"Try it out！"就能实现接口测试，如图 21-2 所示。

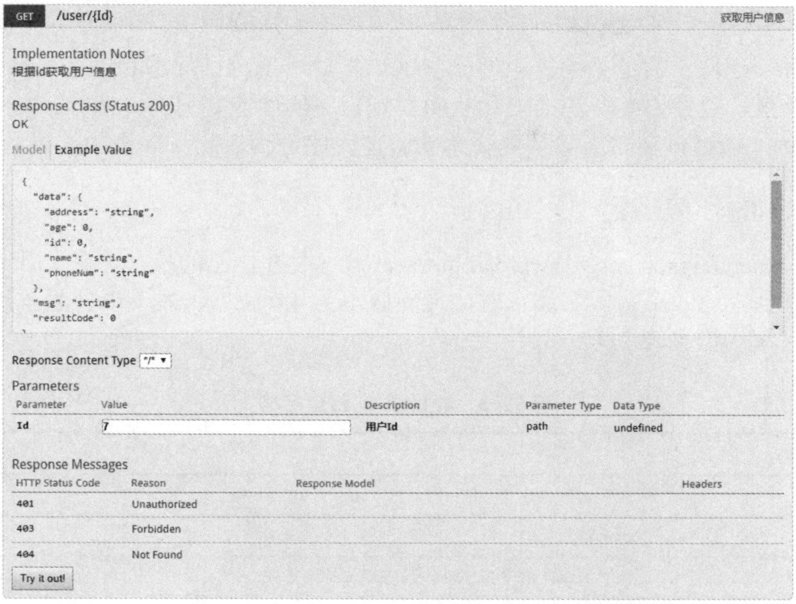

图 21-2　获取用户信息接口

（2）POST 方法测试

点击页面中的 /user/add 方法，同样可以看到相关的接口说明等。由于提交的是一个自定义类型，在页面中"添加参数"的右侧还列出了提交数据的模板，如图 21-3 所示。

图 21-3　添加用户接口

21.2 JMeter

Apache JMeter[一]是一款优秀的开源性能测试工具。用于测试静态和动态资源,通过多线程模拟用户访问场景,监控系统资源的变化从而得到程序的性能。另外,JMeter 能够对应用程序做功能/回归测试,通过创建带有断言的脚本来验证程序是否返回了期望的结果。

21.2.1 JMeter 的环境搭建

登录 http://jmeter.apache.org/download_jmeter.cgi,根据自己的操作系统,下载对应版本的文件,然后解压到本地即可。本书编写时,最新版本为 4.0 版本,本节以此版本进行讲解。

JMeter 的目录结构见表 21-2。

表 21-2 JMeter 的目录结构

目录	作用
bin	可执行文件目录
docs	Jmeter 帮助文档
extras	提供了对 Ant 的支持文件,也可用于持续集成
lib	Jmeter 依赖的 jar 包,同时安装的插件也放于此目录
licenses	软件许可文件

进入 bin 目录,双击 jmeter.bat 即可看到 JMeter 可视化页面,如图 21-4 所示。

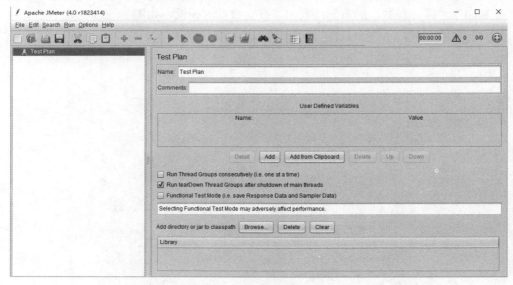

图 21-4 JMeter 可视化页面

软件默认语言为英文,可以通过 "Options->Choose Language->Chinese(Simplified)" 设置成中文。

[一] Jmeter 的官网是 http://jmeter.apache.org/index.html。

21.2.2 测试计划

一个完整的测试计划包括一个或多个元素，例如线程组、逻辑控制器、样品产生控制器、监听器、定时器、断言和配置等。

本节使用之前章节编写的获取用户信息接口的代码，演示接口测试功能。

（1）新建线程组（Thread Group）

在 JMeter 软件的测试计划（Test Plan）上，执行"鼠标右键单击->add->Threads（Users）->Thread Group"。如图 21-5 所示。

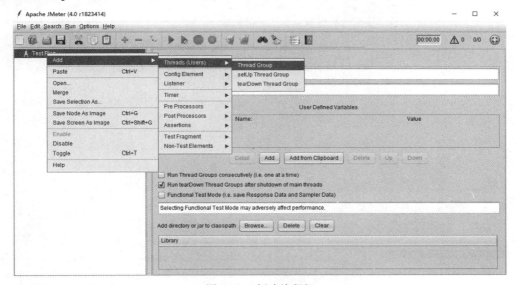

图 21-5　新建线程组

在新建的 Thread Group 页面的 Thread Properites 面板中，保持默认选项即可，即启动一个线程发起一次请求。Thread Properites 中选项的含义如下：

- Number of Threads (Users)：模拟的并发线程数。
- Ramp Up Period (in seconds)：在多长时间内启动所有的线程。例如 Number of Threads 设为 10，Ramp Up Period 设为 1，则 Jmeter 每隔 0.1s 启动 1 个线程。
- Loop Count：单用户任务重复执行的次数。如果设为 Forever，那么 Jmeter 就不会自动停止，需要强制终止。

（2）添加采样器（Sampler）

在新建的线程组节点上，执行"鼠标右键单击->Add->Sampler->HTTP Request"选项，添加 HTTP 请求采样。压力测试获取的用户信息接口是 http://47.95.113.117:18010/ user/7，如图 21-6 所示。

在右边输入页面的 Web Server 页签中，填写请求相关信息，即请求 url 和参数。

（3）添加监听器（Listener）

添加一个监听器，相当于程序的 console 控制台，可以直接查看结果。在 HTTP Request 上执行"鼠标右键单击->Add->Listener->View Results Tree"，添加监听线程组（用户）。如图 21-7 所示。

设置完成后，在菜单栏上点击"Run->Start"执行用例，会弹出对话框让你先保存用例，然后再进行测试。

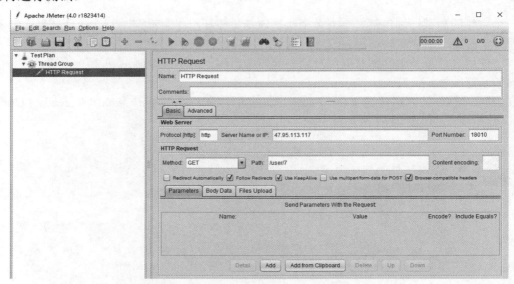

图 21-6　添加采样器

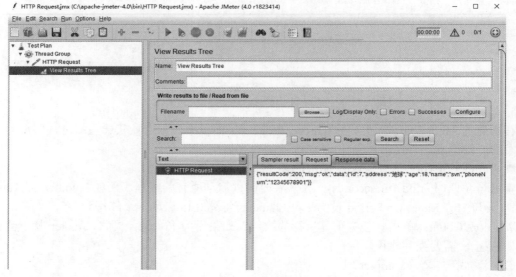

图 21-7　添加监听器

测试完成后，可以看到请求的响应数据。当然在实际测试中，要根据业务需要来设置对应的线程数以及并发测试数。这里演示的是最基本的 JMeter 使用，如果想了解更多功能，可以登录 JMeter 的官网进行学习。

21.3　ab

ab 是 Apache 提供的一款压力测试工具，用来测试服务器负载压力。为了保证测试准确性，ab 需与目标服务器处于同一内网中进行测试。非内网环境对 Web 服务器进行压力测试，会由于网络延时过大或带宽不足，造成测试效果不准确。

本节演示在 Linux 服务器上进行测试。Linux 服务器的版本为 CentOS 7.3。通过客户端登录到服务器上后，执行如下命令安装 ab：

```
$ yum -y install httpd-tools
```

执行如下命令，查看 ab 版本来检测是否安装成功。

```
$ ab -V
```

21.3.1 压力配置

测试之前编写的获取用户信息接口程序，模拟 50 个并发用户，对此接口发送 1000 个请求。

```
$ ab -n1000 -c50 http://127.0.0.1:18010/user/7
```

上面命令中-n 表示总请求数，-c 表示并发数。当然 ab 的常用命令远不止这些，见表 21-3。

表 21-3 ab 工具常用命令

参 数	含 义
-n	请求的总数，默认为 1 次
-c	并发数，同一时间请求数量
-p[注]	POST 请求，文件中包含请求数据，根据数据格式，设置-T 参数
-T	针对 POST/PUT，设置请求头中的 Content-type，例如：application/json。默认是 text/plain
-w	将测试结果打印到 HTML 表格中

21.3.2 结果查看

在执行 ab 测试命令之前，先了解一下 ab 返回参数的含义，见表 21-4。

表 21-4 ab 测试返回参数含义

参 数 名	含 义
Concurrency Level	并发数，等于-c 后面的数值
Time taken for tests	测试总耗时
Complete requests	成功收到返回的数目
Failed requests	请求失败数目
Requests per second	每秒请求数，等于总请求数/测试总耗时
Time per request	每一个请求平均花费时间： 第一个 Time per request 为用户平均请求等待时间； 第二个 Time per request 为服务器平均处理时间； 用户平均请求等待时间 = 服务器平均处理时间×并发数

执行 ab 命令，得到如下输出结果：

```
$ ab -n1000 -c50 http://127.0.0.1:18010/user/7
...//省略
```

[注] 例如 ab -n 1 -c 1 -p 'post.txt' -T 'application/json' 'http://localhost:18010/user/add'命令用于测试 POST 请求。post.txt 为上传的 Json 数据内容的文件。

```
Server Software:
Server Hostname:        127.0.0.1
Server Port:            18010

Document Path:          /user/7
Document Length:        111 bytes

Concurrency Level:      50
Time taken for tests:   1.529 seconds
Complete requests:      1000
Failed requests:        0
Write errors:           0
Total transferred:      289000 bytes
HTML transferred:       111000 bytes
Requests per second:    653.99 [#/sec] (mean)
Time per request:       76.454 [ms] (mean)
Time per request:       1.529 [ms] (mean, across all concurrent requests)
Transfer rate:          184.57 [Kbytes/sec] received

Connection Times (ms)
              min   mean[+/-sd] median   max
Connect:       0     1    2.5      0      24
Processing:    6    74   37.2     65     305
Waiting:       5    73   37.1     65     305
Total:         6    75   36.9     66     305
..//省略
```

从上面的输出结果可见，本次测试总耗时（Time taken for test）1.529s；请求全部完成（Complete requests）共 1000 个，失败请求（Failed requests）个数为 0 个；每秒平均请求数（Requests per second）为 653.99 个；用户平均请求（Time per request）等待时间为 76.454ms，服务器平均请求处理时间为 1.529ms。

21.4 VisualVM

VisualVM 可以对 Java 程序进行运行时监控，此工具提供了图形界面，可以监控程序使用的 CPU、内存、类、线程等，还可以添加插件对程序的其他信息进行监控。VisualVM 随 JDK 同时安装，所在路径为%JAVA_HOME%\bin\jvisualvm.exe。本节演示此工具的基本用法。

21.4.1 查看 CPU

使用之前编写的 SpringBootMybatis 工程，在工程中添加一个简单的 Java 类 JvmCpuTest，此类仅作为测试 VisualVM 工具对 CPU 的监控，不作为业务逻辑。

（1）单个线程的 CPU 使用

在此类中添加如下代码，然后执行此 main 方法。

```
public class JvmCpuTest {
    private static void singleThread(){
        try {
```

```
                    long time = 1;
                    while(time>0) {
                        time = System.currentTimeMillis();
                    }
                } catch (Exception e) {
                    e.printStackTrace();
                }
            }

            public static void main(String[] args) {
                singleThread();
            }
        }
```

在以上代码中编写了一个无限循环。执行此 main 方法后，进入 VisualVM 工具，可以在页面的左侧看到此 Java 服务进程，双击此进程可以看到它的监控页面，选择监控页面的标签可以进入相应的监控项，这里选择"监视"标签，可以看到如图 21-8 所示页面，CPU 使用率大约为 25%（测试程序所运行主机 CPU 为 4 核）。

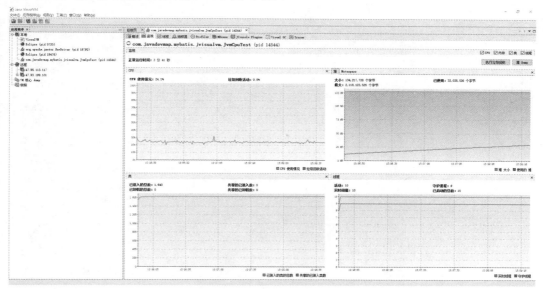

图 21-8　服务监控页面

（2）多线程 CPU 使用

修改以上测试代码，让两个线程同时执行此无限循环，观察 CPU 的使用情况。

```
        public class JvmCpuTest {
            private static void multiThread(){
                ExecutorService eService = Executors.newFixedThreadPool(2);
                for(int i=0;i<2;i++) {
                    eService.execute(new Runnable() {
                        @Override
                        public void run() {
                            try {
                                long time = 1;
                                while(time>0) {
```

```
                            time = System.currentTimeMillis();
                        }
                    } catch (Exception e) {
                        e.printStackTrace();
                    }
                }
            });
        }
        eService.shutdown();
    }

    public static void main(String[] args) {
        multiThread();
    }
}
```

这里使用两个线程执行无限循环，运行程序后，可以看到 CPU 使用情况如图 21-9 所示，CPU 使用率大概为 50%左右。在界面上方的标签中，还可以选择"抽样器"查看 CPU 的具体使用情况。

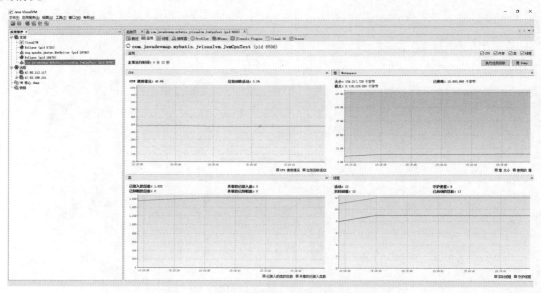

图 21-9　使用两个线程后的服务监控页面

21.4.2　查看线程

在本机启动 SpringBootMybatis 工程，然后在 VisualVM 工具中进入此工程监控中的"线程"标签，可以看到如图 21-10 所示的页面情况，页面的右下方包含了对此进程中各线程状态的说明，在页面中可以看到名为 http-nio-18089-exec*的线程共 10 个。

停止此程序，然后在 yml 文件中添加如下配置，启动工程后，可以在 VisualVM 工具的"线程"标签下看到名为 http-nio-18089-exec*的线程共 30 个，如图 21-11 所示。

```
server:
    port: 18089
```

```
tomcat:
    min-spare-threads: 30
```

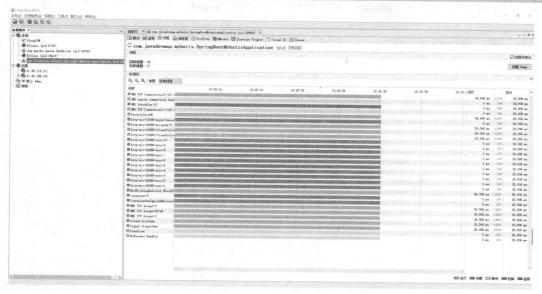

图 21-10 线程监控

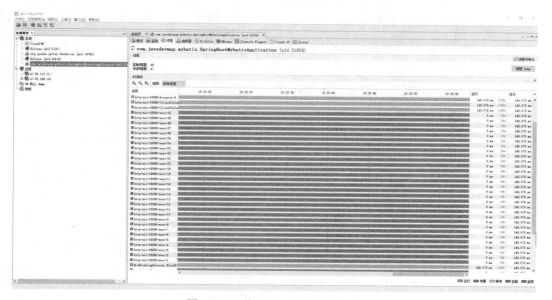

图 21-11 修改配置后的线程监控

21.4.3 监控远程服务

VisualVM 工具除了对本地的 Java 服务进行监控，还可以监控远程服务器上的 Java 服务运行状态。这里简单介绍监控远程服务的方法。

（1）启动远程服务

在远程服务器中，使用如下命令启动服务，此命令包含了远程连接的相关配置。

> $ nohup java -Djava.rmi.server.hostname=47.93.199.101 -Dcom.sun.management.jmxremote -Dcom.sun.management.jmxremote.port=18088 -Dcom.sun.management.jmxremote.rmi.port=18088 -Dcom.sun.management.jmxremote.authenticate=false -Dcom.sun.management.jmxremote.ssl=false -jar -Xmx128m SpringBootMybatis-0.0.1-SNAPSHOT.jar &

（2）在工具中配置连接

在 VisualVM 工具中，选择"远程->添加远程主机"，在输入框中添加远程服务器地址，如图 21-12 所示，然后点击确定，这样就添加了远程服务器。

添加完远程服务器后，鼠标右键单击此服务器，然后选择"添加 JMX 连接"，在弹出的输入框中输入连接地址，如图 21-13 所示。

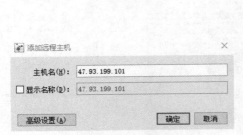

图 21-12　配置远程主机连接

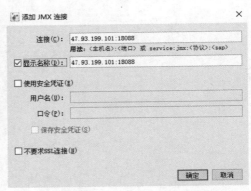

图 21-13　添加 JMX 连接

完成如上设置后，就可以看到要监控的远程程序了，如图 21-14 所示。

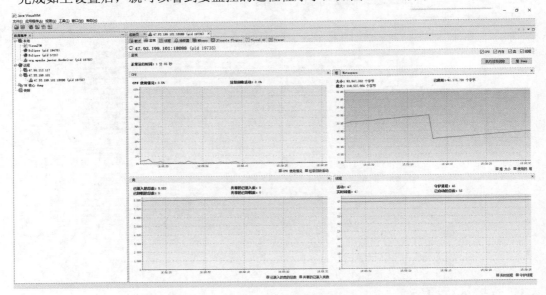

图 21-14　远程服务监控

21.5　JD-GUI

JD-GUI 是一个反编译工具，它可以把已经生成的 Jar 包反编译回代码的形式，没有经过代码混淆的 Jar 包反编译后和实际的源码会有一定的差别，但是不影响阅读。JD-GUI 是可视化的，所以使用起来非常简单，只要引入 Jar 包就可以执行反编译。这里展示 SpringBoot Mybatis 工程的 Jar 包反编译后的效果。如图 21-15 所示。该工具的网址是 http://jd.benow.ca。

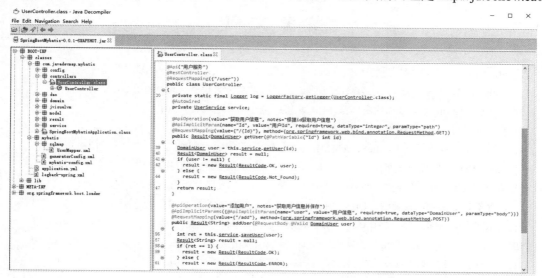

图 21-15　Jar 包反编译

参 考 文 献

[1] Bruce Eckel. Java 编程思想[M]. 4版. 陈昊鹏, 译. 北京: 机械工业出版社, 2007.
[2] Craig Walls. SpringBoot 实战[M]. 丁雪丰, 译. 北京: 人民邮电出版社, 2016.
[3] 翟永超. Spring Cloud 微服务实战[M]. 北京: 电子工业出版社, 2017.
[4] 许晓斌. Maven 实战[M]. 北京: 机械工业出版社, 2011.
[5] Rafal Kuc. 深入理解 ElasticSearch[M]. 张世武, 译. 北京: 机械工业出版社, 2016.
[6] Craig Walls. Spring 实战[M]. 张卫滨, 译. 北京: 人民邮电出版社, 2016.
[7] Josiah L Carlson. Redis 实战[M]. 黄健宏, 译. 北京: 人民邮电出版社, 2015
[8] 杨保华, 戴王剑, 曹亚仑. Docker 技术入门与实战[M]. 北京: 机械工业出版社, 2017.
[9] 姜承尧. MySQL 技术内幕[M]. 北京: 机械工业出版社, 2013.
[10] 饶琛琳. ELK stack 权威指南[M]. 北京: 机械工业出版社, 2015.
[11] 周志明. 深入理解 Java 虚拟机[M]. 北京: 机械工业出版社, 2013.
[12] 鸟哥. 鸟哥的 Linux 私房菜[M]. 北京: 机械工业出版社, 2008.